Phycobiliproteins

Authors

Robert MacColl
and
Deborah Guard-Friar
Wadsworth Center for Laboratories and Research
New York State Department of Health
Albany, New York

CRC Press
Taylor & Francis Group
Boca Raton London New York

CRC Press is an imprint of the
Taylor & Francis Group, an **informa** business

First published 1987 by CRC Press
Taylor & Francis Group
6000 Broken Sound Parkway NW, Suite 300
Boca Raton, FL 33487-2742

Reissued 2018 by CRC Press

Library of Congress Cataloging-in-Publication Data

MacColl, Robert.
Phycobiliproteins.

Bibliography: p.
Includes index.
1. Phycobiliproteins. 2. Algae—Composition.
3. Photosynthesis. I. Guard-Friar, Deborah. II. Title.
QK898.P64M33 1987 589.3'19245 85-32567
ISBN 0-8493-5525-7

A Library of Congress record exists under LC control number: 85032567

Publisher's Note
The publisher has gone to great lengths to ensure the quality of this reprint but points out that some imperfections in the original copies may be apparent.

ISBN 13: 978-1-315-89646-5 (hbk)
ISBN 13: 978-1-351-07556-5 (ebk)

Visit the Taylor & Francis Web site at http://www.taylorandfrancis.com and the
CRC Press Web site at http://www.crcpress.com

TABLE OF CONTENTS

Chapter 1

INTRODUCTION

Biliproteins (also called phycobiliproteins) are photosynthetic antenna pigments found in cyanobacteria (blue-green algae), red algae, and the cryptomonads. These chromoproteins harvest solar energy in regions of the visible spectrum having low chlorophyll absorption and then transfer this excitation energy to chlorophyll in the photosynthetic membrane. The biliproteins obtain their colors from linear tetrapyrrole chromophores which are covalently attached to the apoproteins. The tetrapyrroles are not complexed with metal ions, but being noncyclic, can be readily manipulated by the apoprotein to produce the biologically relevant characteristics.

The several types of biliproteins are distributed in various ways in different organisms. In cyanobacteria and red algae, several types of biliproteins can occur in each organism. The question arises of how the different biliproteins can be organized to achieve efficient energy transfer to chlorophyll *a*. The answer to this is found in the organization of all these biliproteins into discrete organelles that very effectively control the flow of energy. These organelles, the phycobilisomes, were shown by Gantt et al.[1] to be highly ordered protein structures designed to function with high efficiency. A different situation is the case for the cryptomonad biliproteins, since apparently only a single biliprotein occurs in each organism (Table 1). Here phycobilisomes have not been discovered and the question remains to be answered: how are the pigments organized for efficient energy absorption and distribution? In the cryptomonads, chlorophyll *c* is present in addition to chlorophyll *a* which occurs in all three groups.

Biliproteins are named primarily for their colors. Phycoerythrins have absorption maxima between 498 and 568 nm and are red, the blue phycocyanins absorb at lower energies, and allophycocyanins absorb at still lower energies (Table 2). The letter prefixes in their names refer to the types of organisms in which they were discovered. For the cryptomonad biliproteins, the numbers after the names refer to approximate absorption maxima. One of the names used here is not found in the literature: CU-phycoerythrin. CU-Phycoerythrin is the most recently discovered group of phycoerythrins and is prefixed "CU" to indicate its cyanobacterial origins and the presence of a urobilin-like chromophore not found in C-phycoerythrin ("C" indicates cyanobacteria). Allophycocyanin is divided into two functional types: the predominant form which absorbs with a 650-nm maximum and fluoresces at 660 nm and the less abundant form which fluoresces at about 680 nm. This latter type, allophycocyanin 680, is the usual ultimate transfer agent of apparently all the excitation energy from the phycobilisome to the chlorophyll. Allophycocyanin 680 activity is found in two proteins, allophycocyanins I and B.

These absorption and fluorescence characteristics are produced by the chemical nature of the chromophores and by the interactions between chromophores and their surroundings. All biliproteins have either phycocyanobilin or phycoerythrobilin, and one, R-phycocyanin ("R" indicates red alga), has both (Table 3). In addition there are perhaps three minor bilins — phycourobilin, cryptoviolin, and the 697-nm bilin — which occur together with a major type on the same protein. Much less is known about these three chromophores, but all five have unique spectral properties. Even a single type of chromophore can be spectrally flexible, and phycoerythrobilin occurs with a maximum at 565 or 545 nm depending on its interactions within the aproprotein.

What is an antenna pigment? Photosynthesis is mediated by two pigment systems, photosystems I and II, and both systems must be excited to achieve efficient photosynthetic output. Antenna pigments — which depending on the organism may be chlorophylls, car-

Table 1
BILIPROTEIN DISTRIBUTIONS FOR SELECTED CYANOBACTERIA, RED ALGAE, AND CRYPTOMONADS

Classification	Representative species	Biliproteins[a]
Cyanobacteria	*Lyngbya lagerheimii*[2]	C-Phycocyanin, allophycocyanin, allophycocyanin 680[b]
	Tolypothrix tenuis[3]	C-Phycoerythrin, C-phycocyanin, allophycocyanin, allophycocyanin 680
	Gloeobacter violaceus[4]	CU-Phycoerythrin, C-phycocyanin, allophycocyanin
	Anabaena variabilis[5]	Phycoerythrocyanin, C-phycocyanin, allophycocyanin, allophycocyanin 680
Red algae	*Porphyridium cruentum*[2,6,7]	B- and b-Phycoerythrin, R-phycocyanin, allophycocyanin, allophycocyanin 680
	Cyanidium caldarium[8,9]	C-Phycocyanin, allophycocyanin, allophycocyanin 680
	Porphyra perforata[2]	R-Phycoerythrin,[c] R-phycocyanin, allophycocyanin, allophycocyanin 680
	P. tenera[2,3,10,11]	R-Phycoerythrin, C-phycocyanin, allophycocyanin, allophycocyanin 680
	P. naiadum[12,13]	B-Phycoerythrin, C-phycocyanin, allophycocyanin, allophycocyanin 680
	Ceramium rubrum[14,15]	R-Phycoerythrin,[c] R-phycocyanin, allophycocyanin, allophycocyanin 680
Cryptomonads	*Chroomonas* sp.[16]	Phycocyanin 645
	Rhodomonas lens[17,18]	Phycoerythrin 545
	Cryptomonas ovata[16,17]	Phycoerythrin 566
	Hemiselmis virescens (Plymouth)[18]	Phycocyanin 612
	H. rufescens[18]	Phycoerythrin 555
	Cryptomonas magnifica[19]	Phycocyanin 630

[a] The letter prefixes (C, CU, B, b, and R) on the names of the biliproteins originally arose to distinguish the type of organism the biliprotein was found in, e.g., R stood for red algae. It became apparent, however, that some biliproteins originally found in a certain phylum might also occur in another, e.g., C-phycocyanin and allophycocyanin. The prefixes are retained as designations for spectral types. The cryptomonad biliproteins are named for the approximate wavelength of their absorption maxima in nanometers, e.g., phycocyanin 645.

[b] Although allophycocyanin 680 has not been isolated from all of the above organisms, Gantt et al.[20] have proved that the 680-nm emission is characteristic of many phycobilisome-containing cyanobacteria and red algae. It has not yet been detected in cryptomonads.

[c] R-Phycoerythrin has either two or three maxima in its visible absorption spectrum.

otenoids, xanthophylls, or biliproteins — are the vital components that supply the energy to achieve a concerted operation of this dual system. Antenna pigments also may strongly absorb photons having energies that are poorly harvested by chlorophyll *a*.

The efficiency of photosynthesis can be measured as the number of oxygen molecules given off per quantum of light absorbed. Emerson and Lewis[33] first demonstrated that quantum efficiency was not constant across the visible spectrum of a cyanobacterium. Particularly noteworthy was a decline in the efficiency of O_2 evolution in the far-red region where chlorophyll *a* absorbed exclusively ("red-drop" effect). The high yield they found in the region of C-phycocyanin absorption was definite proof that this biliprotein was a photosynthetically active pigment. Haxo and Blinks[34] followed this approach with red algae and likewise demonstrated that in several algae, phycoerythrin and phycocyanin were efficient, whereas chlorophyll and the carotenoids were very much less so. Later, Emerson et al.[35] discovered that the red-drop phenomenon can be overcome and the photosynthetic efficiency improved. This was accomplished when far-red light absorption was coupled with absorption by biliproteins at higher energies. These data provided early evidence for the

Table 2
TYPICAL VISIBLE ABSORPTION AND FLUORESCENCE EMISSION MAXIMA FOR PURIFIED BILIPROTEINS AT pH VALUES NEAR NEUTRAL

Biliprotein[a]	Absorption bands[b] (nm)	Fluorescence emission bands (nm)
C-Phycocyanin	615	647
R-Phycocyanin	555, 617	637
Allophycocyanin	598(sh), 629(sh), 650	660
Allophycocyanin 680	B-618, 671; I-610, 650, 667(sh)	673
Phycoerythrocyanin	535(sh), 575, 595(sh)	625
R-Phycoerythrin	498, 545, 568	578
B-Phycoerythrin	498(sh), 546, 565	578
b-Phycoerythrin	545, 563	570
C-Phycoerythrin	565	577
CU-Phycoerythrin	495, 547, 562	573
Phycocyanin 612	575, 612	634
Phycocyanin 630	583, 630	
Phycocyanin 645	585, 625(sh), 645	660
Phycoerythrin 545	545, 560(sh)	585
Phycoerythrin 555	555	
Phycoerythrin 566	566, 620(sh)	617

[a] Prefixes and suffixes used here are explained in Table 1.
[b] sh is the abbreviation for shoulder.

Table 3
THE DISTRIBUTION OF BILINS AMONG THE BILIPROTEINS

Bilins (chromophores)[a]	Biliproteins
Phycocyanobilin	Allophycocyanin,[21] C-phycocyanin,[22] allophycocyanin 680-tentative[23-25]
Phycoerythrobilin	C-Phycoerythrin,[26] b-phycoerythrin[6]
Phycocyanobilin and phycoerythrobilin	R-Phycocyanin[20,21]
Phycocyanobilin and cryptoviolin	Phycoerythrocyanin,[5] phycocyanin 612[19]
Phycoerythrobilin and cryptoviolin	Phycoerythrin 545[27]
Phycoerythrobilin and phycourobilin	B-Phycoerythrin,[28] R-phycoerythrin,[29] CU-phycoerythrin[30,31]
Phycocyanobilin, cryptoviolin, and 697-nm bilin	Phycocyanin 645[32]

[a] The bilins are the tetrapyrrole chromophores found covalently attached to the biliproteins. The five bilins listed are determined by their visible absorption spectra in acidic urea. Each of these bilins has different absorption spectra under conditions where the solvent has greatly reduced the effect of apoprotein on the spectrum of the bilin. While it is assumed here that the five bilins have different covalent structures, this is unproven and a matter for future experimentation. The 697-nm bilin is found only on a subunit of phycocyanin 645.

existence of two photosystems, both of which must be activated by absorbed light to be maximally effective.

Biliproteins were the most successful of all the antenna pigments in fulfilling the second role, that of shutting the window of low chlorophyll *a* absorption which extended between 450 and 670 nm (Figure 1). The three biliproteins presented in Figure 1 are actual components

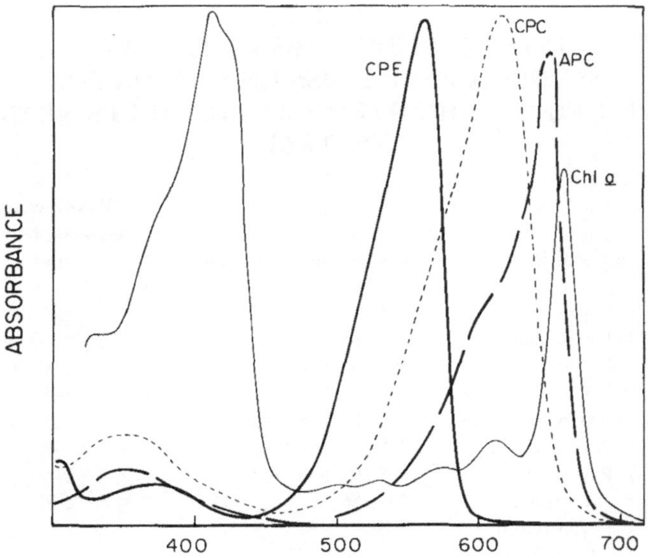

FIGURE 1. Absorption spectra of allophycocyanin (APC), C-phycocyanin (CPC), C-phycoerythrin (CPE), and chlorophyll (Chl *a*). Chlorophyll *a* is dissolved in acetone, and the purified biliproteins are in pH 6 to 7 sodium phosphate buffer.

of phycobilisomes in certain cyanobacteria. Allophycocyanin is almost always a minor component. The amounts of phycoerythrin and phycocyanin vary depending on the organism and in some organisms, on the growth conditions. The amounts of different biliproteins can vary depending on the amount of growth light and its spectral distribution. In some organisms this provides a facile and complex mechanism to adjust to changes in the environment. As noted in Table 2, this range of absorbances (Figure 1) can be extended to even shorter wavelengths for R-, B-, and CU-phycoerythrins which have phycourobilin (498-nm maximum) in addition to phycoerythrobilin.

The structure of the phycobilisome exactly follows the energy levels of the different biliproteins, and the biliproteins of highest-energy absorption are farthest from the photosynthetic membrane. Thus energy absorbed by phycoerythrin, which is located farthest from the membrane, flows through a phycocyanin conduit to the allophycocyanins, which are in direct contact with the membrane.[1]

The principal mechanism of this excitation energy migration is a type of resonance transfer called very weak coupling of dipoles.[36] Energy can be transferred very efficiently by this method. Factors involved in this mechanism include the distance between donor and acceptor chromophores and a modified spectral overlap between the fluorescence emission of the donor and the absorption of the acceptor chromophore. Excitation energy is transferred, not only between heterobiliproteins but also between chromophores on the same biliprotein. Teale and Dale[37] showed that all phycocyanins and phycoerythrins isolated from cyanobacteria and red algae were composed of two functional classes of chromophores, sensitizing and fluorescing. The sensitizing chromophores absorbed photons at higher energies and transferred them to the fluorescing chromophores. When isolated from other members of the excitation energy transfer chain, fluorescing chromophores do three things: receive energy from sensitizing chromophores, absorb photons directly, and fluoresce. When the isolated biliproteins were attached to the next membrane of the chain, the fluorescing chromophores transferred their energy rather than dissipated it via fluorescence. A major criterion used to predict the sensitizing chromophores was fluorescence polarization spectroscopy. For a single type of chromophore a flat polarization spectrum was predicted across the absorption spec-

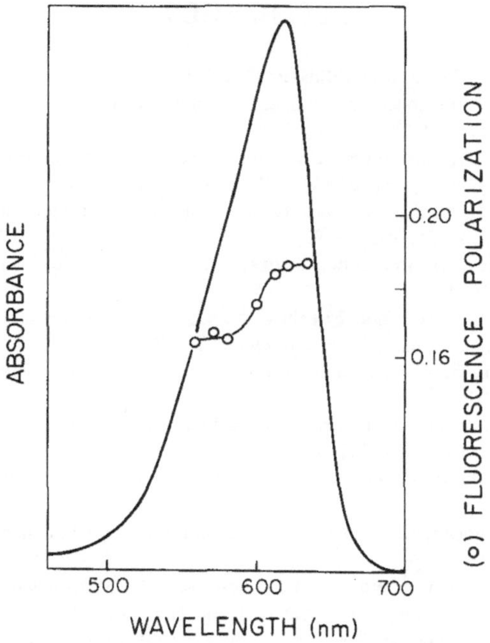

FIGURE 2. Absorption and fluorescence polarization spectra of C-phycocyanin. C-Phycocyanin is from the cyanobacterium *Phormidium luridum* and is in pH 6.0, 0.1-ionic strength sodium phosphate buffer. Experiments are performed at room temperature.

trum of the first excited state. However, even for C-phycocyanin, which had only one type of bilin, a two-plateau polarization spectrum was typically observed, suggesting sensitizing and fluorescing chromophores (Figure 2). The two chromophore types for C-phycocyanin were most likely different conformational states induced by differences in their interactions with the apoprotein.

Biliproteins are complex assemblies of apoproteins (polypeptides) and chromophores (tetrapyrroles or bilins) that interact in several ways to modulate the energy levels of the chromophores. In the phycobilisomes, further interactions were produced by several apparently nonpigmented linker polypeptides which control and stabilize the various assembly stages.[38] Finally, biliproteins must physically attach to and efficiently transfer energy into the thylakoid membranes whose bilayers contain the chlorophyll-protein complexes that complete the photosynthetic matrix of these organisms. In one species of red alga, it has been shown that the light energy absorbed by B-phycoerythrin was transmitted directly to photosystem II, but only 5% of the light absorbed by chlorophyll *a* was sent to photosystem II. Nonetheless, to maintain a proper balance between the two photosystems, there was at least a 55% yield of energy that was subsequently transferred from photosystem II to photosystem I. These energy transfer parameters were sensitive to growth light.[39]

REFERENCES

1. **Gantt, E., Lipschultz, C. A., and Zilinskas, B.,** Further evidence for a phycobilisome model from selective dissociation, fluorescence emission, immunoprecipitation, and electron microscopy, *Biochim. Biophys. Acta,* 430, 375, 1976.
2. **Haxo, F., O'hEocha, C., and Norris, P.,** Comparative studies of chromatographically separated phycoerythrins and phycocyanins, *Arch. Biochem. Biophys.,* 54, 162, 1955.
3. **Hattori, A. and Fujita, Y.,** Spectroscopic studies on the phycobilin pigments obtained from blue-green and red algae, *J. Biochem.,* 46, 903, 1959.
4. **Rippka, R., Waterbury, J., and Cohen-Bazire, G.,** A cyanobacterium which lacks thylakoids, *Arch. Microbiol.,* 100, 419, 1974.
5. **Bryant, D. A., Glazer, A. N.,, and Eiserling, F. A.,** Characterization and structural properties of the major biliproteins of *Anabaena* sp., *Arch. Microbiol.,* 110, 61, 1976.
6. **Gantt, E. and Lipschultz, C. A.,** Phycobilisomes of *Porphyridium cruentum:* pigment analysis, *Biochemistry,* 13, 2960, 1974.
7. **Koch, W.,** Untersuchungen an bakterienfreien Massenkulturen der einzelligen Rotalge *Porphyridium cruentum* Naegeli, *Arch. Mikrobiol.,* 18, 232, 1953.
8. **Allen, M. B.,** Studies with *Cyanidium caldarium,* an anomalously pigmented chlorophyte, *Arch. Mikrobiol.,* 32, 270, 1959.
9. **Nichols, K. E. and Bogorad, L.,** Studies on phycobilin formation with mutants of *Cyanidium caldarium, Nature,* 188, 870, 1960.
10. **Fujiwara, T.,** Studies on chromoproteins in Japanese Nori (*Porphyra tenera*). I. A new method for the crystallization of phycoerythrin and phycocyanin, *J. Biochem.,* 42, 411, 1955.
11. **Lemberg, R.,** Chromoproteide der Rotalgen. II. Splatung mit Pepsin und Sauren. Isolierung eines Pyrrolfarbstoffs, *Liebigs Ann. Chem.,* 477, 195, 1930.
12. **Airth, R. L. and Blinks, L. R.,** A new phycoerythrin from *Porphyra naiadum, Biol. Bull.,* 111, 321, 1956.
13. **French, C. S., Smith, J. H. C., Virgin, H. I., and Airth, R. L.,** Fluorescence-spectrum curves of chlorophylls, pheophytins, phycoerythrins, phycocyanins and hypericin, *Plant Physiol.,* 31, 369, 1956.
14. **Kylin, H.,** Über Phykoerythrin und Phykocyan bei *Ceramium rubrum* (Huds.) Ag., *Hoppe-Seyler's Z. Physiol. Chem.,* 69, 169, 1910.
15. **O'hEocha, C.,** Phycobilins, in *Physiology and Biochemistry of Algae,* Lewin, R. A., Ed., Academic Press, New York, 1962, chap. 25.
16. **Allen, M. B., Dougherty, E. C., and McLaughlin, J. J. A.,** Chromoprotein pigments of some cryptomonad flagellates, *Nature,* 184, 1047, 1959.
17. **Haxo, F. T. and Fork, D. C.,** Photosynthetically active accessory pigments of cryptomonads, *Nature,* 184, 1051, 1959.
18. **O'hEocha, C. and Raftery, M.,** Phycoerythrins and phycocyanins of cryptomonads, *Nature,* 184, 1049, 1959.
19. **O'hEocha, C. O'Carra, P., and Mitchell, D.,** Biliproteins of cryptomonad algae, *Proc. R. Irish Acad.,* 63B, 191, 1964.
20. **Gantt, E., Lipschultz, C. A., Grabowski, J., and Zimmerman, B. K.,** Phycobilisomes from blue-green and red algae, *Plant Physiol.,* 63, 615, 1979.
21. **Chapman, D. J., Cole, W. J., and Siegelman, H. W.,** Chromophores of allophycocyanin and R-phycocyanin, *Biochem. J.,* 105, 903, 1967.
22. **O'hEocha, C.,** Spectral properties of the phycobilins. I. Phycocyanobilin, *Biochemistry,* 2, 375, 1963.
23. **Glazer, A. N. and Bryant, D. A.,** Allophycocyanin B (λmax 671, 618 nm): a new cyanobacterial phycobiliprotein, *Arch. Microbiol.,* 104, 15, 1975.
24. **Lundell, D. J., Yamanaka, G., and Glazer, A. N.,** A terminal energy acceptor of the phycobilisome: the 75,000-dalton polypeptide of *Synechococcus 6301* phycobilisomes — a new biliprotein, *J. Cell Biol.,* 91, 315, 1981.
25. **Redlinger, T. and Gantt, E.,** Phycobilisome structure of *Porphyridium cruentum.* Polypeptide composition, *Plant Physiol.,* 68, 1375, 1981.
26. **O'Carra, P., O'hEocha, C., and Carroll, D. M.,** Spectral properties of the phycobilins. II. Phycoerythrobilin, *Biochemistry,* 3, 1343, 1964.
27. **MacColl, R., Guard-Friar, D., and Csatorday, K.,** Chromatographic and spectroscopic analysis of phycoerythrin 545 and its subunits, *Arch. Microbiol.,* 135, 194, 1983.
28. **O'hEocha, C.,** Biliproteins, in *Biochemistry of Chloroplasts,* Vol. 1, Goodwin, T. W., Ed., Academic Press, New York, 1966, 407.
29. **O'hEocha, C. and O'Carra, P.,** Spectral studies of dentured phycoerythrins, *J. Am. Chem. Soc.,* 83, 1091, 1961.

30. **Bryant, D. A., Cohen-Bazire, G., and Glazer, A. N.,** Characterization of the biliproteins of *Gloeobacter violaceus:* chromophore content of a cyanobacterial phycoerythrin carrying phycourobilin chromophore, *Arch. Microbiol.*, 129, 190, 1981.
31. **Kursar, T. A., Swift, H., and Alberte, R. S.,** Morphology of a novel cyanobacterium and characterization of light-harvesting complexes from it: implications for phycobiliprotein evolution, *Proc. Natl. Acad. Sci. U.S.A.*, 78, 6888, 1981.
32. **Glazer, A. N. and Cohen-Bazire, G.,** A comparison of cryptophytan phycocyanins, *Arch. Microbiol.*, 104, 29, 1975.
33. **Emerson, R. and Lewis, C. M.,** The dependence of the quantum yield of *Chlorella* photosynthesis on the wave length of light, *Am. J. Bot.*, 30, 165, 1943.
34. **Haxo, F. T. and Blinks, L. R.,** Photosynthetic action spectra of marine algae, *J. Gen. Physiol.*, 33, 389, 1950.
35. **Emerson, R., Chalmers, R., and Cederstrand, C.,** Some factors influencing the long-wave limit of photosynthesis, *Proc. Natl. Acad. Sci. U.S.A.*, 43, 133, 1957.
36. **Förster, T.,** Transfer mechanisms of electronic excitation, *Discuss. Faraday Soc.*, 27, 7, 1959.
37. **Teale, F. W. J. and Dale, R. E.,** Isolation and spectral characterization of phycobiliproteins, *Biochem. J.*, 116, 161, 1970.
38. **Tandeau de Marsac, N. and Cohen-Bazire, G.,** Molecular composition of cyanobacterial phycobilisomes, *Proc. Natl. Acad. Sci. U.S.A.*, 74, 1635, 1977.
39. **Ley, A. C. and Butler, W. L.,** Effects of chromatic adaptation on the photochemical apparatus of photosynthesis in *Porphyridium cruentum*, *Plant Physiol.*, 65, 714, 1980.

Chapter 2

PHYCOBILISOMES; A STRUCTURE-FUNCTION MODEL

I. DISCOVERY

Gantt and Conti,[1-3] in a progression of three publications, determined that the granules on the outer surface of the thylakoid membranes of red algae were the locations of the biliproteins. Prior to this discovery, the problem of biliprotein placement was a conflict between two seemingly opposite necessities. In order to be effective antenna pigments and to transfer their excitation energies efficiently into the thylakoid membranes, close proximity to the chlorophyll in these membranes was required. A location inside the thylakoids would thus be ideal; however, biliproteins were highly water soluble and easily extracted from damaged cells (typical nonmembrane protein behavior). The electron micrographs (Figure 1) of thin-sectioned *Porphyridium cruentum* provided Gantt and Conti with a perfect answer to the paradox. The biliproteins of this red alga were assembled into discrete organelles which were on the outer surface of the thylakoids. Their close contact with the membrane allowed for excitation energy transfer, while their lack of penetration into the lipid-protein bilayer allowed for easy removal and was no obstacle to water solubility.

In electron micrographs of *P. cruentum*, the thylakoid membrane appeared to be covered with granules about 32.0 nm in diameter.[1] The granules were regularly spaced and located on the outer (stroma) side of each membrane facing away from a pair of bilayers. The membrane that envelops the chloroplast was free of these granules. These organelles were never observed free and were in close contact with the membranes. Glutaraldehyde-fixed cells showed the granules best. Although other suggestions were also advanced, Gantt and Conti[1] proposed that these granules could be the sites for the biliproteins.

Next, they used electron microscopy in a variety of experiments which clearly suggested the biliprotein content of the granules.[1] The granules varied in size between 30.0 and 40.0 nm in diameter and had a center-to-center spacing along the membrane of 40.0 to 50.0 nm. The granules on adjacent rows were frequently seen as being interdigitated. Algal cells were then disrupted in a French® press and pelleted by centrifugation. The pelleted chloroplast fragments were fixed by glutaraldehyde and osmium tetroxide. Electron micrographs of this material still clearly showed the granules. This experiment proved that the granules and the membranes were connected. The chlorophyll of the red algal cells was then removed by extraction with a solution of 80% methanol and 20% acetone. Glutaraldehyde was used prior to the organic solvent extraction, and osmium tetroxide was applied after. The absorption spectrum of the extracted cells still exhibited the phycoerythrin spectrum, but all the chlorophyll was removed. Likewise, electron micrographs were totally devoid of thylakoid membranes, but the granules were still readily observed, as were the nucleus, chloroplast, and ribosomes. The inverse of this experiment was to remove the granules, while retaining the chloroplast thylakoid membranes. Gantt and Conti[2] reasoned that if the granules contained the biliproteins, the granules could be solubilized by detergent. They then treated the isolated thylakoid membrane with a detergent, deoxycholate. After glutaraldehyde treatment, control membranes not treated by deoxycholate were red, but the detergent-treated sample was green. Electron microscopy demonstrated that the granules were removed by the detergent treatment, but the membrane remained intact. The simultaneous removal of biliproteins and granules by detergent, while both remained after organic solvent extraction, was quite good evidence for the biliprotein being in the granules. Similar granules were also found in four marine red algae, *Spermothamnion, Bangia fusco-purpurea, Nemalion multifidum,* and *Antithamnion glanduliferum,* and in one fresh-water red alga, *Porphyridium aerugineum.* This

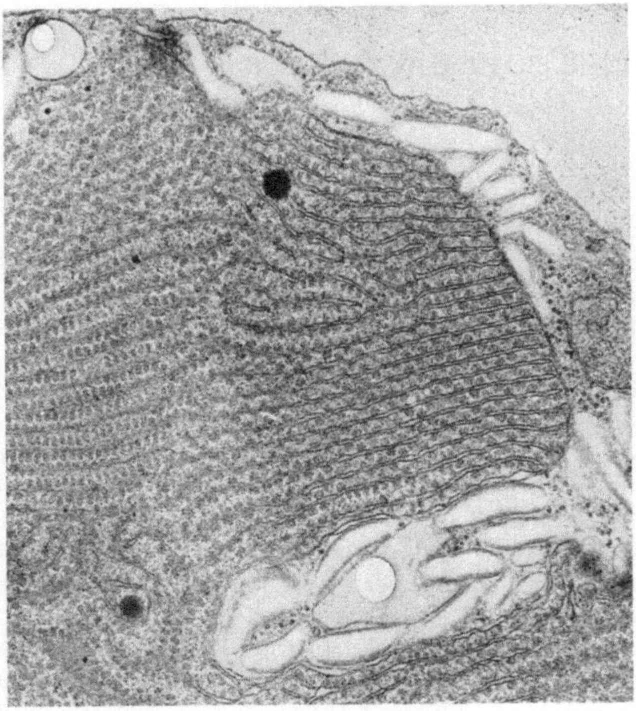

FIGURE 1. Electron micrograph of a thin section of *Porphyridium cruentum*. Face, edge, and grazing (top) views are observed (courtesy of M. R. Edwards). This is the prototype of the hemi-ellipsoidal phycobilisome. (Magnification × 64,000.)

latter alga had no phycoerythrin, and C-phycocyanin and allophycocyanin were its major biliproteins. These experiments pointed toward the granules being composed at least in part of biliproteins.

The final proof for the biliprotein nature of the granules was in their isolation.[3] The *P. cruentum* cells were put through a French® press in the presence of 0.1 *M* phosphate at pH 6.8. They were then treated with deoxycholate and glutaraldehyde. The extract was placed on a sucrose density gradient and centrifuged. A purple band which sedimented into the middle of the gradient had a B-phycoerythrin and R-phycocyanin content similar to intact cells, but chlorophyll was not present. Moreover, this purple band, when stained with phosphotungstic acid, contained particles which electron microscopy showed to have similar dimensions to the granules from the outer thylakoid surface. When glutaraldehyde treatment was omitted, the purple band was absent. The summation of this evidence allowed Gantt and Conti[3] to name the granules, the location of the biliproteins, the phycobilisomes.

II. TWO MAIN TYPES OF PHYCOBILISOMES

Thin sections of *P. aerugineum* exhibited some views of the phycobilisomes very similar to *P. cruentum*,[3] but other aspects were quite distinct. Gantt and Conti[3] demonstrated that these views could be analyzed into two structural types of phycobilisomes. These types are now called hemi-ellipsoidal and hemi-discoidal (Figure 2). These classes of phycobilisomes were derived from different planes of sectioning which reveal different views of the structures. When the plane of sectioning was perpendicular to the thylakoid surface, the phycobilisomes of both classes appeared as semicircles with the flat side on top of the membrane (the face

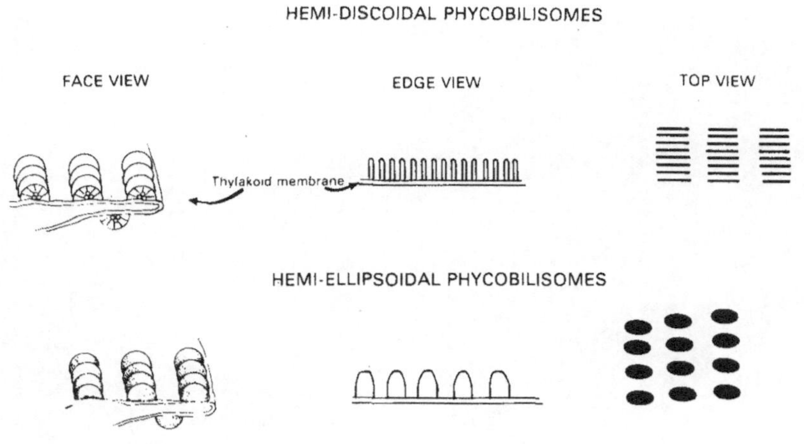

FIGURE 2. Schematic drawing of hemi-ellipsoidal and hemi-discoidal phycobilisomes. The face and edge views are perpendicular to each other. The top view is the projection that the phycobilisomes make on the surface of the thylakoid membrane. The thylakoids are actually entirely closed sacs or vesicles that are quite flattened. The phycobilisomes are located on the stroma or outer sides of the vesicles and do not occur in the intrathylakoid space. In red algae the thylakoids are found inside membrane-bound chloroplasts, but for the cyanobacteria, chloroplasts do not occur.

view). When the plane of sectioning was longitudinal, the phycobilisomes were dissimilar in the two classes. For *P. cruentum* the longitudinal sections showed much thicker phycobilisomes than the *P. aerugineum*. A top view which grazed over the thylakoids showed parallel rows of phycobilisomes similar in appearance to the longitudinal sections.

So far, only phycobilisomes from red algae have been discussed. Lefort[4] and Bourdu and Lefort[5] have produced electron micrographs of cyanobacterial endosymbionts, cyanelles which showed similar granules attached to the thylakoids. The granules in both *Cyanophora paradoxa* and *Glaucocystis nostochinearum* were observed to be of the hemi-discoidal types. By analogy to the work of Gantt and Conti[1-3] on red algal thylakoids, it was clear that the cyanobacterial granules discovered by Lefort[4] were also phycobilisomes. Edwards et al.[6] found phycobilisomes in a free-living cyanobacterium, *Synechococcus lividus*, and Gantt and Conti[7] found them in two cyanobacteria, *Tolypothrix tenuis* and *Fremyella diplosiphon*. A very comprehensive electron microscope study on cyanobacteria by Wildman and Bowen[8] showed that all 15 species examined had phycobilisomes, and all were of the hemi-discoidal class. Other researchers involved with red algae were finding granules now identified as phycobilisomes on the outer surface of the thylakoids.[9-12] Examples of various thin sections of cyanobacteria and red algae clearly showed the phycobilisomes (Figures 3 to 7).

III. ISOLATION OF PHYCOBILISOMES

The first isolation by Gantt and Conti[3] of a phycobilisome used the cross-linking agent glutaraldehyde to preserve the organelles. Although useful at that stage, such chemically altered materials had drawbacks. Gantt and Lipschultz[14] found that phycobilisomes from *P. cruentum* could be prepared without cross-linking. Triton® X-100 was the best detergent for gentle removal of the phycobilisomes from the membrane, and 0.50 *M* phosphate gave better stabilization than lower phosphate concentrations. The isolated phycobilisomes showed a long axis of 40.0 to 50.0 nm and a short axis of 30.0 to 32.0. These isolated preparations measure larger than the phycobilisomes from thin-sectioned cells of 30.0 to 40.0 nm. Gantt and Lipschultz[14] suggested that experimental differences between the treatment of thin-sectioned and detergent-extracted materials account for the discrepancy.

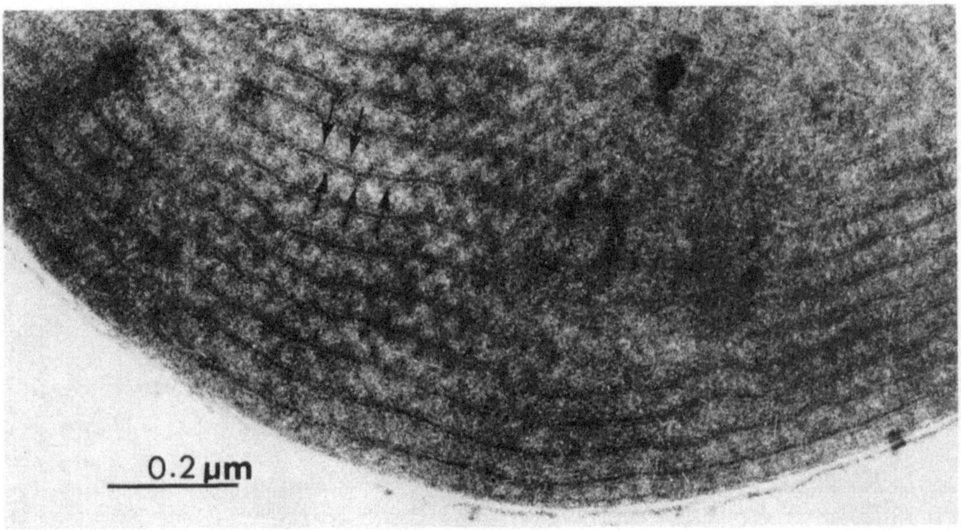

FIGURE 3. Electron micrograph of a thin section of the cyanelle *Cyanophora paradoxa* shows the face view of phycobilisomes (arrows point to some). (From Lefort, M., *C.R.*, 261, 233, 1965. With permission.)

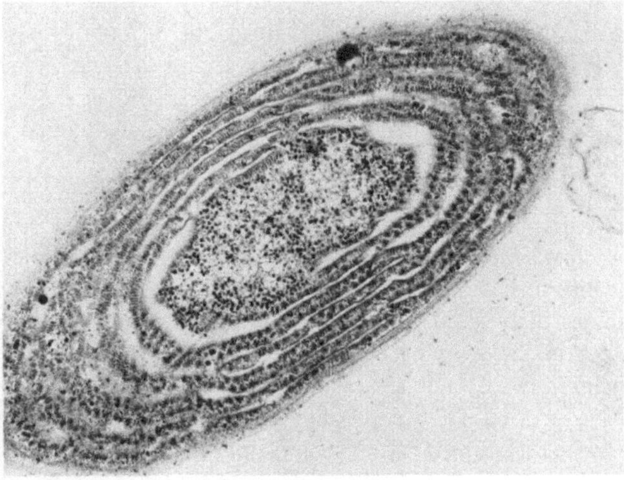

FIGURE 4. Electron micrograph of a thin section of the cyanobacterium *Synechococcus lividus*. Face, top, and edge views are observed. These are typical hemi-discoidal phycobilisomes. (Magnification × 27,000.) (From Edwards, M. R. and Gantt, E., *J. Cell Biol.*, 50, 896, 1971. With permission.)

The first preparation of isolated phycobilisomes from a cyanobacterium was by Gray et al.[15] from *Nostoc* sp. Isolation of these structures confirmed the designation of the granules on the thylakoid surface of cyanobacteria as being phycobilisomes.

Subsequently, Gantt et al.[16] have devised a more general protocol which allowed isolation of intact phycobilisomes from both red algae and cyanobacteria. This new method successfully produced phycobilisomes from 13 species of cyanobacteria and red algae. In addition a sensitive criterion for intactness was demonstrated to be fluorescence emission. Intact phycobilisomes emit with a maximum between 670 and 675 nm at room temperature and 678 to 685 at -196°C. This property of long-wavelength emission is referred to here as

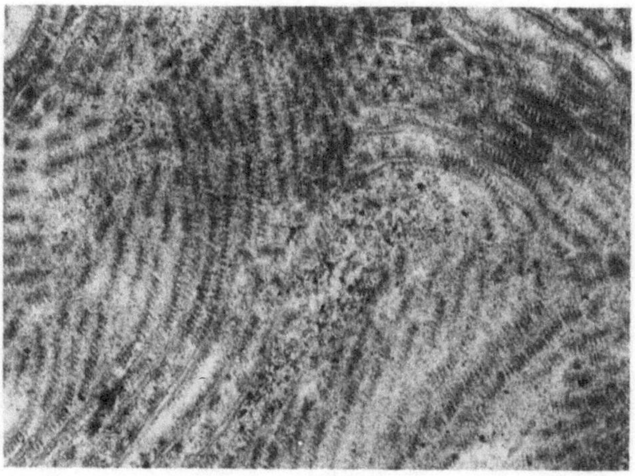

FIGURE 5. Electron micrograph of a thin section of the cyanobacterium *Oscillatoria splendida* showing phycobilisomes. Many views of grazing sections are shown. The micrograph is courtesy of J. C. Thomas. Additional electron micrographs of this organism have been published,[13] and some are shown in Chapter 8.

FIGURE 6. Electron micrograph of a thin section of the cyanobacterium *Microcoleus vaginatus* shows the face view of interdigitating phycobilisomes. The micrograph is courtesy of R. Wildman-Swenson. The bar = 100 nm. Reproduced with permission from the American Society for Microbiology.

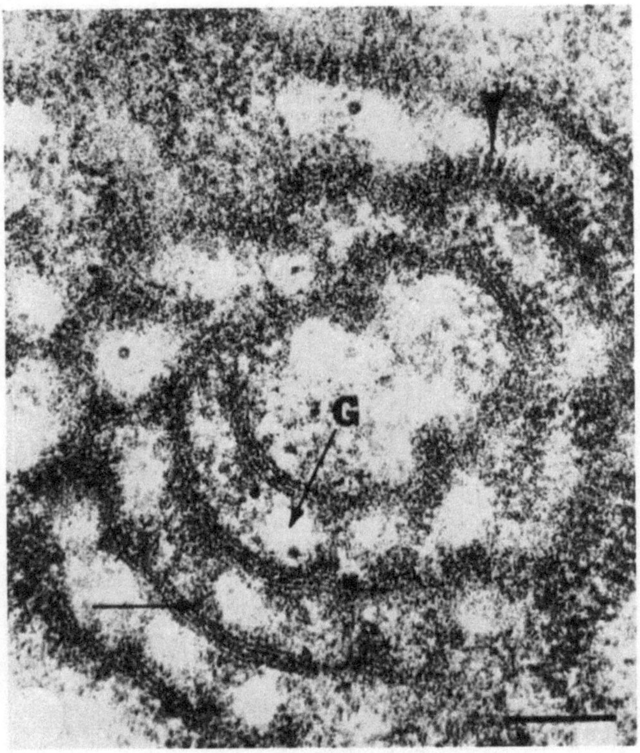

FIGURE 7. Electron micrograph of a thin section of the cyanobacterium *Nostoc muscorum* shows phycobilisomes in edge view (arrow head). The letter G marks a glycogen granule. The micrograph is courtesy of R. Wildman-Swenson. The bar = 100 nm. Reproduced with permission from the American Society for Microbiology.

allophycocyanin 680. When phycobilisomes were induced to dissociate, the emission maximum was shifted to lower wavelengths. This hypochromic shift was caused by uncoupling of the energy transfer chain that linked the biliproteins in each phycobilisome. The isolation procedure performed at 20 to 23°C (Figure 8), since low temperature causes dissociation, was as follows: cells were centrifuged and washed twice with a buffer of 0.75 *M* potassium phosphate at pH 6.8 to 7.0. This buffer was used throughout isolation. Cells (1 to 2 g in 10 mℓ of buffer) were put through a French® press, and Triton® X-100 was added to a final concentration of at least 2%. After being stirred for 20 min, the mixture was centrifuged for 30 min at 25,000 g. Aliquots of the supernatant were layered on a sucrose gradient and centrifuged in a fixed-angle rotor. A sucrose step gradient of 0.25, 0.5, 1.0, and 2.0 *M* buffered sucrose was used, and after 3 hr at 136,000 g the intact phycobilisomes were banded in the 1.0 *M* sucrose layer. The most useful criterion of intact phycobilisomes was their fluorescence emission maximum. If the phycobilisomes were intact, the room temperature emission should be above 670 nm and be relatively independent of excitation wavelength. Dissociation of the phycobilisomes was indicated by increased fluorescence emission from phycoerythrin (575 nm), phycocyanin (640 to 650 nm), and allophycocyanin (660 nm) as well as a loss of the 670- to 680-nm emission.

IV. PHYCOBILISOME ENERGY TRANSFER

The isolated phycobilisomes showed absorption spectra demonstrating the presence of all

Isolation of Phycobilisomes

(Carried out at 20-30°C)

Cells collected by centrifugation
rinsed twice in 0.75 M K-phosphate buffer (pH 6.8-7.0)

↓

1-2g wet weight suspended in 10 ml buffer

↓

disrupted in French pressure cell (10,000 psi)

Triton X-100 added, to final concentration of 2%
stirred 20 min.

↓

centrifuged at 25,000g for 30 min.

pellet
chlorophyll

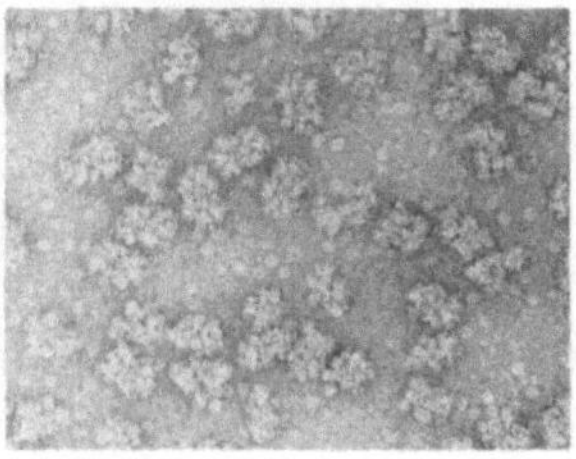

supernatant layered
on sucrose step gradient
(2M, 1M, 0.5M, 0.25M)
centrifuged 3h at
136,000 g

↓

phycobilisome fraction
collected from 1M sucrose
layer

FIGURE 8. Isolation procedure for purification of phycobilisomes. The electron micrograph of phycobilisomes from the red alga *Porphyridium cruentum* is courtesy of M. Kessel. (Magnification × 75,000.)

the biliproteins of the particular cyanobacterium or red alga. Assuming that each phycobilisome had a complete complement of the available biliproteins, the problem remained of how the energy absorbed by a particular biliprotein was delivered to the chlorophyll in the thylakoid membrane. Working with whole red algae and cyanobacteria, Duysens[17] and French and Young[18] found that energy absorbed by phycoerythrin was efficiently transferred to chlorophyll. Phycocyanin acted apparently as an intermediary in this transfer. Although their finding predated the discovery of the phycobilisome, this turned out an important clue in describing its structure-function relationships. Duysens[17] examined the fluorescence spectra of the red algae *Porphyra lacineata* and *Porphyridium cruentum*. The organisms were excited at either 420 or 546 nm, which were the major wavelengths of chlorophyll *a* or phycoerythrin absorbance, respectively. When the phycoerythrin was excited, chlorophyll *a* emission was stronger than it would have been if chlorophyll *a* had been excited directly. The appearance of phycocyanin emission upon phycoerythrin excitation established phycocyanin as a possible intermediate to chlorophyll *a*. Energy transfer from phycocyanin to chlorophyll *a* was also established by exciting the organisms at 600 nm, primarily a wavelength of phycocyanin absorbance. In the cyanobacterium, *Oscillatoria* sp., C-phycocyanin was likewise shown to transfer its excitation energy to chlorophyll *a*.

French and Young[18] excited cells of *P. cruentum* at several wavelengths and recorded the emission spectra. After excitation at wavelengths of 515 and 546 nm, where B-phycoerythrin mainly absorbed, both R-phycocyanin and chlorophyll *a* were observed to fluoresce. The appearance of the action spectra of chlorophyll *a* emission suggested a strong contribution from B-phycoerythrin. It was also noted that excitation of B-phycoerythrin resulted in greater chlorophyll *a* emission than direct excitation of chlorophyll *a* itself. The observation that R-phycocyanin received excitation energy from B-phycoerythrin suggested to French and Young that phycocyanin was an intermediate in excitation energy transfer from B-phycoerythrin to chlorophyll *a*. The function of the antenna pigments to transfer energy to chlorophyll *a* also implied, by these early experiments, that chlorophyll *a* should have a specific role in photosynthesis other than the collection of solar radiation.

Gantt and Lipschultz[19] examined the fluorescence spectra of isolated phycobilisomes from *P. cruentum*. Since the purified phycobilisomes were devoid of chlorophyll *a*, this experiment allowed characterization of the component of the phycobilisome which was the direct transfer agent to chlorophyll *a*. Excitation of B-phycoerythrin at 545 nm resulted in a major emission by the intact phycobilisomes at 675 to 680 nm. The known components of the phycobilisome at this time and their respective emission maxima were B-phycoerythrin, 575 nm; b-phycoerythrin, 570; R-phycocyanin, 636; and allophycocyanin, 660. When the phycobilisomes were placed in a low-phosphate solvent, they started to dissociate. The 680-nm emission from the 545-nm excitation steadily declined, and emission at 575 nm increased correspondingly. This relationship between 680- and 575-nm emission was caused by the coupling of phycoerythrin to the 680-nm final emitter in whole phycobilisome. As the organelles dissociated, the energy formerly transferred from phycoerythrin was now given off as characteristic phycoerythrin emission. Since allophycocyanin was the longest wavelength emitter of the phycobilisome components, Gantt and Lipschultz[19] assigned it the role of final emitter, although its emission maximum was 10 to 20 nm to the blue of the phycobilisome. Gantt and Lipschultz[19] were correct in crediting allophycocyanin as a link in the chain of energy transfer from phycoerythrin to phycocyanin to allophycocyanin to chlorophyll *a*. However, it was later found that special forms of allophycocyanin which occurred in very low amounts were the actual final emitters. Gantt et al.[16] subsequently found this long-wavelength emission to be a general characteristic of intact phycobilisomes.

V. A STRUCTURE-FUNCTION RELATIONSHIP

It seemed plausible that there should be some ordering in the biliproteins of the phycobilisomes that would allow a maximum efficiency in the excitation energy transfer chain. Gantt et al.[20] studied isolated phycobilisomes from the cyanobacterium *Nostoc* sp. which had the useful property of approximately equal amounts of the three major pigments, C-phycoerythrin, C-phycocyanin, and allophycocyanin. Antiserum was prepared against each of the purified proteins. The antiserums of biliproteins were totally specific to the class of biliproteins they were prepared against. Antiserum against b-phycoerythrin will cross-react with all phycoerythrins (B, b, R, and C) but will not react with any phycocyanin or allophycocyanin. Antiserum to phycocyanin or allophycocyanin behaved identically. Intact and dissociated phycobilisomes were tested against each of the three antiserums, and the time of observable precipitation was recorded. For dissociated phycobilisomes, the three antiserums produced visible precipitates within the same time period. This was at least partially caused by the approximately equal concentrations of the three biliproteins. For intact phycobilisomes, however, the experiment yielded considerably different results. Precipitation reactions of phycoerythrin or allophycocyanin antiserum occurred much sooner than those of phycocyanin antiserum. Likewise, reactivity of antiserum against phycocyanin was much slower for the intact phycobilisome than for the dissociated phycobilisome. Gantt et al.[20] interpreted these results to mean that the C-phycocyanin antigenic sites were blocked when the phycobilisomes were intact. The antigenic sites on allophycocyanin and C-phycoerythrin were clearly exposed. One possible explanation for these results would have the C-phycocyanin sandwiched between layers of allophycocyanin on one side and C-phycoerythrin on the other.

Gantt and Lipschultz[21] used a different immunochemical protocol to study this question of biliprotein organization within the phycobilisome. They prepared phycobilisome-containing vesicles and performed a series of immunochemical investigations directly on electron microscope grids. The experimental procedure consisted of reacting the phycobilisome vesicle preparations with rabbit antiserum (to biliproteins). The reaction was marked with ferritin-conjugated goat antirabbit antibodies because the ferritin particles were easily observed in the electron microscope. When antiserum against B-phycoerythrin was used, a strong positive reaction was observed. Antiserum to allophycocyanin and R-phycocyanin gave a much weaker positive reaction with intact phycobilisomes than did antiserum to B-phycoerythrin. When the phycobilisomes were allowed to dissociate and selectively lose their phycoerythrin content, both antiserums to allophycocyanin and R-phycocyanin gave stronger positive reactions. They interpreted these experiments to signify that in the intact phycobilisome vesicle complexes, both R-phycocyanin and allophycocyanin were covered by either the membrane or the phycoeyrthrin. Phycoerythrin was clearly on the outside of the phycobilisomes.

Each of the immunochemical experiments described above[20,21] separately contributed to the understanding of the phycobilisome structure, but taken together their value was synergistic. It was shown by Gantt et al.[20] that the antigenic sites of phycocyanin were blocked in isolated phycobilisomes. An interesting reason would be that one side was coated with phycoerythrin and the other with allophycocyanin, but a mixture of both biliproteins on either side could not be excluded rigorously. The latter explanation was disproven by the immuno-electron microscopy experiments which showed both allophycocyanin and R-phycocyanin to be covered by phycoerythrin.[21] Since phycoerythrin (B and b) was on the outside, it was reasonable to propose that allophycocyanin may be exclusively on the inside in contact with the thylakoid membrane.

As these experiments progressed, Gantt and co-workers gradually formulated a very clear relationship between the structure and the energy transfer function of the phycobilisome. The Gantt phycobilisome model[19,20] proposed that certain biliproteins were located in homologous zones of varying distance from the thylakoid. Allophycocyanin was contiguous to the bilayer, phycoerythrin (or phycoerythrocyanin) was on the periphery, and phycocyanin was between the allophycocyanin and phycoerythrin. This was a perfect structure to foster the excitation energy transfer criterion that energy should flow from high to low. Fluorescence experiments[17-19] endorsed this idea and demonstrated that the energy absorbed by the highest-energy absorber (phycoerythrin) was transferred stepwise first to C- or R-phycocyanin, then to allophycocyanin, and finally to the chlorophyll *a* in the membrane:

$$\text{Phycoerythrin} \rightarrow \text{Phycocyanin} \rightarrow \text{Allophycocyanin} \rightarrow \text{Chlorophyll } a$$

VI. ELECTRON MICROSCOPY OF ISOLATED DISKS, STACKS, AND PHYCOBILISOMES

The primary C-phycocyanin aggregate which was obtained by several methods of purification at low-phosphate concentrations in the pH range of about 4.6 to 6.0 was an 11S hexamer. Berns and Edwards,[22] using potassium phosphotungstate and uranyl acetate as negative stains, found 11S C-phycocyanin to be a disk-like structure composed of six subunits in a ring-like array (Figure 9) with a diameter of about 12.5 nm. Gantt[23] studied a comparable B-phycoerythrin aggregate and found a disk of 10.1-nm diameter and 5.4-nm thickness. These disks were likely candidates for assembly intermediates in the phycobilisomes.

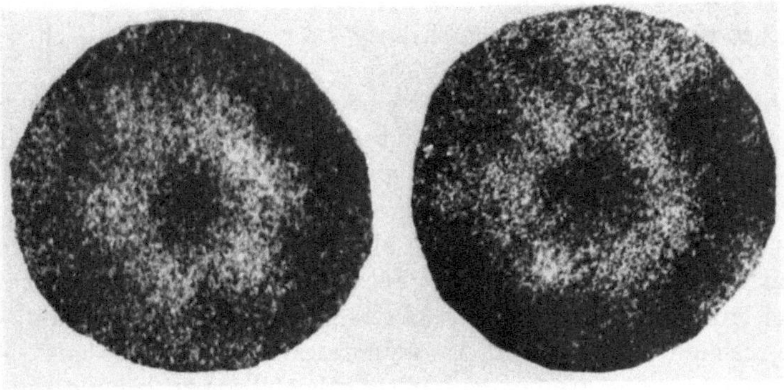

FIGURE 9. Electron micrographs of negatively stained 11S C-phycocyanin. These greatly magnified views show aggregates with six spherical segments. The 11S aggregate is a hexamer ($\alpha_6\beta_6$). (Magnification × 1,900,000.) (From Berns, D. S. and Edwards, M. R., *Arch. Biochem. Biophys.*, 110, 511, 1965. With permission.)

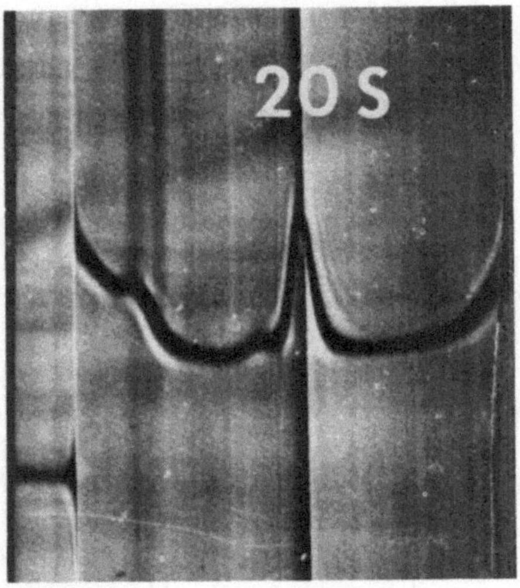

FIGURE 10. Ultracentrifuge pattern of 20S C-phycocyanin. Sedimentation is from left to right. The 20S aggregates are obtained from fresh extracts of red light-grown *Calothrix parietina* into pH 6.0, 0.1-ionic strength sodium phosphate buffer. The 20S aggregates are compiled of three hexameric disks.

Kessel et al.[24] studied the biliproteins of *Calothrix parietina* immediately after they had been extracted into low-phosphate buffer at pH 6.0. The freshly extracted and nonpurified samples of biliproteins from cells grown in red light which greatly depleted the C-phycoerythrin content showed a 20S C-phycocyanin aggregate in the ultracentrifuge (Figure 10) and a stack of three disks in the electron microscope (Figure 11). It is estimated from the sedimentation coefficient that the 20S aggregate was large enough to encompass three hexamers. The isolated stacks showed a specific face-to-face aggregation. The hexamers although clearly linked in the stacks appear to retain some individual identity.

Mörschel et al.[25] isolated phycobilisomes from the red alga, *Rhodella violacea.* Thin sections of this alga in the electron microscope showed typical hemi-discoidal phycobilisomes. Previously, the only isolated phycobilisomes carefully examined by electron mi-

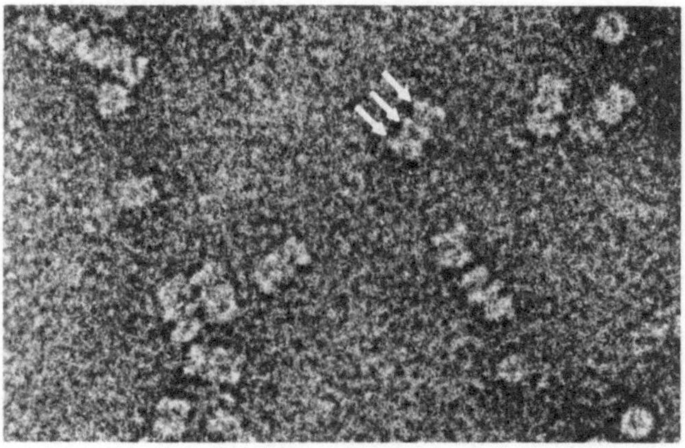

FIGURE 11. Electron micrograph of negatively stained 20S C-phycocyanin. Some views show the 20S aggregate to be a stack of three disks in a face-to-face alignment. (Magnification × 40,000.) (Reproduced with permission of the National Research Council of Canada.)

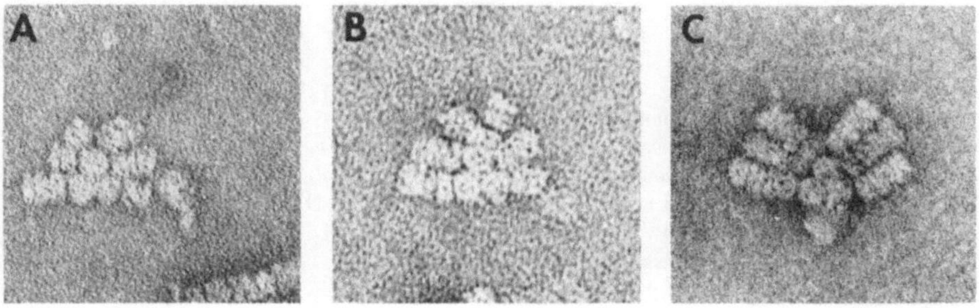

FIGURE 12. Electron micrographs of negatively stained phycobilisomes with triangular cores. Phycobilisomes were isolated from the cyanobacterium *Fremyella diplosiphon* (courtesy of M. C. Ledbetter). Triangular cores with six rod-like projections are seen on each micrograph. (Magnification × 330,000.) (From Rosinski, J., Hainfeld, J. F., Rigbi, M., and Siegelman, H. W., *Ann. Bot.*, 47, 1, 1981. With permission.)

croscopy were the hemi-ellipsoidal variety. The simpler hemi-discoidal organelles showed a highly ordered structure. The phycobilisomes, composed of B-phycoerythrin, C-phycocyanin, and allophycocyanin, were observed to consist of two distinct structural parts. In the center of the long, flat base of the semicircular face were three ring-shaped aggregates in a triangular array, and emanating from the triangle were six stacks or rods. Each stack seemed to consist of three disks assembled face-to-face. The stacks of three disks were highly suggestive of the aggregates observed from extracts of *C. parietina*.[24] Bryant et al.[26] have subsequently found very similar phycobilisomes in several cyanobacteria. Face views showed the cores to have rods attached which are stacks of two to six disks. Although the triangular cores were most common (Figure 12), phycobilisomes with only two circular aggregates in their cores were also detected (Figure 13) depending on the source.[27,28]

Synechocystis sp. 6701 was grown in red or green light to modify the relative amounts of C-phycoerythrin and C-phycocyanin in the phycobilisomes.[26] Red light which was primarily absorbed by C-phycocyanin produced a molar ratio of C-phycocyanin to C-phycoerythrin of 6, and green light mainly absorbed by C-phycoerythrin resulted in a ratio of 0.9. In this cyanobacterium, the ratio of the sum of molar concentration C-phycocyanin plus

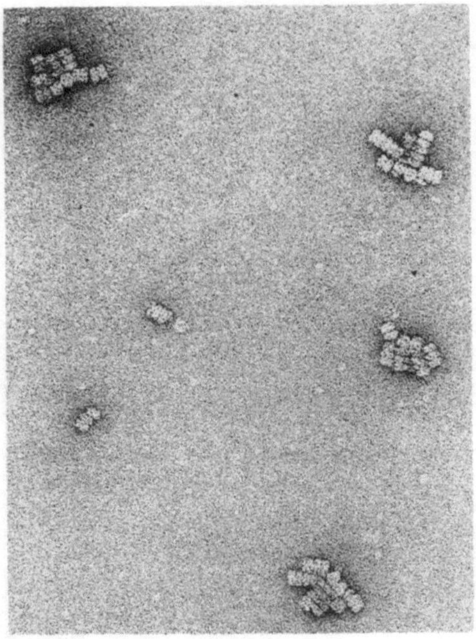

FIGURE 13. Electron micrograph of negatively stained phycobilisomes with a two-cylinder core. Phycobilisomes are isolated from the cyanobacterium *Anacystis nidulans* (courtesy of R. C. Williams). (Magnification × 160,000.) (From Yamanaka, G., Glazer, A. N., and Williams, R. C., *J. Biol. Chem.*, 255, 11,004, 1980. With permission.)

C-phycoerythrin divided by the molar concentration of allophycocyanin was 1.9 for red and 3.7 for green light-grown cells. The method used by the cells to change these allophycocyanin ratios was variable stack length. The ratio of the length of green to red light-grown stacks in these phycobilisomes was 1.6. This suggested that the stacks may not contain allophycocyanin and that the triangular cores, which did not appear to change in red or green light, were adjacent to the membrane and were very logical locations for the allophycocyanin.

Bryant et al.[26] obtained a few views of phycobilisomes with their bottom edge up. The middle of this membrane-adjacent edge showed two of the three circular structures of the triangular core. The circular units of the triangular core were each about 12.0 nm in diameter and were actually stacks of two disks, each about 6.0 to 7.0 nm thick. In an additional characterization of the core, Anderson et al.[29] obtained a UV mutant of *Synechocystis* sp. 6701, which was defective in phycobilisome assembly. Only a small amount of C-phycocyanin was made, and none was observed attached to the triangular core. The stacks and the triangular cores were obtained entirely unattached. It was therefore an excellent system for core study. Electron micrographs of both the face and edge views were obtained (Figure 14). The basic measurements for the triangle were a 22.0-nm-long edge and a thickness of 14.0 nm. This was in excellent agreement with Bryant et al.[26] Mörschel et al.[30] have studied purified allophycocyanin trimers by electron microscopy and found the aggregates to be 10.0 × 3.0 nm. There was space for 12 trimers of this size in each triangular core if significant space for other components was neglected. The electron micrographs of the core (Figure 14) suggested that these 12 trimers may be paired into 6 hexamers. Bryant et al.[26] pointed out that a phycobilisome of 6 hexamers in the core and 6 stacks on the periphery suggested a possible one-to-one relationship. The stacks in the Bryant et al.[26] model were staggered around the core, while for Mörschel et al.[25] the stacks were coplanar.

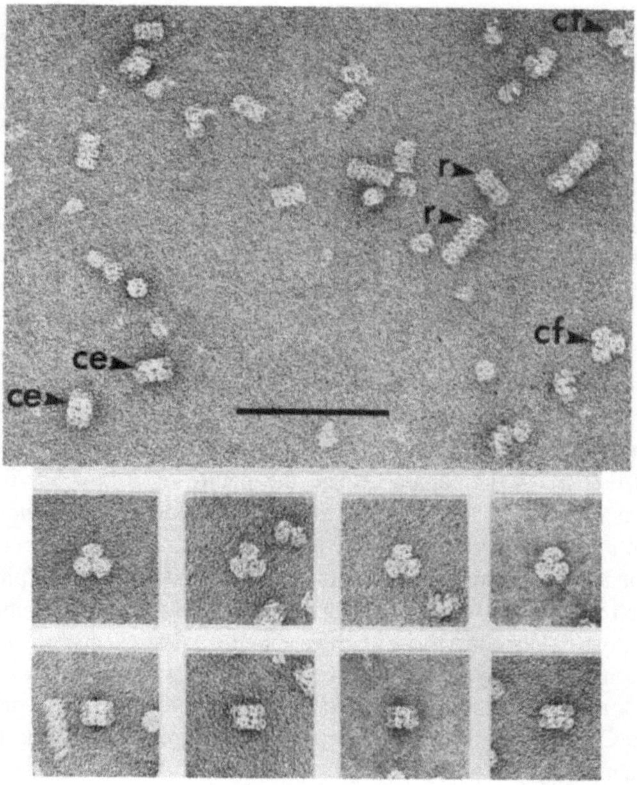

FIGURE 14. Electron micrographs of separated stacks and cores. These particles are a fraction from a sucrose density gradient of a preparation from an UV-induced mutant of the cyanobacterium *Synechocystis* sp. No intact phycobilisomes are obtained from this mutant which is defective in C-phycocyanin synthesis. The letters in the top field signify rods, r; core face view, cf; and core edge view, ce. The galleries are face and edge views (bottom row) of the cores (courtesy of L. K. Anderson). Bar = 100 nm. (From Anderson, L. K., Rayner, M. C., and Eiserling, F. A., *Arch. Microbiol.*, 138, 237, 1984. With permission.)

VII. A MODEL

It has been suggested previously that the core was an obvious location for allophycocyanin and the stacks for phycocyanin and phycoerythrin. This, of course, needed to be verified. Also, was each stack a hetero-rod of phycocyanin and phycoerythrin? Or are some of the rods pure phycocyanin and others pure phycoerythrin? Koller et al.[31] performed a series of experiments designed to answer these questions. They prepared purified stacks of the phycobilisomes from the red alga *R. violacea*. These stacks were about 2:1 in their B-phycoerythrin to C-phycocyanin ratio. The stacks were prepared by passing the phycobilisomes through a Sephadex G-25 column in distilled water and then adjusting the biliprotein fractions to 1 m*M* phosphate at pH 7.0. The samples were kept at 4°C for about 12 hr. The samples were then centrifuged at 14°C through a continuous sucrose gradient in 0.65 *M* phosphate. Of the four biliprotein bands obtained, the fastest sedimenting was shown by electron microscopy to be a stack of three face-to-face disks having a length of 15.0 nm and a diameter of 8.7 nm. The integrity of the individual hexamers was at least partially retained, since a clear line with no apparent protein was observed between each pair of disks in the stacks.

These purified stacks showed a loss of the absorbance at 652 nm which was characteristic of allophycocyanin. The fluorescence emission that occurred at 644 nm indicated that C-phycocyanin and not allophycocyanin was the final emitter.[31] Measurements of the spectra of these preparations indicated a molar ratio of two B-phycoerythrins to one C-phycocyanin, and Koller et al. suggested that two disks in the stack of three were B-phycoerythrins. The stacks were reacted with the cross-linking reagent glutaraldehyde, and the partially dissociated stacks were separated on gel electrophoresis. Seven bands were observed and three of these contained both phycocyanin and phycoerythrin. The coexistence of both phycocyanin and phycoerythrin in the same stack was therefore proven. The ratio of phycoerythrin to phycocyanin was 61:23% in the phycobilisomes and 71:29% in the stacks of three disks. These measurements suggested but were not sufficiently close to prove that all the phycocyanin and phycoerythrin were in the stacks. On the basis of C-phycocyanin emission being observed after B-phycoerythrin excitation, Koller et al.[31] suggested that the hetero-stacks were polar, having C-phycocyanin on one end and B-phycoerythrin on the other.

Lipschultz and Gantt[32] further characterized the nature of B-phycoerythrin and C-phycocyanin as hetero-aggregates. A complex from the red alga *Porphyridium sordidum* was shown by electron microscopy to be composed of two disks. The pigment content suggested the stacks might be an $\alpha_6\beta_6$-B-phycoerythrin disk connected to an $\alpha_6\beta_6$-C-phycocyanin disk. The complex, for reasons that were not clear, dissociated in a 0.10 *M* phosphate sucrose gradient at 4°C into a B-phycoerythrin aggregate and a wide C-phycocyanin zone. When the B-phycoerythrin and C-phycocyanin were recombined, the complex reformed. When first mixed, the fluorescence emission of the individual components was observed. However, within 30 min the C-phycocyanin emission had greatly increased, and the B-phycoerythrin emission had concurrently fallen. This kinetics linkage of the two emissions clearly demonstrated the polar nature of the complex and proved excitation energy transfer from phycoerythrin to phycocyanin within the stacks. The percentage of reassociation was greater when larger molecular weight C-phycocyanin fractions from the gradient were used. These reassociative fractions in addition to C-phycocyanin contained a 30,000-mol wt colorless polypeptide. The functions of these colorless polypeptides will be discussed later (see Chapter 5).

The Gantt phycobilisome model[19,20] discussed above was confirmed, and many pertinent details were added from the studies of the electron microscopy of hemi-discoidal phycobilisomes[25,26] and the properties of isolated stacks.[31,32] Since phycocyanin and phycoerythrin were found in the stacks, while allophycocyanin was not, Koller et al.[31] suggested the triangular core for the site of allophycocyanin. Since electron micrographs of thin-sectioned cells and isolated phycobilisomes indicated the triangular cores to be adjacent to the thylakoid membrane, the proposal of the Gantt model[19,20] that allophycocyanin was closer to chlorophyll than phycocyanin or phycoerythrin was proven. A schematic model (Figure 15) put forward by Bryant et al.[26] and Mörschel et al.[25] and now confirmed by the work on isolated stacks[31,32] showed a structural arrangement of the three major biliproteins in a typical hemi-discoidal arrangement that was energetically sound. No single model, of course, was a real representation of phycobilisomes. Their shapes varied from discoidal to ellipsoidal, their stack lengths were influenced by conditions of growth, and their sizes differed from organism to organism. Their diameters may vary from 33.0 nm for *R. violacea* to 58.0 nm for a *Lyngbya-Plectonema-Phormidium* group 7409. Hemi-discoidal organelles were about 7.0 to 8.0 nm thick, and the hemi-ellipsoidal variety was two to four times thicker. In addition, there were two cyanobacteria that had phycobilisomes which did not conform to the hemi-discoidal or hemi-ellipsoidal classes.[33,34] The electron micrograph of *Porphyridium cruentum* (Figure 1) clearly showed the three characteristic views of a hemi-ellipsoidal phycobilisome, and the electron micrograph[35] of *Synechococcus lividus* (Figure 4) illustrated the face, edge, and the top views of a typical hemi-discoidal phycobilisome.

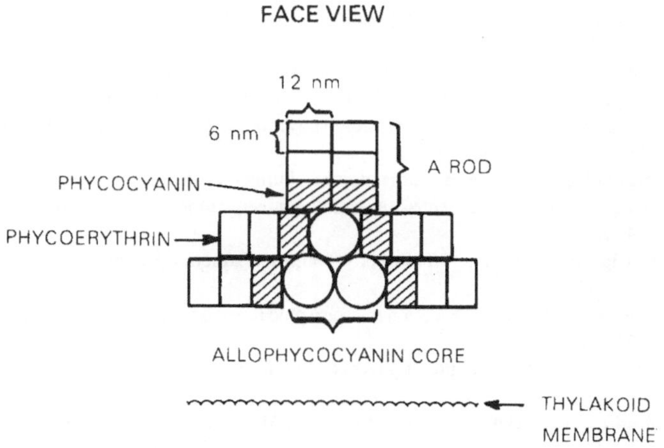

FIGURE 15. Schematic model of a hemi-discoidal phycobilisome. Each disk is a hexameric aggregate, usually of phycocyanin or phycoerythrin. In many representations in the literature the hexameric disks which comprise the phycobilisome rods are subdivided in half, but this is not shown here. The core is composed usually of three cylinders which include allophycocyanin and allophycocyanin 680. Colorless polypeptides are distributed in both rods and the core.

REFERENCES

1. **Gantt, E. and Conti, S. F.**, The ultrastructure of *Porphyridium cruentum*, *J. Cell Biol.*, 26, 365, 1965.
2. **Gantt, E. and Conti, S. F.**, Granules associated with the chloroplast lamellae of *Porphyridium cruentum*, *J. Cell Biol.*, 29, 123, 1966.
3. **Gantt, E. and Conti, S. F.**, Phycobiliprotein localization in algae, *Brookhaven Symp. Biol.*, 19, 393, 1966.
4. **Lefort, M.**, Sur le chromatoplasma d'une Cyanophycée endosymbiotique: *Glaucocystis nostochinearum* Itzigs, *C.R.*, 261, 233, 1965.
5. **Bourdu, R. and Lefort, M.**, Structure fine, observée en cryodècapage, des lamelles photosynthétiques des Cyanophycées endosymbiotiques: *Glaucocystis nostochinearum* Itzigs, et *Cyanophora paradoxa* Korschikoff, *C.R.*, 265, 37, 1967.
6. **Edwards, M. R., Berns, D. S., Ghiorse, W. C., and Holt, S. C.**, Ultrastructure of the thermophilic blue-green alga, *Synechococcus lividus* Copeland, *J. Phycol.*, 4, 283, 1968.
7. **Gantt, E. and Conti, S. F.**, Ultrastructure of blue-green algae, *J. Bacteriol.*, 97, 1486, 1969.
8. **Wildman, R. B. and Bowen, C. C.**, Phycobilisomes in blue-green algae, *J. Bacteriol.*, 117, 866, 1974.
9. **Myers, A., Preston, R. D., and Ripley, G. W.**, Fine structure in red algae. I. X-ray and electronmicroscope investigation of *Griffithsia flosculosa*, *Proc. R. Soc. London Ser. B*, 144, 450, 1956.
10. **Bisalputra, T. and Bisalputra, A.-A.**, The occurrence of DNA fibrils in chloroplasts of *Laurencia spectabilis*, *J. Ultrastruct. Res.*, 17, 14, 1967.
11. **Peyrière, M.**, Les problèmes cytologiques de *Rytiphlea tinctoria* algue rouge à floridorubine, *C.R.*, 266, 2253, 1968.
12. **Lichtlé, C. and Giraud, G.**, Ètude ultrastructurale de la zone apicale du thalle du *Polysiphonia elongata* (Harv.) Rhodophycée, Floridée. Évolution des plastes, *J. Microsc.*, 8, 867, 1969.
13. **Lichtlé, C. and Thomas, J. C.**, Étude ultrastructuale des thylacoides des algues à phycobiliproteines, comparaison des résultats obtenus par fixation classique et cryodècapage, *Phycologia*, 15, 393, 1976.
14. **Gantt, E. and Lipschultz, C. A.**, Phycobilisomes of *Porphyridium cruentum*. I. Isolation, *J. Cell Biol.*, 54, 313, 1972.
15. **Gray, B. H., Lipschultz, C. A., and Gantt, E.**, Phycobilisomes from a blue-green alga *Nostoc* species, *J. Bacteriol.*, 116, 471, 1973.
16. **Gantt, E., Lipschultz, C. A., Grabowski, J., and Zimmerman, B. K.**, Phycobilisomes from blue-green and red algae, *Plant Physiol.*, 63, 615, 1979.

17. **Duysens, L. N. M.,** Transfer of light energy within the pigment systems present in photosynthesizing cells, *Nature,* 168, 548, 1951.
18. **French, C. S. and Young, V. K.,** The fluorescence spectra of red algae and the transfer of energy from phycoerythrin to phycocyanin and chlorophyll, *J. Gen. Physiol.,* 35, 873, 1952.
19. **Gantt, E. and Lipschultz, C. A.,** Energy transfer in phycobilisomes from phycoerythrin to allophycocyanin, *Biochim. Biophys. Acta,* 292, 858, 1973.
20. **Gantt, E., Lipschultz, C. A., and Zilinskas, B.,** Further evidence for a phycobilisome model from selective dissociation, fluorescence emission, immunoprecipitation, and electron microscopy, *Biochim. Biophys. Acta,* 430, 375, 1976.
21. **Gantt, E. and Lipschultz, C. A.,** Probing phycobilisome structure by immuno-electron microscopy, *J. Phycol.,* 13, 185, 1977.
22. **Berns, D. S. and Edwards, M. R.,** Electron micrographic investigations of C-phycocyanin, *Arch. Biochem. Biophys.,* 110, 511, 1965.
23. **Gantt, E.,** Properties and ultrastructure of phycoerythrin from *Porphyridium cruentum, Plant Physiol.,* 44, 1629, 1969.
24. **Kessel, M., MacColl, R., Berns, D. S., and Edwards, M. R.,** Electron microscope and physical chemical characterization of C-phycocyanin from fresh extracts of two blue-green algae, *Can. J. Microbiol.,* 19, 831, 1973.
25. **Mörschel, E., Koller, K. P., Wehrmeyer, W., and Schneider, H.,** Biliprotein assembly in the disc-shaped phycobilisomes of *Rhodella violacea.* I. Electron microscopy of phycobilisomes in situ and analysis of their architecture after isolation and negative staining, *Cytobiologie,* 16, 118, 1977.
26. **Bryant, D. A., Guglielmi, G., Tandeau de Marsac, N., Castets, A.-M., and Cohen-Bazire, G.,** The structure of cyanobacterial phycobilisomes: a model, *Arch. Microbiol.,* 123, 113, 1979.
27. **Rosinski, J., Hainfeld, J. F., Rigbi, M., and Siegelman, H. W.,** Phycobilisome ultrastructure and chromatic adaptation in *Fremyella diplosiphon, Ann. Bot.,* 47, 1, 1981.
28. **Yamanaka, G., Glazer, A. N., and Williams, R. C.,** Molecular architecture of a light-harvesting antenna. Comparison of wild type and mutant *Synechococcus* 6301 phycobilisomes, *J. Biol. Chem.,* 255, 11,004, 1980.
29. **Anderson, L. K., Rayner, M. C., and Eiserling, F. A.,** Ultra-violet mutagenesis of *Synchocystis* sp. 6701: mutations in chromatic adaptation and phycobilisome assembly, *Arch. Microbiol.,* 138, 237, 1984.
30. **Mörschel, E., Koller, K.-P., and Wehrmeyer, W.,** Biliprotein assembly in the disc-shaped phycobilisomes of *Rhodella violacea* electron microscopical and biochemical analyses of C-phycocyanin and allophycocyanin aggregates, *Arch. Microbiol.,* 125, 43, 1980.
31. **Koller, K.-P., Wehrmeyer, W., and Mörschel, E.,** Biliprotein assembly in the disc-shaped phycobilisomes of *Rhodella violacea.* On the molecular composition of energy-transferring complexes (tripartite units) forming the periphery of the phycobilisome, *Eur. J. Biochem.,* 91, 57, 1978.
32. **Lipschultz, C. A. and Gantt, E.,** Association of phycoerythrin and phycocyanin: in vitro formation of functional energy transferring phycobilisome complex of *Porphyridium sordidum, Biochemistry,* 20, 3371, 1981.
33. **Guglielmi, G., Cohen-Bazire, G., and Bryant, D. A.,** The structure of *Gloeobacter violaceus* and its phycobilisomes, *Arch. Microbiol.,* 129, 181, 1981.
34. **Isono, T. and Katoh, T.,** Cylindrical phycobilisomes from a blue-green alga, *Anabaena variabilis, Plant Cell Physiol.,* 2, 1347, 1982.
35. **Edwards, M. R. and Gantt, E.,** Phycobilisomes of the thermophilic blue-green algae *Synechococcus lividus, J. Cell Biol.,* 50, 896, 1971.

Chapter 3

BILIPROTEINS: COVALENT STRUCTURES

I. INTRODUCTION

Phycobilisomes are composed of proteins: biliproteins and linker polypeptides. Here the covalent chemical structures of the biliproteins isolated from these phycobilisomes will be considered. The composition of biliprotein subunits, the distribution of bilins (chromophores) among the subunits, the tetrapyrrole structures of the various bilins, the covalent bonding between tetrapyrrole and apoprotein, and the sequences of the amino acids in these biliproteins will be discussed.

Individual biliproteins are readily obtained from the phycobilisomes by either suspending the latter directly or by suspending lysed cells in a buffer having a low-phosphate ion concentration. The dissociated phycobilisomes are composed of various size protein aggregates which are then purified.

II. SUBUNIT STRUCTURE

Sodium dodecyl sulfate (SDS) gel electrophoresis experiments first clearly demonstrated the possibility that all the purified biliproteins were composed of at least two subunits (Table 1). Generally, the lower molecular weight subunit was designated α and the larger β. Amino acid analysis of the α and β subunits of C-phycocyanin showed that they were distinct gene products.[14] B-Phycoerythrin and R-phycoerythrin had in addition a still larger subunit or group of subunits called γ.[7,10-13] An earlier gel filtration experiment using Sephadex G-200 under denaturing conditions in 8 *M* urea at pH 3.0 found an elution pattern for R-phycoerythrin that agreed extremely well with the presence of an α, β, and γ subunit structure.[15] Although it was noted that there are multiple types of γ subunits, the stoichiometry of isolated B- and R-phycoerythrins was $\alpha_6\beta_6\gamma$ so that the various γ subunits must occur on different protein aggregates. Much less is currently known concerning the γ subunits compared to those of α and β.

Bennett and Bogorad[4] have shown that the α and β subunits occur in equal molar amounts in these proteins. The subunits were not joined by disulfide bonds and were invariably pigmented (Table 2). The chromophore compositions of the cryptomonad biliproteins, phycocyanins 612 and 645, are shown in Table 2 for comparison purposes and will not be discussed here in detail. So far, phycocyanin 645 is the only biliprotein to have the 697-nm bilin. This bilin was named for its absorption maximum in acidic urea, and little else is known concerning its properties.

III. CHROMOPHORE CONTENT

The characteristic absorption maxima and absorptivities in acidic urea were used to obtain the number and the type of bilin on a biliprotein or its subunits. Bennett and Bogorad[4] first used urea to estimate the ratio of phycoerythrobilins on C-phycoerythrin subunits. Glazer and Fang[16] subsequently refined the technique to estimate the actual number of chromophores on C-phycocyanin and allophycocyanin. They noted that the absorption of C-phycocyanin is 1.5 times that of an equal amount of allophycocyanin in acidic urea. These data were used to propose that there are three phycocyanobilins on C-phycocyanin and two on allophycocyanin and to estimate the absorptivity of phycocyanobilin in acidic urea. The spectroscopic data on acidic urea of the separated subunits agreed with those for whole protein.

Table 1
SUBUNIT STRUCTURE OF BILIPROTEINS FROM SDS POLYACRYLAMIDE GEL ELECTROPHORESIS

Biliproteins	Group	Subunits	Mol wt	Ref.
Allophycocyanin	Cyanobacteria	α	17,000	1
		β	19,000	
	Red algae	α	16,300	2
		β	17,900	
C-Phycocyanin	Cyanobacteria	α	15,900—18,500	3—6
		β	18,300—20,000	
	Red algae	α	17,000	2
		β	19,500	
C-Phycoerythrin	Cyanobacteria	α	17,600—20,000	3—6
		β	19,000—22,000	
R-Phycocyanin	Red algae	α	16,400	3, 7
		β	18,400	
CU-Phycoerythrin	Cyanobacteria	α	20,500	8
		β	21,700	
Phycoerythrocyanin	Cyanobacteria	α	16,000	9
		β	20,200—20,500	
R-Phycoerythrin	Red algae	α	19,000—21,000	10,
		β	19,000—21,600	11
		γ_1	30,000—31,000	
		γ_2	33,000	
B-Phycoerythrin	Red algae	α	17,500	12,
		β	17,500	13
		γ_1	30,000	
		γ_2	29,500	
		γ_3	29,000	
b-Phycoerythrin	Red algae	α	17,500	13
		β	17,500	

Table 2
CHROMOPHORE DISTRIBUTION ON THE BILIPROTEIN SUBUNITS

Biliproteins	Chromophores[a] per subunit α	β	γ_1	γ_2	Ref.
C-Phycocyanin	1 PCB	2 PCB			16
Allophycocyanin	1 PCB	1 PCB			16
C-Phycoerythrin	2 PEB	3 or 4 PEB			17—19
R-Phycoerythrin	2 PEB	2 or 3 PEB, 1 PUB	3 PEB, 2 PUB[b]	1 or 2 PEB, 1 PUB[b]	10, 19
B-Phycoerythrin	2 PEB	3 PEB	2 PEB, 2 PUB		12, 13, 18, 20
R-Phycocyanin	1 PCB	1 PCB, 1 PEB			17
Phycoerythrocyanin	1 CV	2 PCB			9
CU-Phycoerythrin	3 PEB	3 PEB, 1 PUB			8, 21
Phycocyanin 645	1 697B	2 PCB, 1 CV			22
Phycocyanin 612	1 PCB	2 PCB, 1 CV			23

[a] PCB, phycocyanobilin; PEB, phycoerythrobilin; PUB, phycourobilin; CV, cryptoviolin; and 697B, 697-nm bilin.

[b] For R-phycoerythrin from *Gastroclonium coulteri*, two subunits were reported to occur, and the average bilin content was one PEB and three PUB per subunit.[19]

Table 3
ABSORPTIVITIES OF BILINS[a] IN ACIDIC UREA

	Molar absorptivity (M^{-1} cm^{-1})			
Wavelength (nm)	Phycoerythrobilin[18,19]	Phycourobilin[19]	Phycocyanobilin[16]	Cryptoviolin[9]
495	13,400	94,000		
550	45,500—53,700[b]	0[c]	6,000	
590			16,200	43,000
662	0		35,500	5,700

[a] The 697-nm bilin occurs only in cryptomonad biliproteins, and there is no direct measurement of its molar absorptivity. An assumption is usually made that its absorptivity at its maximum is the same as phycocyanobilin but red shifted 35 nm in acidic urea.

[b] The 53,700 M^{-1} cm^{-1} value is the most recent estimate[19].

[c] Since phycourobilin only occurs together with phycoerythrobilin, this is an estimate.

The subunits of C-phycocyanin were separated by chromatography on a column of Bio-Rex® 70, a cation ion exchange resin. The pH of the solvents which contained 0.01 *M* 2-mercaptoethanol and formic acid was 1.9. The α and β subunits were separated using a stepwise urea gradient. The α-subunit eluted at 8 *M* urea and the β subunit at 9 *M*. The assignment of two phycocyanobilins to β and one to α was established by the approximately double absorptivity of the chromophore absorption band of the β to the α subunit.

A typical acidic urea solvent used for the spectroscopic measurements was 8.0 *M* urea at pH 3.0 containing 0.01 *M* 2-mercaptoethanol. Estimates of the absorptivities of phycocyanobilin, phycoerythrobilin, cryptoviolin, and phycourobilin in acidic urea are reported in Table 3. This technique brought rapid progress and good results (Table 2). There are dangers, however, in its indiscriminate use due to bleaching of bilins in denatured biliproteins and erroneous absorptivities.

Glazer and Hixson[13] used the acidic urea and spectral techniques on B-phycoerythrin isolated from *Porphyridium cruentum*. A phycoerythrobilin content for the protein ($\alpha_6\beta_6\gamma$) was found to be 38 with 2 phycourobilins by their calculations. They argued that in order to achieve 38 phycoerythrobilins, each α had 2 phycoerythrobilins and each β had 4, leaving 2 phycoerythrobilins and 2 phycourobilins for γ. However, Muckle and Rüdiger[18] determined a different absorptivity for phycoerythrobilin. They recalculated the data of Glazer and Hixson to be 5.25 total phycoerythrobilins per αβ unit or 2 per α and 3 per β as the better fit. Finally, it was found by Lundell et al.[20] that only 5 chromopeptides could be isolated from the α and β subunits of B-phycoerythrin, and the distribution of 2 per α and 3 per β was assigned to B-phycoerythrin (Table 2). A second problem arose with the γ subunit of this same protein. Glazer and Hixson[13] reported 2 phycoerythrobilins and 2 phycourobilins on γ, while Redlinger and Gantt[12] found only phycourobilin on γ. Following these discoveries, Klotz and Glazer[19] have evaluated the absorptivities of phycoerythrobilin and phycourobilin. They suggested that the formerly reported chromophore contents[13,17] of B-, R-, and C-phycoerythrins were incorrect.

Guard-Friar and MacColl[24] have analyzed the possible pitfalls of the acidic urea method for proteins containing phycocyanobilin and phycocyanobilin together with cryptoviolin. Protein concentration, pH, and mercaptoethanol were all shown to affect the bleaching of bilins in acidic urea. High biliprotein concentrations and pH values near 2 promoted retention of chromophore absorbance. Reducing agents like mercaptoethanol or dithiothreitol promoted loss of bilin absorbance. These effects were fairly slow, and it would be expected that after a few hours it would still be possible to obtain the correct chromophore content. The case of phycocyanobilin plus cryptoviolin was quite difficult, since the mercaptoethanol reduction of phycocyanobilin produced a modified tetrapyrrole which overlapped spectrally with cryptoviolin. The combined effect of phycocyanobilin absorption decreasing and cryptoviolin absorption apparently increasing due to the formation of a new chemically produced bilin could readily produce an erroneous ratio of phycocyanobilin to cryptoviolin. Glazer and Hixson[13] pointed out that measurements on whole protein can be made rapidly after dissolving lyophilized biliprotein in acidic urea. Therefore this measurement was intrinsically more reliable than measurements on subunits which have been exposed to acidic urea during chromatographic separation. This argument was valid, but the caveat must be made that if results from whole protein differ from separated subunits, it would be advantageous to search for an experimental protocol that gave comparable data for both. Methods of subunit separation deleting mercaptoethanol and operating near pH 2 were optimal.[24] In addition, the current absorptivities were only estimates, especially for cryptoviolin, phycourobilin, and the 697-nm bilin.

The chromophore content of some biliproteins is variable. The CU-phycoerythrins vary as a class in their phycoerythrobilin to phycourobilin composition. Bryant et al.[8] found a 6:1 ratio of phycoerythrobilin to phycourobilin in CU-phycoerythrin from *Gloeobacter violaceus* (Table 2). Kursar et al.[21] found a 4:1 ratio for the biliprotein from *Synechococcus* sp. strain DC2. Fujita and Shimura[25] reported a CU-phycoerythrin spectrum from yet a third cyanobacterium *Trichodesmium thiebautii*. Because it differed from the two already discussed, it probably represented a third ratio for the two bilins. Ong et al.[26] reported a CU-phycoerythrin that was totally unique because the minor bilin surpassed the amount of the major bilin (phycourobilins to phycoerythrobilin >1). R-Phycoerythrin from *Callithamnion roseum* was another example of bilin diversity for a phycoerythrin. Yu et al.[11] reported that this biliprotein possessed a phycoerythrobilin to phycourobilin ratio that varied with the intensity of growth light. A unique observation requiring additional study was that C-phycoerythrin had a species-dependent phycoerythrobilin composition. Muckle and Rüdiger[18] showed that C-phycoerythrin from *Pseudanabaena* W1173 and *Tolypothrix tenuis* had β subunits with three phycoerythrobilins, while C-phycoerythrin from *Pseudanabaena catenata, Fremyella diplosiphon, Phormidium persicinum, Phormidium autumnale,* and *Nostoc muscorum* had four phycoerythrobilins on β. All the α subunits appeared to have two phycoerythrobilins.

However, no phycocyanin (except for the R- or C-phycocyanin type) or allophycocyanin has yet been found to have either a species-dependent chromophore content or one that varies with growth conditions. There are two mutually compatible arguments for why the chromophore compositions of phycoerythrins can be more flexible than those of other biliproteins. The first possibility is that allophycocyanin and C- and R-phycocyanin are located in strategic places in the phycobilisomes. Any tampering with their chromophores, especially for allophycocyanin, could result in a protein that would slightly unbalance the interactions needed to build a structurally competent phycobilisome. Another answer may concern the functional role of phycoerythrin, since this biliprotein has a major responsibility for the adaptability of the light harvesting system to different environments. It is advantageous for this biliprotein to exist in several different forms (C, CU, R, B, and b), and to have in some cases variability within a single species.

IV. BILIN STRUCTURE

A. Bilin Cleavage from Apoprotein

Bilins are linear tetrapyrroles.[27-29] The structures of phycocyanobilin and phycoerythrobilin are well established. The first methods to cleave phycocyanobilin from the apoprotein used concentrated HCl. O'hEocha[30] improved on the Lemberg protocol by using gentler acid treatment: 12 *N* HCl for short periods at 25°C. It was found that acid produces structural changes in the cleaved bilin, and O'hEocha[30] noted three different forms of the bilin cleaved from C-phycocyanin.

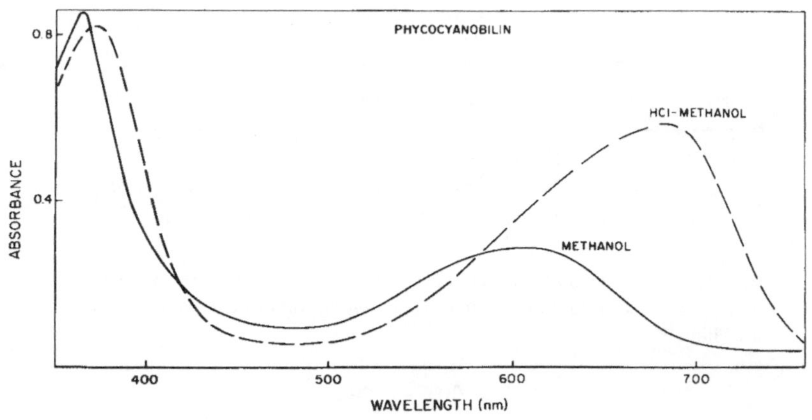

FIGURE 1. Spectra of chromophore cleaved from C-phycocyanin (*Phormidium luridum*). The protein is refluxed in methanol overnight, and the insoluble protein residue is removed. The spectrum is measured for methanol and 8% HCl in methanol. The addition of acid causes both hyperchromic and bathochromic shifts.

Experiments by Fujita and Hattori[31] indicated that a milder method of obtaining phycocyanobilin was refluxing the protein in methanol. O'Carra and O'hEocha[32] improved the methanol method and found that the liberated chromophores of phycocyanin and phycoerythrin had one unprotonated pyrrole nitrogen and two nonesterified carboxylic acid groups. In concentrated HCl, these bilins converted to spectrally different forms which resembled those of acid-released bilins. The methanol-cleaved phycoerythrobilin was further converted to a urobilin on longer treatment with concentrated HCl.

B. Structure of Phycocyanobilin

Two groups, Cole et al.[33,34] and Crespi et al.,[35,36] used this methanol technique to obtain phycocyanobilin for structural studies. In general, denatured C-phycocyanin was refluxed in absolute methanol overnight, and the chromophore obtained from the methanol solution was called phycocyanobilin.

The spectra of the cleaved chromophore of C-phycocyanin from *Phormidium luridum* showed the result of the methanol technique (Figure 1). Crespi et al.[35,36] purified the methanol-cleaved bilin by extraction into other organic solvents. Cole et al.,[33,34] however, first esterified the two free-acid groups of the bilin by adding boron trifluoride to the methanol extract and refluxing further. The resulting dimethyl ester was purified by thin-layer chromatography. Cole et al.[33,34] then used nuclear magnetic resonance (NMR) spectroscopy and mass spectroscopy to determine the structure of phycocyanobilin. They found a molecular weight by mass spectroscopy of 614 for the molecular ion of the dimethyl ester. Crespi et al.[35,36] found a molecular weight of 588 for the molecular ion of the diacid. The molecular weight of 614 for the dimethyl ester minus 28 equals 586 for the molecular weight of the diacid. This showed a difference of 2 from the Cole et al.[33,34] study.

Two possibilities exist for the different results. Either the diacid picked up two protons in the mass spectrograph and was too high, or the dimethyl ester lost two protons during the process and was too low in molecular weight. Using phycocyanobilin cleaved from protein by hydrogen bromide in trifluoroacetic acid, Schram and Kroes[37] determined the mass spectrum of both the free acid and the dimethyl ester. They found an m/e of 588 for the diacid and an m/e of 614 for the dimethyl ester. While Schram and Kroes[37] argued for acceptance of the 588-mol wt structure for phycocyanobilin, other experiments have confirmed the validity of the 586 structure.

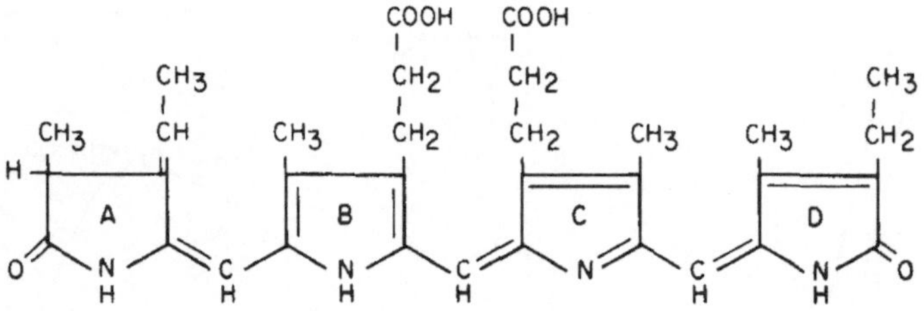

FIGURE 2. Structure of phycocyanobilin diacid. This structure and those following are not meant to give information on the various configurations of the bilins.

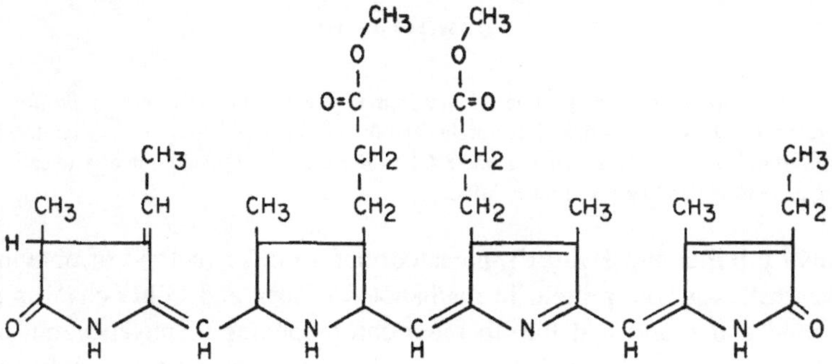

FIGURE 3. Structure of phycocyanobilin dimethyl ester.

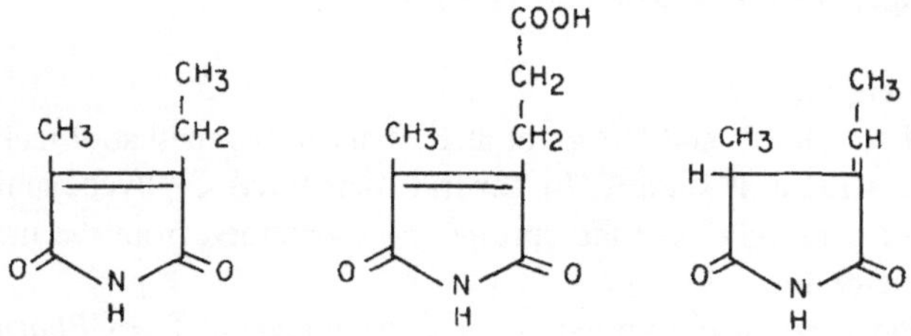

FIGURE 4. Structure of phycocyanobilin imides. The imides are, from left to right, methylethylmaleimide, hematinic acid, and methylethylidenesuccinimide.

Support for acceptance of the 586-mol wt structure (Figures 2 and 3) at least as the structure of the methanol-cleaved bilin came first from Rüdiger et al.,[38] who studied the structure of phycocyanobilin by chromic acid cleavage into imides (Figure 4). Three imides, methylethylmaleimide, hematinic acid, and methylethylidenesuccinimide, were identified by thin-layer chromatography. Since phycocyanobilin was known to be a diacid, it was concluded that the imide with the propionic acid side chain occurred twice in the tetrapyrrole. The structure they deduced from these imides was the same as reported by Cole et al.[33,34] from mass spectral and NMR data. However, certain problems exist with the chromic acid technique. Rüdiger et al.[38] studied C-phycocyanin itself and two types of acid-cleaved bilin from C-phycocyanin, but all three gave the same imides. At least one of the three was expected to have a different structure to explain the different absorption spectra of the two acid-released forms.

Fu et al.[39] reopened the mass spectral study of methanol- and *n*-butanol-cleaved phycocyanobilin from C-phycocyanin using new technology which enabled unambiguous exam-

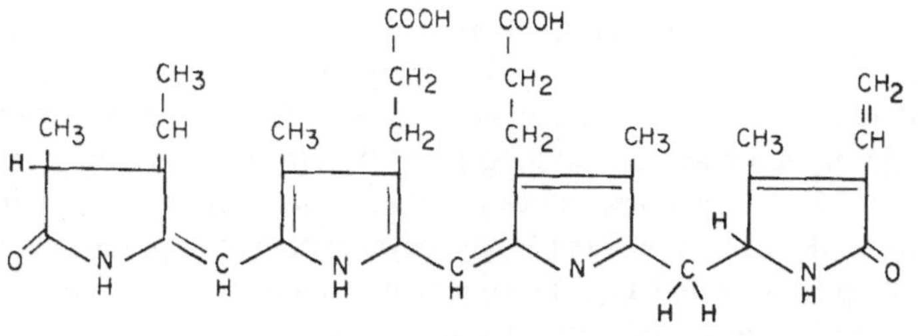

FIGURE 5. Structure of phycoerythrobilin.

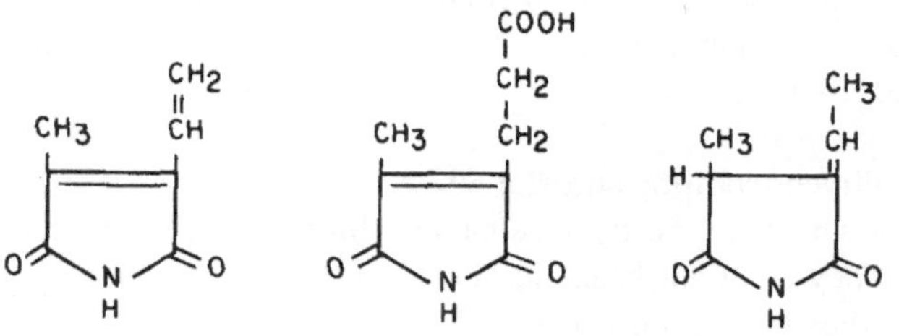

FIGURE 6. Structure of phycoerythrobilin imides. The imides are, from left to right, methylvinylmaleimide, hematinic acid, and methylethylidenesuccinimide.

ination of diacids directly. Previously, purification by chromatography was possible only after esterification, but now the diacids could be purified by reverse-phase liquid chromatography. Furthermore, a volatility-enhancement technique improved the opportunity for direct study of the diacid by proton-transfer chemical ionization mass spectroscopy. This yielded a protonated molecular ion at m/e 587 for phycocyanibilin which gives a nonprotonated species at m/e 586. The agreement among the mass spectral results from esterified and diacid phycocyanobilin together with the chromic acid results confirmed that the pigment with an m/e of 586 molecular ions was correct for phycocyanobilin (Figure 2). However, the difficulties in performing these experiments and in validating the conclusions should not be ignored.

Killilea et al.[40] have continued these inquiries into the structure of phycocyanobilin. Inspection of the proposed structure for methanol-cleaved phycocyanobilin showed an ethylidene group which is an exocyclic component of ring A (Figure 2). Killilea et al.[40] presented spectroscopic evidence and an analogy to the methanol-released bilin that indicated that the acid-cleaved phycocyanobilin obtained by them was almost identical to the methanol-produced product. The only difference was the substitution of a vinyl group for the ethylidene group on ring A. They maintained that the isomer with the vinyl group was the natural form of the bilin. It must be stressed that the ethylidene or its isomer was involved in covalent attachment to the apoprotein and that this double bond does not exist in the native protein-bound apoprotein. This renders any discussion of the position of this double bond not germane to the question of phycocyanobilin structure, although it will be an important point in the study of bilin biosynthesis.

C. Structure of Phycoerythrobilin

Chapman et al.,[41] using the same protocols as for C-phycocyanin, obtained the structure of phycoerythrobilin from C-phycoerythrin (Figure 5). A study of both the exchangeable hydrogens on the tetrapyrrole[42] and the imides from chromic acid degradation (Figure 6) showed good agreement with their proposed structure.[41] It is seen that phycocyanobilin and phycoerythrobilin are isomers. A comparison of the structure of phycocyanobilin and phycoerythrobilin shows that there are more conjugated double bonds in phycocyanobilin, since the bridge between rings C and D is saturated for phycoerythrobilin. Since the longer chains of conjugated double bonds yield excited states of lower energy, phycocyanobilin should have its visible absorption maximum at longer wavelengths than phycoerythrobilin. This is true, but the structure of the chromophores is only the first step in determining the spectra of the biliproteins. Allophycocyanin and C-phycocyanin both have phycocyanobilins as their sole bilins, but allophycocyanin has its maximum at 650 nm, while C-phycocyanin is at 615 to 620 nm. The interactions of the bilins within the apoprotein refine the electronic energy levels to produce the biologically relevant electronic spectra. The same principle applies to phycoerythrobilin which, depending on its protein environment, displays a chromophore absorption band at 545 or 565 nm. Gossauer and Hirsch[43,44] and Gossauer and Weller[45] have synthesized both phycocyanobilin (Figure 2) and phycoerythrobilin (Figure 5) corresponding to the methanol-released bilins.

D. Cryptoviolin, Phycourobilin, and the 697-nm Bilin

Much less is known about the three remaining bilins: phycourobilin, cryptoviolin, and the 697-nm bilin. They have not been cleaved from their apoproteins by methanol, and no direct structural studies have been performed on them. Phycourobilin is found on R-, B-, and CU-phycoerythrins. Its absorption maximum is at 498 nm. It is named after the urobilin class of tetrapyrroles produced in man which all absorb in this region. Urobilins have fewer double bonds in their conjugated system and therefore absorb at shorter wavelengths than phycoerythrobilins. Since they always occur with phycoerythrobilin in phycoerythrins, they function as sensitizing chromophores (see Chapter 6) in these biliproteins. Cryptoviolin occurred in only one protein from the cyanobacteria, phycoerythrocyanin,[9] and not at all in the red algae. It appears to be present in all of the phycocyanins[46] and phycoerythrins[47] from cryptomonads. It was named for its spectral similarity to the violin class of tetrapyrroles.[48] Its spectra suggested that it had more conjugated double bonds than phycoerythrobilin but less than phycocyanobilin. It might have served as a sensitizing chromophore when coupled with phycocyanobilin (phycoerythrocyanin, phycocyanin 615 and 645) but as a fluorescing chromophore when it was present with phycoerythrobilin (phycoerythrin 545). The 697-nm bilin only occurred on phycocyanin 645. Its native spectrum is unknown, since it overlaps substantially with the phycocyanobilins on this protein.

Vaughan[15] has cleaved phycourobilin from R-phycoerythrin (*Ceramium rubrum*) by the following method: the protein was first acetylated with acetic anhydride in the presence of sodium acetate at 0°C and HCl gas was passed through an acetic acid solution of the acetylated protein, lyophilized to dryness, dissolved in 1.0 *M* HCl, and extracted with chloroform. The urobilin-like tetrapyrrole was purified and shown to possess an absorption and IR spectra similar but not identical to known urobilins.

Chapman et al.[49] considered the possibility that phycourobilin was a protein-induced phycoerythrobilin absorption band. They also pointed out that phycoerythrobilin can isomerize to a urobilin when cleaved. An argument for a chemically distinct protein-linked urobilin was that when phycourobilin-containing phycoerythrins were denatured in acid urea the 498-nm band was retained.[48] Other protein-induced absorption differences, e.g., those of allophycocyanin, C-phycocyanin, and phycoerythrobilin 545 and 565, were eliminated when the influences of apoproteins were minimized by denaturing the protein. They also found tryptic peptides that contained phycourobilin and not phycoerythrobilin. The urobilin-like absorption occurred in cells, purified native protein, denatured protein, and cleaved chromophores. A structural study of phycourobilin would be of great value.

Cryptoviolin occurs in the cyanobacterial protein, phycoerythrocyanin, and cryptomonad biliproteins. As in the case of phycourobilin, its characteristic absorption spectrum survived acidic urea treatment,[9] but also like phycourobilin, careful studies of its cleavage from apoprotein and its structure are needed.

V. CHROMOPHORE-APOPROTEIN LINKAGE

A. An Overview

Perhaps the most intensely studied problem in biliprotein chemistry has been the covalent linkage of tetrapyrrole to apoprotein. There is now substantial evidence concerning phycocyanobilin and phycoerythrobilin to state that ring A of these bilins is covalently attached to the sulfur atom of a cysteine residue on the apoprotein, and there are data substantiating the suggestion that the ethylidene side chain is the place of attachment on ring A. Suggestions of a second site of covalent linkage are controversial or not fully proven.

B. Phycocyanobilin Linkage to Cysteine

Evidence for the attachment of bilins in the biliproteins to the sulfur atom of cysteine will be outlined first. Fujiwara[50] performed a peptic digest of R-phycoerythrin (*Porphyra tenera*) and isolated several chromopeptides. All the chromopeptides contained cysteine. She noted that in the case of cytochrome *c*, where the apoprotein was linked to the prosthetic group by two thioether bonds, such a bond also seemed probable for the attachment of phycoerythrobilin in R-phycoerythrin.

Crespi and Smith,[51] by digesting C-phycocyanin with the broad spectrum protease Nagarse, found indirect evidence of bilin attachment through a sulfur atom of cysteine. Furthermore, they found metal ions, Hg^{+2}, Ag^{+}, and Cd^{+2}, that formed insoluble sulfides and promoted the liberation of phycocyanobilin.

Strong evidence for the linkage of a phycocyanobilin to cysteine was found in the enzymatic digest studies of C-phycocyanin (*Mastigocladus laminosus*) by Byfield and Zuber.[52] They studied both the α and β subunits. However, more conclusive results were obtained for the α subunit which had only one chromophore. A 20-residue chromopeptide of the α subunit was obtained with pepsin. This peptide was then treated by trypsin, and a green color was found on the resulting decapeptide. The amino acid analysis of this decapeptide showed that one cysteine was present. The sequence of these ten amino acids was determined by the Edman degradation. The third Edman cycle removed the green color from the peptide, but the identity of the amino acid cleaved with the chromophore was not determined at this step. The Edman procedure was continued for a short sequence beyond the chromophore-amino acid step, and then the remainder of the peptide apparently was analyzed to determine the identities of the remaining amino acids. All the amino acids seemed to be accounted for except cysteine, and it was concluded that the phycocyanobilin must be attached to this amino acid, since the amino acid attached to the bilin was the one that was unidentified.

Further evidence for the involvement of cysteine in the covalent attachment of phycocyanobilin to apoprotein came from the studies of Williams and Glazer[53] on the cysteine content of biliproteins. They determined the number of free cysteines on C-phycocyanin (*Anacystis nidulans*), then cleaved off some chromophores, and redetermined the number of unattached cysteines. The chemicals employed for cleavage were Ag^{+}, Hg^{+2}, and HBr which had all been previously shown to cleave some of the chromophores.[51] Typical results for free-cysteine sulfurs were for unmodified C-phycocyanin, 1.2 cysteines; for Ag_2SO_4 treated, 3.1; and for HBr treated, 3.3. Similar results were shown for the α and β subunits. Therefore there was a clear demonstration of some blockage of cysteines on the native protein, and this was indirect support for the blockage being phycocyanobilins. The Crespi and Smith[51] studies showing that Ag^{+} and Hg^{+2} promoted chromophore liberation provided the main support for this idea.

There was substantial evidence of cysteine being the amino acid residue that covalently attached to phycocyanobilin based on the experiments of Brown et al.[54] with C-phycocyanin (*Cyanidium caldarium*). Radioactive sulfur was incorporated in the growth medium of this red alga, and the purified C-phycocyanin subunits were then digested with trypsin or cyanogen bromide. Chromopeptides were then obtained with radioactivity present, and Edman degradation was then used to sequence the peptides. Both chromophore color and radioactivity from these peptides were simultaneously cleaved and recovered during Edman degradation. Further analysis of their data indicated that in fact cysteine was the amino acid to which all three phycocyanobilins were attached.

C. Phycoerythrobilin Linkage to Cysteine

Phycoerythrins were studied for chromophore-protein linkage at the same time as C-phycocyanin. Köst-Reyes et al.[55] used pepsin to degrade B-phycoerythrin (*Porphyridium cruentum*) and obtain chromopeptides. Their use of pepsin treatment produced a chromotripeptide consisting of chromophore, cysteine, valine, and leucine. Edman degradation of the tripeptide showed that the red color was associated with cysteine.

Muckle et al.[56] studied enzymatic digests of C-phycoerythrins from two cyanobacteria, *Pseudanabaena* W1173 and *Phormidium persicinum*. The chromopeptides were purified and subjected to Edman degradation. As previously mentioned, no amino acid was detected by usual methods when the amino acid bearing the chromophore was cleaved from the peptides in the Edman degradation. This was because the amino acid remained attached to the tetrapyrrole and was not readily identified. In this study the loss of chromophore color was also accompanied by nondetection of an amino acid. The chromophore complex deleted from the peptide by the Edman technique was extracted and hydrolyzed, and cysteine was identified as the amino acid component.

For C-phycocyanin, a very critical demonstration of cysteine linkage to phycocyanobilin began with radioactive sulfur being incorporated into the biliprotein and ended with a direct demonstration of a cysteine-bilin complex. The same year that this was accomplished, Köst-Reyes and Köst[57] presented a very similar study for B-phycoerythrin (*Porphyridium cruentum*). The alga, from which B-phycoerythrin was purified, grew in media containing the same radioactive sulfur compound used by Brown et al.[54] The biliprotein was digested with trypsin and pepsin and the chromopeptides purified. The red chromophore and radioactive sulfur were recovered together in the organic phase of the Edman degradation. This organic soluble product was purified and hydrolyzed with acid. The radioactive sulfur was found to be on a cysteine. The yield was low, however, and precautions were taken to provide protection for the cysteine. These experiments demonstrated that cysteine was the linking agent for most and probably all the phycoerythrobilins of B- (α and β subunits) and C-phycoerythrin.

D. Thioether Bond

The point of interest, then, is the nature of the sulfur attachment to the tetrapyrroles. What is the chemical composition and location of the linkage? Chromic acid degradation of C-phycocyanin and phycoerythrin showed that ring A and one of the two middle rings (B or C) of the tetrapyrroles cleaved more slowly than the other two rings (B or C and D) of the tetrapyrrole. This was interpreted to mean that these rings (A and B or C) could be the protein-binding positions.[38] Crespi and Smith[51] suggested a thioether bond to the ethylidene group on ring A together with an ester bond also to ring A. The thioether bonding was suggested as analogous to the situation in cytochrome *c*. Several investigators have pointed out that the exocyclic ethylidene on ring A was a good candidate for the bonding to cysteine. When C-phycocyanin was denatured in acidic urea it had an absorption maximum at 662 nm, but methanol-cleaved phycocyanobilin had a maximum at 683 nm in acidic methanol (Figure 1). Investigators have interpreted this type of data to mean that the conjugated double-bond system was longer in the free bilin. They suggested that this was evidence for the ethylidene group being formed upon chromophore cleavage. More substantial data on this came from mass spectroscopy studies. In addition to phycocyanobilin (Figure 2), some methanol or

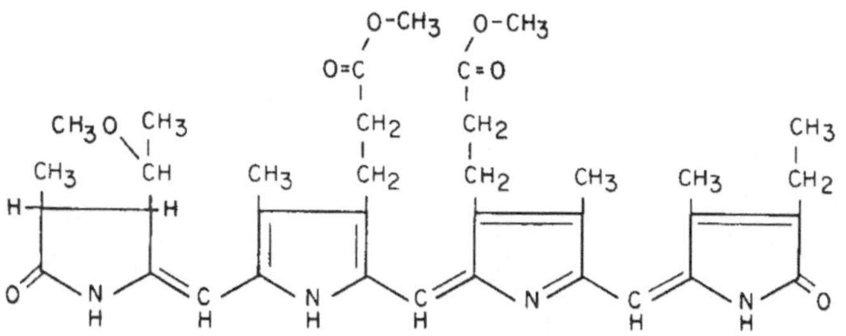

FIGURE 7. Structure of phycocyanobilin adduct to methanol.

hydrochloric acid adducts were observed by mass spectroscopy depending on whether methanol or HCl was used as the chromophore-cleaving agent.[58] Beuhler et al.[58] then proceeded to obtain data on the site of adduct formation. Cleaved phycocyanobilin can react with methanol and 7% boron trifluoride to form in 10% yield the methanol adduct after 3 min. When mesobiliverdin, which had a saturated (ethyl) side chain instead of the ethylidene but was otherwise a similar tetrapyrrole, was studied in the same reaction, no adduct was detected. The lack of reactivity of mesobiliverdin suggested that the ethylidene group on ring A of phycocyanobilin was the site of adduct formation (Figure 7) during methanol cleavage. Gossauer et al.[59] have proved by total synthesis that the methanol binds to the bilin at the ethylidene group. Beuhler et al.[58] suggested that these data supported the concept that the ethylidene group may be the binding site of bilin to cysteine, and adduct formation occurred during the cleavage. Of course, it could also be argued that the ethylidene group was present on protein-bound phycocyanobilin, and certain of the amino acids acted to catalyze the adduct formation.

Noting that chromic acid treatment at 20°C did not produce cleavage of ring A from the protein, Schoch et al.[60] found that in model compounds a thioether bonding to a pyrrole was oxidized to a sulfone by chromic acid. Treatment of this sulfone with ammonia liberated methylethylmaleimide. Studies by Klein and Rüdiger[61] on C-phycocyanin and C-phycoerythrin using the chromic acid/ammonia treatment demonstrated that the imide from ring A was structurally consistent with the presence of a thioether bond to ring A at the position of the exocyclic ethylidene group of the free chromophore.

An NMR spectroscopy study by Lagarias et al.[62] has added substantially to the confidence in the thioether linkage through ring A. They used cyanogen bromide to obtain a blue heptapeptide from C-phycocyanin (*A. nidulans*). The same heptapeptide was then synthesized without the tetrapyrrole. A study of both peptides was then performed using 360-MHz ^1H NMR spectroscopy. All the NMR signals of the synthetic peptide were assigned, and a comparison was made to the chromopeptide. The resonances of two hydrogens on cysteine and a hydrogen on alanine were the only significant amino acid shifts. A chemical shift in the chromopeptides was noted for a hydrogen that was not found in free chromophore but was predicted by the thioether linkage to ring A. This and other details of the NMR were very compatible with the thioether bonding to the ethylidene group of ring A. Schoenleber et al.[63] have used NMR spectroscopy to study purified chromopeptides from the α and β subunits of B-phycoerythrin. As with the chromopeptide from C-phycocyanin, they reported that the NMR resonances were in agreement with a thioether bond between the side chain on ring A of phycoerythrobilin and cysteine. All the above results supported the validity of a thioether bond between ring A of C-phycocyanin and phycoerythrin and a cysteine residue (Figures 8 and 9).

E. Second Covalent Linkage

No agreement can be put together for the existence of a second covalent linkage from the

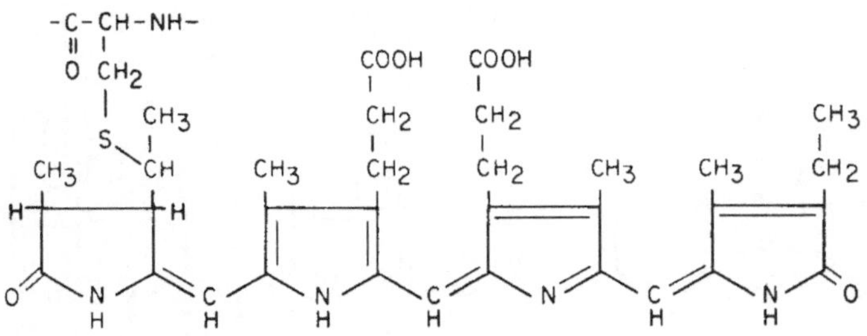

FIGURE 8. Structure of phycocyanobilin thioether linkage.

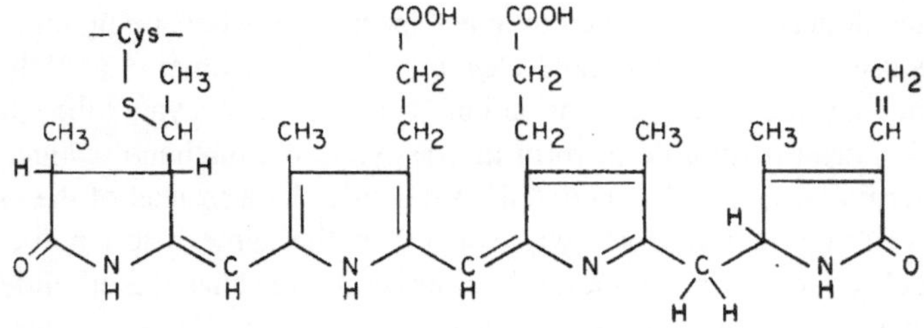

FIGURE 9. Structure of phycoerythrobilin thioether linkage.

apoprotein to the tetrapyrroles. The most widely discussed example of a linkage other than the thioether to ring A was an additional covalent bond to ring C. Rüdiger and O'Carra[64] obtained evidence for this second linkage during their chromic acid degradation studies on various biliproteins. The chromic acid degradation of R-phycoerythrin (*Rhodymenia palmata*) and C-phycocyanin were carried out at two temperatures, 20 and 100°C. The proteins were treated with chromic acid in sulfuric acid and after 1 hr were extracted with ether. The resulting imides (Figures 4 and 6) were separated on thin-layer chromatography and their identities confirmed by IR spectroscopy. All the biliproteins examined showed similar results. Ring D and one of the two middle rings were converted to free imides at 20°C, but 100°C was required to cleave ring A and the second propionic acid ring. This suggested that two degradation-resistant rings were directly bound to the apoprotein. The possibility that only one was bound and the second was tightly attached to this bound ring and therefore cleaved slowly was discounted because the two rings seemed to cleave independently in certain experiments. Rüdiger and O'Carra[64] pointed out that the propionic acid side chain of ring C was the most likely attachment site and could form an ester linkage to a hydroxyl group of the apoprotein. Support for this type of linkage came from the study of a tetrapyrrole isolated from the defensive ink of the sea hare *Aplysia*.[65] These animals feed on red algae and cleave chromophores from the biliproteins for their ink. One component of the ink has been identified as phycoerythrobilin monomethyl ester.[65] Perhaps this monomethyl ester derivative was a result of cleavage of an ester bond to the apoprotein.[58] Köst et al.[66] performed experiments on several different biliproteins and maintained that in all cases there were two covalent linkages to the apoprotein.

A serious challenge to the second ester linkage concept came out of the experiments on the chromic acid degradation of C-phycocyanin (*C. caldarium*).[67] It was established that 5-aminolevulinic acid can be incorporated into the chromophores but not the apoprotein. Using this technique to label the phycocyanobilins with radioactivity, Troxler et al.[67] studied quantitatively the chromic acid conversion of phycocyanobilin to imides. They found that for C-phycocyanin, the cleavage of the propionic acid-containing rings (B or C) to hematinic acid increased markedly at 100°C as compared to 25°C. This observation was in agreement with Rüdiger and O'Carra.[64] However, for phycocyanobilin already cleaved from the apo-

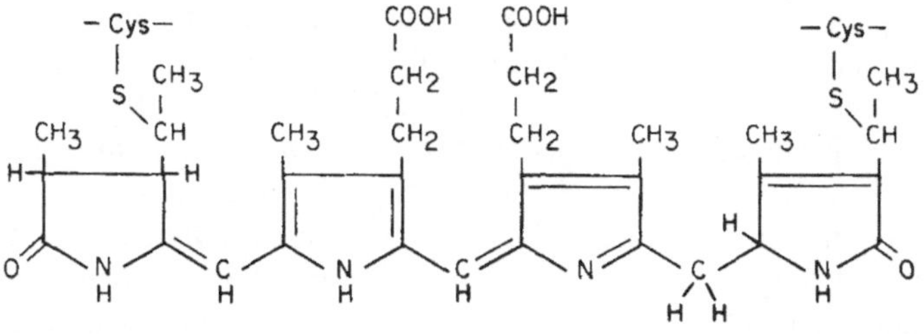

FIGURE 10. Dual linkage of phycoerythrobilin through rings A and D.

protein they found the same pattern of radioactivity as for protein-linked phycocyanobilin at 25 vs. 100°C. Therefore, the argument used by Rüdiger and O'Carra[64] which explained that the increase in hematinic acid obtained at 100°C was a result of its being linked to apoprotein was questionable. Also in the various Edman degradation studies reported on so far, little evidence for this ester linkage has appeared. Of course, this could be a consequence of the Edman protocols breaking the ester bond.

Killilea et al.[40] have obtained data that supported the ester linkage. Using a mixture of mammalian proteinases, they obtained a phycoerythrobilin-containing peptide which was then subjected to chromic acid degradation. A peptide was isolated that on hydrolysis yielded hematinic acid, serine, glycine, and alanine. Killilea et al.[40] considered this to be evidence for a serine to ring C ester linkage as discussed above; although this type of second linkage has been widely sought after, there is a lack of critical consensus for such bonding. Without confirmatory findings, such a linkage must be viewed skeptically.

Muckle et al.[56] also looked for an ester linkage using C-phycoerythrin (*Pseudanabaena* W1173). One tryptic chromopeptide had a different absorption spectrum, absorbing with a 559-nm maximum instead of the 542 to 550 nm found for the others. Edman degradation of this 559-nm peptide produced an unusual result. Some chromophores were extracted along with cysteine at an Edman step, but the rest extracted with serine at the next step. As with the Killilea et al.[40] results, serine was implicated with chromophore linkage to a phycoerythrobilin. Muckle et al.[56] found no evidence for a second linkage when they studied C-phycoerythrin from *Phormidium persicinum*. They suggested that this serine linkage may be genuine or it may be an artifact of their protocols. Kufer et al.[68] have used milder chromophore-degradation methods to study the problem. Definitive data are needed on this type of covalent linkage, and its validity clearly is unproven.

Glazer et al.[69] studied free and bound cysteines in various biliproteins. They concluded that all phycocyanobilins and phycoerythrobilins were linked through a single thioether bond to the apoprotein. Other types of linkage were not measured in their assay. Phycourobilins were perhaps linked via two thioether bonds. Research on B-phycoerythrin (*Porphyridium cruentum*) may have uncovered a different situation.[20,70] Lundell et al.[20] and Schoenleber et al.[70] have studied the five tryptic chromopeptides from the α and β subunits of this biliprotein. One of these chromopeptides was quite unique. Part of its sequence was as follows: Ala-Ser-Cys-Met-Val-Ser-Asp-Val-Ser-Gly-Met-Ile-Cys-Glu with both Cys attached to the *same* phycoerythrobilin to form a loop of amino acids and chromophores. Thermolysin further digested the chromopeptide, and this second chromopeptide retained the dual linkage but had a severed loop (Figure 10). As with dual-linked phycoerythrobilin from C-phycoerythrin,[57] this unusual bilin had an absorption maximum at 559 nm which was red shifted compared to the monolinked peptides. NMR and mass spectroscopy have confirmed the dual binding. The first cysteine had the well-established thioether bonding to ring A. The second linkage was to ring D and was also most likely a thioether bond. This type of second linkage cannot occur for phycocyanobilin, since the side chain on ring D was saturated. The C-[56] and B-phycoerythrin[20,70] studies suggested that a diversity of bilin attachment to the apoprotein was possible. In each case, one of the chromophores on a biliprotein was unique in this property of dual linkage. Both the Lundell et al.[20] and the Muckle et al.[56] studies used trypsin to produce the chromopeptide. Nagy et al.[71] have also found a doubly linked phycourobilin on the β subunit of R-phycoerythrin from *Gastroclonium coulteri*.

Another approach to learning more about the structure of bilins is to employ different types of chemical modification. Kufer and Scheer[72,73] have used sodium dithionite and sodium borohydride to study C-phycocyanin and C-phycoerythrin. Klein and Rüdiger[74] studied the formation of a thioether bond to phycocyanobilin through a reaction with ethanethiol.

VI. PRIMARY STRUCTURE

Major progress has been made on the amino acid sequencing of biliproteins. Frank et al.[75] completed the first entire sequence of a biliprotein. They studied a cyanobacterial C-phycocyanin (*M. laminosus*) and completed the sequences of both the α and β subunits. The subunits were separated either by chromatography on hydroxylapatite in the presence of SDS or on DEAE-Sephadex in 8 *M* urea. Various rather large peptides were prepared by cleavage, purified, and sequenced. The α chain had 162 amino acid residues, and the β had 172. When the sequences were lined up matching the N terminus of each, 41 positions were identical for both subunits. A gap of two amino acids had to be introduced in the β sequence to allow for best homology between α and β. The two subunits were quite distinct, but the 25% homology overall and the even greater homology when calculated in certain segments suggested that they may have evolved from a common ancestor. One cysteine on the β subunit did not carry a chromophore, and there was no homology between the region of the second β chromophore and a comparable α region. The first sequence of C-phycocyanin from a red alga was from *C. caldarium* by Offner et al.[76] and Troxler et al.[77] The α subunit had 162 amino acids, and β had 172; this was the same result as for the cyanobacterial C-phycocyanin. By allowing two gaps, one in α and one in β, a 30% homology was obtained between them. There were certain short stretches of excellent homology, and between residues 10 and 26, Troxler et al.[77] reported 71% agreement of sequence. The sequences of the α and β subunits of the cyanobacterial and red algal C-phycocyanins were quite similar. For example, there were only three differences in sequences for the first 25 amino acids of the two α subunits. It was quite interesting that the α subunit of one C-phycocyanin was much closer in homology to the α from a very different organism than it was to β of the same organism.

The first complete sequence of allophycocyanin was also from the cyanobacterium *M. laminosus*.[78] Sidler et al.[78] found 160 amino acid residues in α and 161 in β. They reported a 37% homology between α and β subunits. When the sequences were aligned to achieve maximum homology, it was shown that the first phycocyanobilin-bearing cysteines were in the same position from the N terminus for α and β of C-phycocyanin and α and β for allophycocyanin. While the first chromophore was located near the middle of the sequence, the second phycocyanobilin of the β subunit of C-phycocyanin was nearer the C terminus. The first complete sequence of a red algal allophycocyanin was established by Offner and Troxler[79] using protein isolated from *C. caldarium*. Again as for C-phycocyanin, a great deal of homology, over 80%, was found to exist between allophycocyanins from different sources. The great degree of homology between the amino acid sequences of different C-phycocyanins and different allophycocyanins was in agreement with the finding that no variation occurred in the chromophore contents for all the organisms so far studied. Offner and Troxler[79] proposed that the demands of energy transfer and the arrangement of proteins in the phycobilisomes must be similar for both red algae and cyanobacteria.

Phycoerythrocyanin from the cyanobacterium *M. laminosus* has also been sequenced.[80] The α subunit had 162 amino acids, and the β had 171. The chromophore position could

Table 4
AMINO ACID SEQUENCES AROUND CYSTEINES OF C-PHYCOCYANIN

Organism	Subunit	Residue from N terminus	Sequence
		Chromophore-Binding Cysteines	
Mastigocladus laminosus	α	84	ADARGKSKCARDIGHYL
Cyanidium caldarium	α	84	SSAVGKAKCARDIGYYL
M. laminosus	β	82	TRNGTMAACLRDMEIIL
	β	153	PNGITKGDCSALISEVA
C. caldarium	β	82	DRRM--AACLRDMEIIL
	β	153	PSGITTGDCSALMAEVG
		Nonchromophore-Binding Cysteines	
M. laminosus	β	109	DASILDDRCLNGLRETY
C. caldarium	β	109	DSSILDDRCLNGLRETY
	α	98	YYLRMVTYCLVVGGTGP

be observed by loss of the color from the sequencer cup. Two blue chromophores were associated with the β subunit and one red chromophore (cryptoviolin) with α. This agreed with the chromophore composition obtained by absorption spectroscopy in acidic urea.[9] The identity of the amino acid which binds the cryptoviolin chromophore was indirectly found to be cysteine.

The homology between the α subunit of phycoerythrocyanin and the α subunit of C-phycocyanin (*M. laminosis*) and between the β subunits was outstanding. For the two α subunits it was 63%, and for the β subunits it was 67%. This was much higher than the corresponding homologies between subunits of C-phycocyanin and allophycocyanin or allophycocyanin and phycoerythrocyanin. For example, the homology between the α subunits of allophycocyanin and C-phycocyanin was 32%. Such close sequence homology almost suggests that phycoerythrocyanin might be best viewed as an exotic type of phycocyanin.

The cysteine residues for these sequences are especially interesting because of their function in covalently linking the chromophores to the apoprotein. The amino acid sequences surrounding both the chromophore-linked and the chromophore-free cysteines are compared for both C-phycocyanins (Table 4), both allophycocyanins (Table 5), and phycoerythrocyanin (Table 6). The sequences proceed in a N terminal to C terminal direction. For C-phycocyanin, one α and two β chromophore-binding cysteines occur at identical positions from the N terminus in the sequence of the proteins for both microbial sources (α, 84 and β 82 and 153). The amino acids around these cysteines were well conserved in comparing the two sources. There was also homology of the second and third residues to the C terminal side of the cysteine-binding residues in positions 82 and 84 of both β and α. However, the sequence around the cysteine at position 153 did not have any consistent homology to the other chromophore-binding cysteines in C-phycocyanin. Thus no single amino acid in the sequence region around the tetrapyrrole-cysteine was essential. In addition, homology in itself was not an exclusive hallmark of the chromophore-binding cysteines, since cysteine on β at position 109 was well conserved for both C-phycocyanin sources. For phycoerythrocyanin, the α subunit contained the unusual cryptoviolin chromophore. Its position at residue 84 was identical to the C-phycocyanin result. Also the sequence around this residue showed a striking homology to the regions around the same cysteine for C-phycocyanin (Tables 4 and 6). Evidently, the cryptoviolin chromophore simply replaced phycocyanobilin without requiring a major realignment of the amino acids. In reporting these sequences the

Table 5
AMINO ACID SEQUENCES AROUND CYSTEINES OF ALLOPHYCOCYANIN

Organism	Subunit	Residue from N terminus	Sequence
		Chromophore-Binding Cysteines	
Mastigocladus laminosus	α	80	QEMTAT--CLRDLDYYL
Cyanidium caldarium	α	80	EEMT--ATCLRDLDYYL
M. laminosus	β	81	TR--RYAACIRDLDYYL
C. caldarium	β	81	TRRY--AACIRDLDYYL
		Nonchromophore-Binding Cysteines	
M. laminosus	β	157	MGVYFDYICSGLS
C. caldarium	β	157	MGIYFDYICSGLS

Table 6
AMINO ACID SEQUENCE AROUND CYSTEINES OF PHYCOERYTHROCYANIN

Organism	Subunit	Residue from N terminus	Sequence
		Chromophore-Binding Cysteines	
Mastigocladus laminosus	α	84	STPEGKAKCVRDIDHYL
	β	84	FHHRNQAACIRDLGFIL
	β	155	PNGITKGDCSQLMSELA
		Nonchromophore-Binding Cysteines	
M. laminosus	α	98, 99	HYLRTISYCCVVGGTGP
	β	75	FNPGGP--CFHHRNQAA
	β	110	DTSVMDDRCLNGLRETY

dashes (Table 5) were put in to show gaps that establish maximum homologies of the entire sequences of biliproteins from a given organism. Table 5 shows that these gaps were not most advantageously placed to show the local homologies between organisms in the particular region under scrutiny. Ignoring the gaps, Table 5 shows that the sequences around the chromophore-binding cysteines were completely homologous out to five amino acids or more when the α and the β subunits of allophycocyanins from the two organisms were compared.

REFERENCES

1. **Brown, A. S., Foster, J. A., Voynow, P. V., Franzblau, C.,, and Troxler, R. F.,** Allophycocyanin from the filamentous cyanophyte, *Phormidium luridum, Biochemistry*, 14, 3581, 1975.
2. **Brown, A. S. and Troxler, R. F.,** Properties and N-terminal sequence of allophycocyanin from the unicellular rhodophyte *Cyanidium caldarium, Biochem. J.*, 163, 571, 1977.
3. **O'Carra, P.,** Algal biliproteins, *Biochem. J.*, 119, 2P, 1970.
4. **Bennett, A. and Bogorad, L.,** Properties of subunits and aggregates of blue-green algal biliproteins, *Biochemistry*, 10, 3625, 1971.
5. **O'Carra, P. and Killilea, S. D.,** Subunit structures of C-phycocyanin and C-phycoerythrin, *Biochem. Biophys. Res. Commun.*, 45, 1192, 1971.
6. **Glazer, A. N. and Cohen-Bazire, G.,** Subunit structure of the phycobiliproteins of blue-green algae, *Proc. Natl. Acad. Sci. U.S.A.*, 68, 1398, 1971.
7. **Gantt, E. and Lipschultz, C. A.,** Phycobilisomes of *Porphyridium cruentum:* pigment analysis, *Biochemistry*, 13, 2960, 1974.
8. **Bryant, D. A., Cohen-Bazire, G., and Glazer, A. N.,** Characterization of the biliproteins of *Gloeobacter violaceus*. Chromophore content of a cyanobacterial phycoerythrin carrying phycourobilin chromophore, *Arch. Microbiol.*, 129, 190, 1981.
9. **Bryant, D. A., Glazer, A. N., and Eiserling, F. A.,** Characterization and structural properties of the major biliproteins of *Anabaena* sp., *Arch. Microbiol.*, 110, 61, 1976.
10. **Stadnichuk, I. N., Odinzova, T. I., and Strongin, A. Y.,** Molecular organization and chromophore composition of R-phycoerythrin from red alga *Callithamnion rubosum, Mol. Biol.*, 18, 343, 1984.
11. **Yu, M.-H., Glazer, A. N., Spencer, K. G., and West, J. A.,** Phycoerythrins of the red alga *Callithamnion*. Variation in phycoerythrobilin and phycourobilin content, *Plant Physiol.*, 68, 482, 1981.
12. **Redlinger, T. and Gantt, E.,** Phycobilisome structure of *Porphyridium cruentum*. Polypeptide composition, *Plant Physiol.*, 68, 1375, 1981.
13. **Glazer, A. N. and Hixson, C. S.,** Subunit structure and chromophore composition of Rhodophytan phycoerythrins. *Porphyridium cruentum* B-phycoerythrin and b-phycoerythrin, *J. Biol. Chem.*, 252, 32, 1977.
14. **Binder, A., Wilson, K., and Zuber, H.,** C-phycocyanin from the thermophilic blue-green alga *Mastigocladus laminosus*. Isolation, characterization, and subunit composition, *FEBS Lett.*, 20, 111, 1972.
15. **Vaughan, M. H., Jr.,** Structural and Comparative Studies of the Algal Protein Phycoerythrin, Ph.D. thesis, Massachusetts Institute of Technology, Cambridge, 1964.
16. **Glazer, A. N. and Fang, S.,** Chromophore content of blue-green algal phycobiliproteins, *J. Biol. Chem.*, 248, 659, 1973.
17. **Glazer, A. N. and Hixson, C. S.,** Characterization of R-phycocyanin. Chromophore content of R-phycocyanin and C-phycoerythrin, *J. Biol. Chem.*, 250, 5487, 1975.
18. **Muckle, G. and Rüdiger, W.,** Chromophore content of C-phycoerythrin from various cyanobacteria, *Z. Naturforsch.*, 32c, 957, 1977.
19. **Klotz, A. V. and Glazer, A. N.,** Characterization of the bilin attachment sites in R-phycoerythrin, *J. Biol. Chem.*, 260, 4856, 1985.
20. **Lundell, D. J., Glazer, A. N., DeLange, R. J., and Brown, D. M.,** Bilin attachment sites in the α and β subunits of B-phycoerythrin. Amino acid sequence studies, *J. Biol. Chem.*, 259, 5472, 1984.
21. **Kursar, T. A., Swift, H., and Alberte, R. S.,** Morphology of a novel cyanobacterium and characterization of light-harvesting complexes from it: implications for phycobiliprotein evolution, *Proc. Natl. Acad. Sci. U.S.A.*, 78, 6888, 1981.
22. **MacColl, R. and Guard-Friar, D.,** Phycocyanin 645. The chromophore assay of phycocyanin 645 from the cryptomonad protozoa *Chroomonas* species, *J. Biol. Chem.*, 258, 14,327, 1983.
23. **MacColl, R. and Guard-Friar, D.,** Phycocyanin 612: a biochemical and photophysical study, *Biochemistry*, 22, 5568, 1983.
24. **Guard-Friar, D. and MacColl, R.,** Spectroscopic properties of tetrapyrroles on denatured biliproteins, *Arch. Biochem. Biophys.*, 230, 300, 1984.
25. **Fujita, Y. and Shimura, S.,** Phycoerythrin of the marine blue-green alga *Trichodesmium thiebautii, Plant Cell Physiol.*, 15, 939, 1974.
26. **Ong, L. J., Glazer, A. N., and Waterbury, J. B.,** An unusual phycoerythrin from a marine cyanobacterium, *Science*, 224, 80, 1984.
27. **Lemberg, R.,** Die Chromoproteide der Rotalgen. I, *Liebigs Ann. Chem.*, 461, 46, 1928.
28. **Lemberg, R.,** Chromoproteide der Rotalgen. II. Spaltung mit Pepsin und Säuren. Isolierung eines Pyrrolfarbstoffs, *Liebigs Ann. Chem.*, 477, 195, 1930.
29. **Lemberg, R. and Bader, G.,** Die phycobiline der rot-algen. Überführung in Mesobilirubin und Dehydro-Mesobilirubin, *Liebigs Ann. Chem.*, 505, 151, 1933.

30. **O'hEocha, C.,** Spectral properties of the phycobilins. I. Phycocyanobilin, *Biochemistry*, 2, 375, 1963.
31. **Fujita, Y. and Hattori, A.,** Preliminary note on a new phycobilin pigment isolated from blue-green algae, *J. Biochem.*, 51, 89, 1962.
32. **O'Carra, P. and O'hEocha, C.,** Bilins released from algae and biliproteins by methanolic extraction. *Phytochemistry*, 5, 993, 1966.
33. **Cole, W. J., Chapman, D. J., and Siegelman, H. W.,** The structure of phycocyanobilin, *J. Am. Chem. Soc.*, 89, 3642, 1967.
34. **Cole, W. J., Chapman, D. J., and Siegelman, H. W.,** The structure and properties of phycocyanobilin and related bilatrienes, *Biochemistry*, 7, 2929, 1968.
35. **Crespi, H. L., Boucher, L. J., Norman, G. D., Katz, J. J., and Dougherty, R. C.,** Structure of phycocyanobilin, *J. Am. Chem. Soc.*, 89, 3642, 1967.
36. **Crespi, H. L., Smith, U., and Katz, J. J.,** Phycocyanobilin. Structure and exchange studies by nuclear magnetic resonance and its mode of attachment in phycocyanin. A model for phytochrome, *Biochemistry*, 7, 2232, 1968.
37. **Schram, B. L. and Kroes, H. H.,** Structure of phycocyanobilin, *Eur. J. Biochem.*, 19, 581, 1971.
38. **Rüdiger, W., O'Carra, P., and O'hEocha, C.,** Structure of phycoerythrobilin and phycocyanobilin, *Nature*, 215, 1477, 1967.
39. **Fu, E., Friedman, L., and Siegelman, H. W.,** Mass-spectral identification and purification of phycoerythrobilin and phycocyanobilin, *Biochem. J.*, 179, 1, 1979.
40. **Killilea, S. D., O'Carra, P., and Murphy, R. F.,** Structures and apoprotein linkages of phycoerythrobilin and phycocyanobilin, *Biochem. J.*, 187, 311, 1980.
41. **Chapman, D. J., Cole, W. J., and Siegelman, H. W.,** The structure of phycoerythrobilin, *J. Am. Chem. Soc.*, 89, 5976, 1967.
42. **Crespi, H. L. and Katz, J. J.,** Exchangeable hydrogen in phycoerythrobilin, *Phytochemistry*, 8, 759, 1969.
43. **Gossauer, A. and Hirsch, W.,** Synthesen von Gallenfarbstoffen. II. Totalsynthese des Phycocyanobilins, *Tetrahedron Lett.*, 17, 1451, 1973.
44. **Gossauer, A. and Hirsch, W.,** Synthesen von Gallenfarbstoffen. IV. Totalsynthese des Racemischen Phycocyanobilins (Phycobiliverdins) Sowie eines "Homophycobiliverdins", *Liebigs Ann. Chem.*, 1974, 1496, 1974.
45. **Gossauer, A. and Weller, J.-P.,** Chemical total synthesis of (+)-(2R,16R)- and (+)-(2S,16R)-phycoerythrobilin dimethyl ester, *J. Am. Chem. Soc.*, 100, 5928, 1978.
46. **O'hEocha, C., O'Carra, P., and Mitchell, D.,** Biliproteins of cryptomonad algae, *Proc. R. Ir. Acad.*, 63B, 191, 1964.
47. **MacColl, R., Guard-Friar, D., and Csatorday, K.,** Chromatographic and spectroscopic analysis of phycoerythrin 545 and its subunits, *Arch. Microbiol.*, 135, 194, 1983.
48. **O'Carra, P., Murphy, R. F., and Killilea, S. D.,** The native forms of the phycobilin chromophores of algal biliproteins. A clarification, *Biochem. J.*, 187, 303, 1980.
49. **Chapman, D. J., Cole, W. J., and Siegelman, H. W.,** A comparative study of the phycoerythrin chromophore, *Phytochemistry*, 7, 1831, 1968.
50. **Fujiwara, T.,** Studies on chromoproteins in Japanese Nori (*Porphyra tenera*). III. Chromopeptides derived from phycoerythrin by peptic digestion, J. *Biochem.*, 44, 723, 1957.
51. **Crespi, H. L. and Smith, U. H.,** The chromophore-protein bonds in phycocyanin, *Phytochemistry*, 9, 205, 1970.
52. **Byfield, P. G. H. and Zuber, H.,** Chromophore-containing peptide sequences in C-phycocyanin from *Mastigocladus laminosus, FEBS Lett.*, 28, 36, 1972.
53. **Williams, V. P. and Glazer, A. N.,** Structural studies on phycobiliproteins. I. Bilin-containing peptides of C-phycocyanin, *J. Biol. Chem.*, 253, 202, 1978.
54. **Brown, A. S., Offner, G. D., Ehrhardt, M. M., and Troxler, R. F.,** Phycobilin-apoprotein linkages in the α and β subunits of phycocyanin from the unicellular rhodophyte, *Cyanidium caldarium.* Amino acid sequences of ^{35}S-labeled chromopeptides, *J. Biol. Chem.*, 254, 7803, 1979.
55. **Köst-Reyes, E., Köst, H.-P., and Rüdiger, W.,** Bonding between chromophore and protein in biliproteins. II. Detection of cysteine as binding amino acid in B-phycoerythrin, *Liebigs Ann. Chem.*, 1975, 1594, 1975.
56. **Muckle, G., Otto, J., and Rüdiger, W.,** On the linkages between chromophore and protein in biliproteins. VIII. Amino acid sequence in the chromophore regions of C-phycoerythrin from *Pseudanabaena* W1173 and *Phormidium persicinum, Hoppe-Seyler's Z. Physiol. Chem.*, 359, 345, 1978.
57. **Köst-Reyes, E. and Köst, H.-P.,** The protein-chromophore bond in B phycoerythrin from *Porphyridium cruentum.* Radiosulfur labeling experiments, *Eur. J. Biochem.*, 102, 83, 1979.
58. **Beuhler, R. J., Pierce, R. C., Friedman, L., and Siegelman, H. W.,** Cleavage of phycocyanobilin from C-phycocyanin. Separation and mass spectral identification of the products, *J. Biol. Chem.*, 251, 2405, 1976.

59. **Gossauer, A., Hinze, R.-P., and Kutschan, R.,** Total synthesis and elucidation of the relative configuration of two epimeric methanol adducts of phycocyanobilin dimethyl ester, *Chem. Ber.*, 114, 132, 1981.
60. **Schoch, S., Klein, G., Linsenmeier, U., and Rüdiger, W.,** On the linkages between chromophore and protein in biliproteins. III. Elimination reactions with model compounds, *Liebigs Ann. Chem.*, 1976, 549, 1976.
61. **Klein, G. and Rüdiger, W.,** On the linkages between chromophore and protein in biliproteins. V. Stereochemistry of model imides, *Liebigs Ann. Chem.*, 1978, 267, 1978.
62. **Lagarias, J. C., Glazer, A. N., and Rapoport, H.,** Chromopeptides from C-phycocyanin. Structure and linkage of a phycocyanobilin bound to the β subunit, *J. Am. Chem. Soc.*, 101, 5030, 1979.
63. **Schoenleber, R. W., Lundell, D. J., Glazer, A. N., and Rapoport, H.,** Bilin attachment sites in the α and β subunits of B-phycoerythrin. Structural studies on the singly linked phycoerythrobilins, *J. Biol. Chem.*, 259, 5485, 1984.
64. **Rüdiger, W. and O'Carra, P.,** Studies on the structures and apoprotein linkages of the phycobilins, *Eur. J. Biochem.*, 7, 509, 1969.
65. **Rüdiger, W.,** Über die Abwehrfarbstoffe von Aplysia-Arten. II. Die Struktur von Aplysioviolin, *Hoppe-Seyler's Z. Physiol. Chem.*, 348, 1554, 1967.
66. **Köst, H.-P., Rüdiger, W., and Chapman, D. J.,** Bonding between chromophore and protein in biliproteins. I. Degradation experiments and spectral investigations on biliproteins, *Liebigs Ann. Chem.*, 1975, 1582, 1975.
67. **Troxler, R. F., Brown, A. S., and Köst, H.-P.,** Quantitative degradation of radiolabeled phycobiliproteins. Chromic acid degradation of C-phycocyanin, *Eur. J. Biochem.*, 87, 181, 1978.
68. **Kufer, W., Krauss, C., and Scheer, H.,** Two mild, regioselective methods of degrading biliprotein chromophores, *Angew. Chem. Int. Ed. Engl.*, 21, 446, 1982.
69. **Glazer, A. N., Hixson, C. S., and DeLange, R. J.,** Determination of the number of thioether-linked cysteine residues in cytochromes *c* and phycobiliproteins, *Anal. Biochem.*, 92, 489, 1979.
70. **Schoenleber, R. W., Lundell, D. J., Glazer, A. N., and Rapoport, H.,** Bilin attachment sites in the α and β subunits of B-phycoerythrin. Structural studies on a doubly peptide-linked phycoerythrobilin, *J. Biol. Chem.*, 259, 5481, 1984.
71. **Nagy, J. O., Bishop, J. E., Klotz, A. V., Glazer, A. N., and Rapoport, H.,** Bilin attachment sites in the α, β, and γ subunits of R-phycoerythrin. Structural studies on singly and doubly linked phycourobilins, *J. Biol. Chem.*, 260, 4864, 1985.
72. **Kufer, W. and Scheer, H.,** Studies on the plant bile pigments. VII. Preparation and characterization of phycobiliproteins with chromophores chemically modified by reduction, *Hoppe-Seyler's Z. Physiol. Chem.*, 360, 935, 1979.
73. **Kufer, W. and Scheer, H.,** Rubins and rubinoid addition products from phycocyanin, *Z. Naturforsch.*, 37c, 179, 1982.
74. **Klein, G. and Rüdiger, W.,** Thioether formation of phycocyanobilin: a model reaction of phycocyanin biosynthesis, *Z. Naturforsch.*, 34c, 192, 1979.
75. **Frank, G., Sidler, W., Widmer, H., and Zuber, H.,** The complete amino acid sequence of both subunits of C-phycocyanin from the cyanobacterium *Mastigocladus laminosus*, *Hoppe-Seyler's Z. Physiol. Chem.*, 359, 1491, 1978.
76. **Offner, G. D., Brown-Mason, A. S., Ehrhardt, M. M., and Troxler, R. F.,** Primary structure of phycocyanin from unicellular rhodophyte *Cyanidium caldarium*. I. Complete amino acid sequence of the α subunit, *J. Biol. Chem.*, 256, 12,167, 1981.
77. **Troxler, R. F., Ehrhardt, M. M., Brown-Mason, A. S., and Offner, G. D.,** Primary structure of phycocyanin from the unicellular rhodophyte *Cyanidium caldarium*. II. Complete amino acid sequence of the β subunit, *J. Biol. Chem.*, 256, 12,176, 1981.
78. **Sidler, W., Gysi, J., Isker, E., and Zuber, H.,** The complete amino acid sequence of both subunits of allophycocyanin, a light harvesting protein-pigment complex from the cyanobacterium *Mastigocladus laminosus*, *Hoppe-Seyler's Z. Physiol. Chem.*, 362, 611, 1981.
79. **Offner, G. D. and Troxler, R. F.,** Primary structure of allophycocyanin from the unicellular rhodophyte, *Cyanidium caldarium*. The complete amino acid sequences of the α and β subunits, *J. Biol. Chem.*, 258, 9931, 1983.
80. **Füglistaller, P., Suter, F., and Zuber, H.,** The complete amino-acid sequence of both subunits of phycoerythrocyanin from the thermophilic cyanobacterium *Mastigocladus laminosus*, *Hoppe-Seyler's Z. Physiol. Chem.*, 364, 691, 1983.

Chapter 4

BILIPROTEINS: SOME PHYSICAL PROPERTIES

I. NOMENCLATURE

Biliproteins were named for their colors; phycocyanins were blue and phycoerythrins were red. Prefixes and suffixes were used to differentiate between types. To distinguish between biliproteins from red algae and cyanobacteria the prefixes R and C were introduced, e.g., R-phycocyanin, C-phycocyanin, R-phycoerythrin, and C-phycoerythrin.[1] Clearly, the rationale for these prefixes is tenuous, since C-phycocyanin occurs in both cyanobacteria and red algae, but the system has survived mainly because the names can be arbitrarily related to different unique spectral characteristics.

When a distinctively different spectral form of phycoerythrin was isolated from the red alga *Porphyra naiadum*, it was named B-phycoerythrin to distinguish it from R-phycoerythrin. B was chosen because this alga belongs to the order of red algae known as Bangiales.[2] At the same time, B-phycoerythrin was also isolated from *Porphyridium cruentum*.[3] B-Phycoerythrin had much less 490-nm absorbance than did R-phycoerythrin, the result of a much smaller phycourobilin content. A closely related biliprotein, b-phycoerythrin, was completely devoid of phycourobilin.[4]

Phycoerythrocyanin was named because it has a combination of red and blue characteristics.[5] It is a confusing name because it suggests that the protein has both phycocyanobilin and phycoerythrobilin, while in fact phycoerythrocyanin is composed of phycocyanobilin and cryptoviolin.

An unusual group of cyanobacterial biliproteins were first isolated in 1974.[6,7] Unlike the C-phycoerythrins which have only phycoerythrobilin, this type of phycoerythrin has both phycoerythrobilin and phycourobilin. Its bilin content is therefore similar to the red algal phycoerythrins. Here, to denote both its cyanobacterial origin and its phycourobilin content, it is named CU-phycoerythrin.

Allophycocyanin arose as a word in 1933[8] but only much later as a name for a distinct biliprotein group.[3] Allophycocyanin has the same chromophore as C-phycocyanin but a much different spectrum and a unique function.

Allophycocyanin has an absorption maximum at about 650 nm and a fluorescence emission maximum at 660 nm. Isolated phycobilisomes have an emission maximum near 675 nm. This characteristic will be referred to as allophycocyanin 680. There are two different polypeptide complexes that have the allophycocyanin 680 property: allophycocyanin I[9] and allophycocyanin B.[10]

Cryptomonad biliproteins will not be discussed below in detail. When several cryptomonad biliproteins were examined, it was noted that their visible absorption spectra fell roughly into six groups: phycocyanins 645, 630, and 612 and phycoerythrins 566, 555, and 545.[11]

An entire overhaul of this nomenclature would seem appropriate.

II. BILIPROTEINS

Phycobilisomes are composed of a variety of biliproteins which are easily dissociated from each other in the proper solvent. The individual biliproteins can be purified and studied. Surveys have been published listing the distributions of various biliproteins among red algae and cyanobacteria.[12,13]

A. C-Phycocyanin

C-Phycocyanin, which has a visible absorption maximum at 615 to 620 nm, occurs as the major biliprotein in many cyanobacteria and some red algae. Even prior to the realization that large aggregates of C-phycocyanin are needed to produce phycobilisomes, its aggregation properties were being studied. Svedberg and Katsurai[1] found that C-phycocyanin (*Aphanizomenon flos aquae*) had a pH-dependent reversible aggregation ability. At pH 4.6 the protein was noted to have a uniform 223,000 mol wt, while at pH 6.8 it was a mixture of two aggregates of 208,000 and 104,000 mol wt. The corresponding sedimentation coefficients were 11.2 and 5.8S. Much later, sodium dodecyl sulfate (SDS) gel electrophoresis established that C-phycocyanin has two chromophore-bearing subunits: α of about 17,000 mol wt and β of about 19,000 in an equimolar ratio.[14-17] Thus the interconvertible aggregates studied by Svedberg and Katsurai[1] were hexamers ($\alpha_6\beta_6$) and trimers ($\alpha_3\beta_3$). The $\alpha\beta$ unit will be called a monomer. The subunits of C-phycocyanin could be separated for example by gel filtration after the protein was denatured by SDS.[18] Since 1929 several measurements of the molecular weight of C-phycocyanin have appeared; an example was a study by Kato et al.[19] in which the sedimentation coefficient ($s_{20,w}$) was found to be 10.2S and the diffusion coefficient was determined by intensity fluctuation spectroscopy to be 4.73×10^{-7} cm²/sec. Using the Svedberg equation the molecular weight can be found:

$$M = \frac{RT\,s}{D\,(1 - \bar{v}\rho)} \qquad (1)$$

where M is the molecular weight, R is the ideal gas constant, T is the absolute temperature, s is the sedimentation coefficient extrapolated to zero concentration and calculated for the conditions of water and 20°C, D is the diffusion coefficient under the same conditions, $\bar{v}$ is the partial specific volume of the protein, and ρ is the density of the solution. A result of 209,000 was obtained for the molecular weight. They further developed a model in which each monomer of the hexamer had a 4.4-nm diameter. The diameter of the hexamer was 13.2 nm and was in excellent agreement with the measurements from electron microscopy.[20] Also, it appeared that colorless linker polypeptides were not part of these purified hexamers, since the sum of the α and β subunit molecular weights closely equals the experimental value for the hexamers. The hexamers were important structures in the rods of the phycobilisomes. The chromophore content was known from spectroscopy in acidic urea and amino acid sequence to be 18 per hexamer. Brody and Brody,[21] a decade prior to these measurements using physical measurements on native protein, calculated 22 chromophores per 273,000 mol wt. Using a more accurate 220,000 for the hexamer molecular weight, their analysis also gave 18 chromophores per hexamer.

B. R-Phycocyanin

The visible absorption spectra of C- and R-phycocyanin differ in that the latter has a second absorption maximum at 555 nm along with the 615-nm maximum they share (Figure 1). The R-phycocyanin absorption spectrum was first published in 1910.[22] The protein had the same 11.4 and 6.2S aggregates as C-phycocyanin.[23] Below pH 6.2 the 11S aggregate predominated, and above pH 7 only the 6S aggregate was found. As with C-phycocyanin, the tendency of this biliprotein to associate or dissociate depending on the experimental situation made an exact measurement of molecular weight difficult. Using sedimentation equilibrium methods on the model E analytical ultracentrifuge, the molecular weight of 103,000 was determined.[24] Noting the molecular weights of α and β from SDS gel electrophoresis to be about 16,000 and 18,000,[14] respectively, the subunit structure of R-phycocyanin was $\alpha_3\beta_3$ and the 11S form most likely was $\alpha_6\beta_6$. Diffusion coefficients for both R-phycocyanin and R-phycoerythrin have been determined.[25]

C. C-Phycoerythrin

The absorption spectrum of C-phycoerythrin shows a visible maximum at 565 nm. A

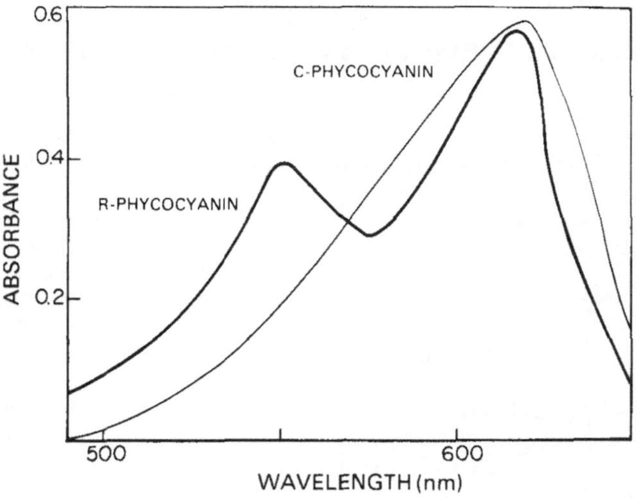

FIGURE 1. Absorption spectra of C-phycocyanin and R-phycocyanin. All absorption spectra are taken at room temperature. The C-phycocyanin is from the cyanobacterium *Phormidium luridum*, and the R-phycocyanin is from the red alga *Ceramium rubrum*. Unless noted otherwise, all the spectra shown in this chapter are for proteins in buffers near neutral pH.

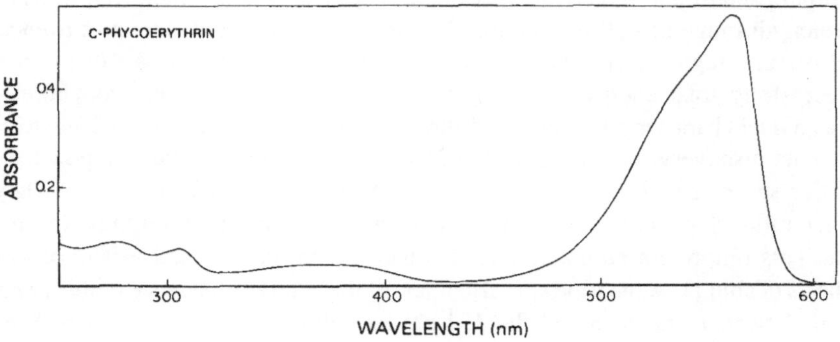

FIGURE 2. Absorption spectrum of C-phycoerythrin. The protein is isolated from the cyanobacterium *Phormidium persicinum*.

small UV maximum is observed at 308 nm which is characteristic of phycoerythrobilin (Figure 2). C-Phycoerythrin can sometimes be obtained with two visible maxima at 542 and 565 nm.[3] The double-peaked C-phycoerythrin was an apparently denatured and dissociated form of C-phycoerythrin.[26,27] The 565- and 542-nm maxima appeared to represent different types of protein-bilin interactions. Hattori and Fujita[28] purified C-phycoerythrin from *Tolypothrix tenuis*. Its sedimentation coefficient ($s_{20,w}$) was 10.2S and its diffusion coefficient 4.4×10^{-7} cm^2/sec. These data in the Svedberg equation (Equation 1) gave a molecular weight of 226,000. Since SDS gel electrophoresis gave molecular weights of about 18,800 for α and 21,000 for β,[15-17] the subunit structure of C-phycoerythrin was $\alpha_6\beta_6$. Its isoelectric point was 4.4.[28] Alberte et al.[29] have purified a C-phycoerythrin from *Synechococcus* sp. WH 8018 with an unusual absorption maximum at 551 nm.

D. CU-Phycoerythrin

The properties of the CU-phycoerythrins are a topic of current investigation. The group, simply called phycoerythrin in the literature, is very heterogeneous in its spectroscopic and

Table 1
PROPERTIES OF CU-PHYCOERYTHRINS FROM CYANOBACTERIA

Organism	Absorption maximum (nm)	Ratio of phycoerythrobilin to phycourobilin	Subunits	Ref.
Synechococcus sp., DC-2	500, 542	4:1	α — 17,000 β — 19,500	29, 30
Synechococcus sp. WH 8103	492, 543	0.6:1	α — 19,500 β — 20,000 γ — 29,000	31
Trichodesmium thiebautii	500, 547, 565(sh)[a]			6, 32
Gloeobacter violaceus	501, 564	6:1	α — 20,500 β — 21,700	7, 33

[a] sh indicates shoulder.

subunit properties (Table 1). Discovered in 1974 by Fujita and Shimura[6] in marine *Trichodesmium thiebautii* and by Rippka et al.[7] in freshwater *Gloeobacter violaceus,* these biliproteins were the first observation of phycourobilin in the cyanobacteria. The 500-nm absorption band of phycourobilin was observed in the intact cells, and it was noted that it enhanced photosynthetic efficiency at 500 nm more than a cyanobacterium having only C-phycoerythrin.[6] Shimura and Fujita[32] noted that beneath the ocean surface, available light was limited to the wavelength range of 450 to 550 nm. *T. thiebautii* was very common in tropical and subtropical oceanic regions, and its ecological success may be due in part to its ability to harvest successfully solar energy at 500 nm through its phycourobilin chromophores.

There was a dual characterization of two distinct examples of this new type of biliprotein[30,33] 7 years after its discovery. Kursar et al.[30] found a phycourobilin-containing phycoerythrin in a particular strain (DC-2), of the marine cyanobacterium *Synechococcus* sp. They calculated a 4:1 ratio of phycoerythrobilin to phycourobilin from its absorption spectrum in acidic urea. This biliprotein completely lacked the 565-nm band characteristic of most C-phycoerythrins (Table 1). A mention was also made of the natural abundance of this organism. Bryant et al.[33] further characterized the CU-phycoerythrin from *G. violaceus*. The phycoerythrobilin to phycourobilin ratio was a very different 6:1. The α and β subunits of this CU-phycoerythrin were then separated by chromatography on a Bio-Rex® 70 ion exchange resin at pH 3.0 using a linear urea gradient in the presence of 2-mercaptoethanol. The α subunit had only phycoerythrobilin, and the β subunit possessed both phycoerythrobilin and phycourobilin. The ratio of phycoerythrobilin to phycourobilin on the β subunit was 3:1. Although Bryant et al.[33] regarded the complete description of the chromophore assignment as tentative, they suggested three phycoerythrobilins for both α and β. No evidence for a chromophore-bearing γ subunit was detected. The undenatured CU-phycoerythrin from this cyanobacterium was separated into two fractions by different phosphate ion concentrations on a column of hydroxylapatite. These two fractions had similar spectra and differed in that one type had a small quantity of colorless polypeptides associated with it. Both fractions of CU-phycoerythrin were composed of a mixture of different aggregates.

Recently, two more studies on CU-phycoerythrins have appeared.[29,31] Alberte et al.[29] have carefully examined the CU-phycoerythrin from *Synechococcus* sp. and have found no evidence for a γ subunit. They used SDS gel electrophoresis, and neither overloading a gel with protein nor using the extremely sensitive silver-staining protocol showed a trace of a γ. Ong et al.,[31] however, have uncovered yet another CU-phycoerythrin in another strain (WH 8103) of a marine *Synechococcus* sp. in which a red-colored γ subunit was present (Table 1). This particular CU-phycoerythrin was a truly unique biliprotein in its chromophore content. The phycoerythrobilin to phycourobilin ratio was 0.6:1! Obviously, the native protein had a very high ratio of absorption at 492 nm compared to 543 nm. This appeared to be the first biliprotein discovered in which a minor bilin (phycourobilin, cryptoviolin, or 697-nm bilin) occurred in excess of the two principal bilins, phycocyanobilin and phycoerythrobilin.

These results demonstrated that the CU-phycoerythrins were very diverse in both chromophore content and subunit structure. They also suggested the wide distribution and probable importance of this protein despite its recent discovery. Perhaps an increased emphasis on the research of marine life forms will result in a continuation of similarly unexpected discoveries.

E. R-Phycoerythrin

R-Phycoerythrins had at least two spectroscopic varieties. One variety had three prominent bands at 568, 545, and 498 nm (*Ceramium rubrum*),[22] and the other had a variation in which the 545-nm band was only a shoulder (*Porphyra perforata*).[3] Haxo et al.[3] noted that R-phycoerythrin from *P. tenera* was a two-peak variety, but others have isolated it with three peaks. They suggested that the spectrum may vary with growth conditions. This speculation was verified for R-phycoerythrin from *Callithamnion roseum*.[34] The intensity of growth light affected the appearance of the band near 540 nm. Increased light intensity produced more 540-nm absorbance relative to the 564- and 496-nm bands. The chromophore content of the low 540-nm form of R-phycoerythrin (*C. roseum*) was compared to that of a three-peak variety (*C. byssoides*). The two-peak variety had a much smaller ratio of phycoerythrobilin to phycourobilin (24.1:10.9) than did the three-peak type (27.6:7.3), although the total number of bilins seemed invariant.

The molecular weight of R-phycoerythrin was reported by several groups to be in the range of 214,000 to 290,000.[1,23,34] The protein had the usual α and β subunits each of about 20,000 mol wt plus an additional γ subunit at about 30,000. Yu et al.[34] calculated a molar ratio of $\alpha+\beta$ to γ of about 12:1 from densities on stained SDS gels. Although the range of native molecular weights was a problem, the subunit molecular weights of approximately 20,000 and the 12:1 ratio of subunits suggested an $\alpha_6\beta_6\gamma$ structure which would have a calculated molecular weight of about 270,000.

The chromophore-bearing γ subunit is much less studied that α or β. Yu et al.[34] found γ subunits of 30,000 and 33,000 mol wt for R-phycoerythrin from *C. byssoides* and γ subunits of 31,000 and 33,000 for the *C. roseum* protein. Stadnichuk et al.[35] found two γ subunits of 31,000 and 33,000 from *C. rubosum* which have different bilin contents. If the $\alpha_6\beta_6\gamma$ subunit structure was valid, then there are at least two different aggregates with different γ subunits for any particular organism. Yu et al.[34] proved this to be correct by separating R-phycoerythrin into two types by electrophoresis on nondenaturing gels. One type had the 33,000 mol wt γ and the second had 30,000. R-Phycoerythrin was also isolated as a lower molecular weight aggregate.[36,37]

F. B-Phycoerythrin

Gantt and Lipschultz[4] have studied B-phycoerythrin from *Porphyridium cruentum*. They found a molecular weight of 260,000, and two bands on SDS gel electrophoresis of 17,300 and 30,000 mol wt. Glazer and Hixson[38] showed that the 17,300-mol wt band contains both the α and β subunits in equal amounts. Gantt and Lipschultz[4] measured a 6:1 ratio (bound dye concentrations) of the 17,300- to the 30,000-mol wt band. They noted that the γ subunit has all the 490-nm (phycourobilin) absorbance bands. Considering the relative molecular weights of α and β to γ, i.e., $\alpha + \beta$ nearly equal to γ, the 265,000-mol wt aggregate is best described as $\alpha_6\beta_6\gamma$ in order to satisfy the subunit ratios [$\alpha/\beta = 1$, $(\alpha + \beta)/\gamma = 6$].

Koller and Wehrmeyer[39] have isolated two forms of B-phycoerythrin (*Rhodella violacea*)

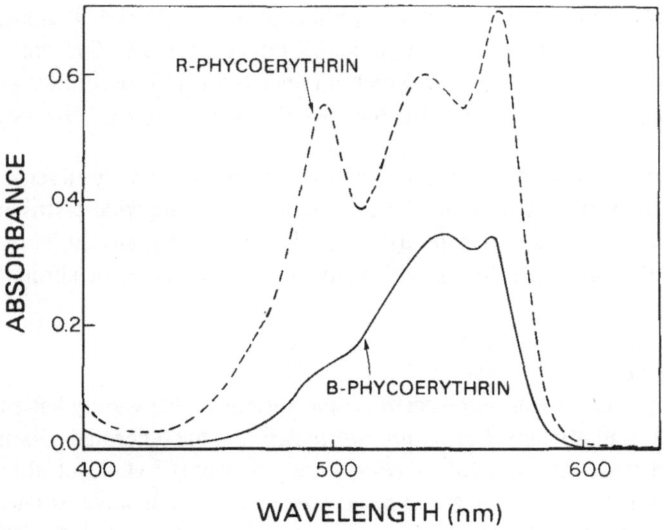

FIGURE 3. Absorption spectra of B-phycoerythrin and R-phycoerythrin. The B-phycoerythrin is from the red alga *Porphyridium cruentum*, and the R-phycoerythrin is isolated from the red alga *Ceramium rubrum*.

by nondenaturing gel electrophoresis. Their physical properties were similar, but they differed slightly in amino acid composition. They referred to these forms as isoproteins, but two other possible explanations can be advanced. Either the two proteins differ in their γ subunits, or one has some colorless linker polypeptide attached. They determined the isoelectric points of the B-phycoerythrins to be 4.4 and 4.2. Van der Velde[36] reported two forms of B-phycoerythrin (*Acrochaetium virgatulum*) which differed in molecular weight.

The visible absorption spectra of B- and R-phycoerythrin have very similar wavelength maxima, but the relative intensities of these bands differed (Figure 3).

G. b-Phycoerythrin

A low-molecular weight form of phycoerythrin from *P. cruentum* can be isolated which lacked a γ subunit. Since the γ subunit had the phycourobilin chromophores, b-phycoerythrin lacked the 490-nm band of B-phycoerythrin. b-Phycoerythrin was found to be composed of two different molecular weight aggregates.[4] Additional studies on this biliprotein showed that its subunits were α and β probably with identical molecular weights and in general very similar properties to the subunits of B-phycoerythrin. The protein showed a protein concentration and solvent-dependent aggregation.[38] B-Phycoerythrin from *Porphyra naiadum* also can be isolated with two forms.[40]

H. Phycoerythrocyanin

In 1976, Bryant et al.[5] isolated and characterized a new biliprotein which was named phycoerythrocyanin from *Anabaena variabilis* and *Anabaena* sp. 6411. Its absorption spectrum had a maximum at 575 nm and shoulders at 595 and 535 nm (Figure 4). The maximum was then between phycoerythrin and phycocyanin in energy. The unique spectrum of phycoerythrocyanin was produced from a mixture of bilins: phycocyanobilin and cryptoviolin. The cryptoviolin chromophore had a characteristic UV band at 332 nm. The protein was purified by three chromatographic procedures: the crude biliproteins were chromatographed on DEAE-cellulose DE52 at pH 7.0 in a linear phosphate gradient; one fraction was then rechromatographed on DEAE-cellulose DE52 at pH 5.5 in a linear acetate gradient; and finally pure phycoerythrocyanin was obtained by selecting a 570-nm rich fraction and placing

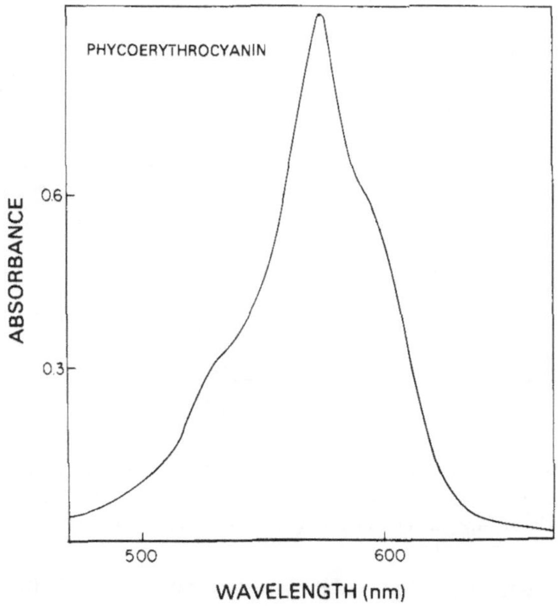

FIGURE 4. Absorption spectrum of phycoerythrocyanin. The phycoerythrocyanin is purified from the cyanobacterium *Anabaena variabilis*.

it on a hydroxylapatite column and eluting with a phosphate gradient. The molecular weight of the protein was determined to be 101,000 to 104,000, and it was composed of two subunits α (16,000 mol wt) and β (20,200 to 20,500 mol wt). Bryant et al. proposed an $\alpha_3\beta_3$ subunit structure. The principal isoelectric point was found at 5.16 to 5.17.

Almog and Berns[41] reported a novel method for the extraction of the α subunit of phycoerythrocyanin (containing the cryptoviolin chromophore) from a mixture of biliproteins. This subunit was quite useful, since it contained a single cryptoviolin chromophore and allowed accurate measurement of the spectral properties of this chromophore.

Phycoerythrocyanin was the only biliprotein from the cyanobacteria that had the cryptoviolin chromophore. This bilin was common on cryptomonad biliproteins and was not detected on any red algal protein. It was the second example of a cyanobacterial biliprotein having two different bilins, since CU-phycoerythrin had phycoerythrobilin and phycourobilin. In both cases the minor bilin (phycourobilin or cryptoviolin) was at a higher energy than the principal bilin with which it was associated.

I. Allophycocyanin

Haxo et al.[3] purified allophycocyanin (Figure 5) by chromatography from four different organisms: *Porphyridium cruentum, Porphyra tenera, Fremyella diplosiphon,* and *Lyngbya lagerheimii.* Although allophycocyanin and C-phycocyanin had identical chromophores, their absorption spectra were quite different, and the 660-nm fluorescence of allophycocyanin was about 20 nm to the red of C-phycocyanin. Although allophycocyanin was a minor component of the biliprotein extracts, Haxo et al.[3] suggested that it was not an artifact. Craig and Carr[42] noted that the C-phycocyanin to allophycocyanin ratio remained fairly constant throughout the growth cycle of *A. variabilis.* This ruled out the possibility of allophycocyanin being an artifact that accumulated at the end of the cycle.

Following earlier unsuccessful efforts to obtain the subunits of allophycocyanin, Brown et al.[43] used SDS gel electrophoresis to demonstrate two subunits, α (17,300 mol wt) and β (19,000 mol wt), for allophycocyanin from the cyanobacterium *Phormidium luridum.* The

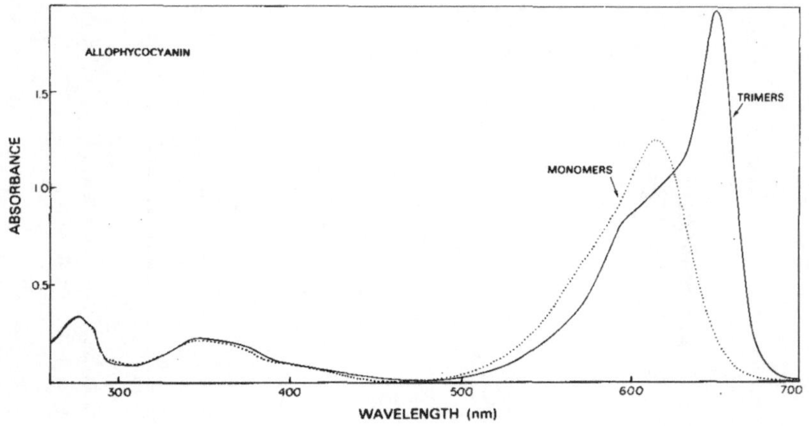

FIGURE 5. Absorption spectra of allophycocyanin monomers and trimers. The trimers are in pH 7.0 sodium phosphate buffer, and the monomers are in 1.0 *M* $NaClO_4$ at pH 7.0. The protein is isolated from the cyanobacterium *Phormidium luridum*.

two subunits were shown to be equimolar. Brown and Troxler,[44] likewise, showed two subunits in a 1:1 *M* ratio for allophycocyanin from the red alga *Cyanidium caldarium*. Cohen-Bazire et al.[45] reported the molecular weight of allophycocyanin from four different sources to be in the range of 95,000 to 105,000. Noting the subunit molecular weights and the 1:1 ratio of subunits,[44] they proposed an $\alpha_3\beta_3$ structure for isolated allophycocyanins. The isoelectric point of allophycocyanin (*A. variabilis*) was 4.95.

Gysi and Zuber[46] developed a chromatography method on DEAE Sephadex A-50 to separate the α and β subunits of allophycocyanin (*Mastigocladus laminosus*) with a linear KCl gradient in the presence of 8 *M* urea.

Gysi and Zuber[47] were able to separate two very similar types of allophycocyanin by ion-exchange chromatography. Zilinskas et al.[9] isolated two forms of allophycocyanin, types II and III, which differed in their 650- to 620-nm ratios. They had similar sedimentation coefficients, molecular weights, and subunits. Allophycocyanin III had a small colorless polypeptide associated with it, and it was possible that this polypeptide induced the absorption change. Both these forms (II and III) fluoresced at 660 to 662 nm. Troxler et al.[48] showed that the N terminal amino acid sequences of the α and β subunits of allophycocyanins II and III were identical. They determined the sequence as far as 30 amino acids from the N-terminus.

J. Allophycocyanin B

Glazer and Bryant[10] isolated allophycocyanin B for the first time from two cyanobacteria *Synechococcus* sp. and *A. variabilis*. Its primary point of interest was a fluorescence emission maximum near 680 nm. Since intact phycobilisomes also fluoresced at this wavelength, allophycocyanin B was an immediate selection as the pigment responsible for transferring excitation energy from the phycobilisomes to chlorophyll *a* in the photosynthetic membrane. This biliprotein had two visible absorbance maxima at 671 and 618 nm. It had a molecular weight of 89,000, and they assumed an $\alpha_3\beta_3$ subunit structure. One phycocyanobilin was associated with each subunit. The yield of purified allophycocyanin B was estimated to be 0.5% of the C-phycocyanin of *Synechococcus* sp. 6301.

Ley et al.[49] purified allophycocyanin B from a red alga *Porphyridium cruentum*. The final step in this purification was an isoelectric focusing experiment in which two bands were obtained. The band with an isoelectric point of 5.45 was shown to be allophycocyanin B. At 20°C the absorption spectrum of this protein had maxima at 619 and 669 nm and a fluorescence emission at 673 nm, and at − 196°C the emission maximum shifted to 679 nm. They suggested that allophycocyanin B is a widespread constituent of phycobilisomes.

Lundell and Glazer[50] have determined from a comparison of several properties that the β subunit of the 650-nm absorption allophycocyanin is probably identical to the β subunit of allophycocyanin B. The α subunits of the two biliproteins were quite distinctive. In acidic urea the α subunit of allophycocyanin B was identical to other phycocyanobilin-containing subunits. However, when the denatured α subunit of allophycocyanin B was returned to pH 6.5 ammonium acetate buffer it had an absorption maximum at 645 nm; whereas the α subunit of allophycocyanin in the same situation had a maximum at 617 nm. The α subunit, therefore, was reponsible for the long-wavelength emission of allophycocyanin B. Another interesting finding was that when allophycocyanin and allophycocyanin B were mixed the subunits exchanged and formed heterologous trimers.[50]

K. Allophycocyanin I

Zilinskas et al.,[9] as mentioned above, have isolated two forms of allophycocyanin (II and III) which have an emission at 660 to 662 nm. They also purified a third form of allophycocyanin with a 678-nm emission maximum from the cyanobacterium *Nostoc* sp. This form, called allophycocyanin I, had a molecular weight of 145,000 and showed three subunits: α, β, and a colored 35,000-mol wt subunit. It had an absorption maximum at 654 nm, slightly to the red of allophycocyanin. A key observation was that the allophycocyanin I fractions originally had two polypeptides, the larger being 73,000 mol wt, which disappeared during further purification. Following this important discovery of a second form of allophycocyanin in addition to allophycocyanin B which had the 680-nm fluorescence emission, two reports appeared that greatly enhanced the characterization of allophycocyanin I. The critical importance of the 35,000-mol wt polypeptide[9] was made clear when Lundell et al.[51] isolated a 75,000-mol wt polypeptide from the cyanobacterium which had absorption maxima at 610 and 665 nm and most importantly an emission maximum at 676 nm. Lundell et al.[52] had isolated this 75,000-mol wt polypeptide earlier but found it to be colorless. Obviously, this first isolation had bleached the chromophore. The second isolation method used a selective precipitation of the 75,000-mol wt polypeptide with 0.5 *M* NH_4SCN. This material was then partially purified by gel filtration chromatography on a Sephacryl® S-200 column. Fractions rich in the 75,000-mol wt polypeptide and showing 660-nm absorbance were selected. When denatured the visible absorption spectrum of this polypeptide closely resembled but was not identical to phycocyanobilin-containing subunits. Digestion of the 75,000-mol wt polypeptide with trypsin resulted in its cleavage to a blue-colored 40,000-mol wt polypeptide. Returning to the findings of Zilinskas et al.,[9] they obtained a chromophore-bearing polypeptide which seemed to be identical in origin to the 40,000-mol wt trypsin digest of Lundell et al.[51] At the same time, from the red alga *P. cruentum,* Redlinger and Gantt[53] obtained a blue-colored 95,000-mol wt polypeptide. Its absorption spectrum showed bands at 610, 650, and 667 nm. At − 196°C, its fluorescence emission maximum was at 680 nm. A 95,000-mol wt polypeptide was also found in thylakoid membrane that had been washed free of biliproteins. Redlinger and Gantt[53] suggested that the same polypeptide can occur in both phycobilisomes and the membrane and therefore was a link between them. Clearly, the 75,000- and 95,000-mol wt polypeptides served the same function in different organisms.

Rusckowski and Zilinskas[54] continued the investigation of the allophycocyanin I from the cyanobacterium *Nostoc* sp. in the presence of a protease inhibitor phenylmethylsulfonyl fluoride. They obtained 95,000- and 80,000-mol wt polypeptides, both of which were colored. Both of these polypeptides were found in complex with other polypeptides as allophycocyanin I. The proteolysis of the 95,000-mol wt polypeptide to the 35,000 form was confirmed. The 80,000-mol wt polypeptide was also a product of proteolysis. In addition to phenylmethylsulfonyl fluoride, diisopropylfluorophosphate also inhibited proteolysis, but toluenesulfonylphenylalanine, *p*-toluenesulfonyllysine chloromethylketone, and soybean trypsin inhibitor were ineffectual. Zilinskas,[55] also using *Nostoc* sp., isolated an aggregate of the 95,000-mol wt polypeptide together with an allophycocyanin trimer. Its fluorescence emission maximum was at 675 nm. The 95,000-mol wt polypeptide was removed from the other components of the allophycocyanin I by gel filtration chromatography with Sepharose® 6B in the presence of 8 *M* urea. The urea was partially removed by a short-duration dialysis into pH 7.0 sodium phosphate buffer containing phenylmethylsulfonyl fluoride to inhibit proteolysis. Complete removal of urea resulted in precipitation of most of the 95,000-mol wt polypeptide. The partially renatured polypeptide had absorption maxima at 660 and 610 nm and an emission maximum at 675 nm. SDS gel electrophoresis showed that the 95,000-mol wt polypeptide was purified.

A comparison of allophycocyanins I and B showed their activities to be associated with quite different polypeptides, although perhaps both had identical phycocyanobilin chromophores. Allophycocyanin B had its important 680-nm fluorescence emission emanating from a specialized α subunit (15,000 to 16,000 mol wt), while the key ingredient of allophycocyanin I was a 75,000- to 95,000-mol wt polypeptide. Since the fluorescence emission of both corresponded very nicely to that of intact phycobilisomes, both were good choices for the phycobilisome final emitter. Canaani and Gantt[56] have obtained fluorescence emission spectra of allophycocyanins I and B and intact phycobilisomes from *Nostoc* sp. and demonstrated that all three have identical emissions of 688 nm at $-196°C$. Moreover, since both I and B occurred in the same organism,[9,54] there was a strong possibility that both were present in the same phycobilisome. Then there would be dual pathways to funnel excitation energy out of each phycobilisome and to two possibly distinct receptors.

The ratio of allophycocyanin I to B and their distribution among various organisms were not generally known. Phycobilisomes from *G. violaceus* were a unique case and did not fluoresce with a 680-nm maximum nor was allophycocyanin B (or I) isolated from this organism.[33,57] The phycobilisomes of this cyanobacterium were unusual and did not correspond to either the hemi-discoidal or the hemi-ellipsoidal classes.

L. Cryptomonad Biliproteins

Cryptomonads do not have phycobilisomes and will not be discussed here in detail. Although the biliproteins from cyanobacteria and red algae were typically isolated as trimers ($\alpha_3\beta_3$) and hexamers ($\alpha_6\beta_6$ and $\alpha_6\beta_6\gamma$), cryptomonad biliproteins were isolated as dimers ($\alpha_2\beta_2$).[58] Since the trimers and hexamers (disk-like structures) were engineered as basic building blocks of the unique phycobilisome structure, finding a different aggregate for biliproteins from nonphycobilisome-containing organisms was not surprising.

III. C-PHYCOCYANIN AGGREGATION

The assembly properties of C-phycocyanin have been examined in more detail than the other biliproteins. The aggregation of C-phycocyanin from cyanobacterial sources has been studied as a function of temperature, ionic strength, pH, and protein concentration. The great majority of these experiments were performed before the discovery of the colorless linker polypeptides. It is not possible, therefore, to assess fully the role these polypeptides play in each step of the aggregation process. The correlation that is noted between the degree of purification of a particular biliprotein and its aggregation is probably caused by the removal of colorless linker polypeptides during the purification. In general, biliproteins that were purified mainly through reliance on ammonium sulfate fractionation aggregated differently and more extensively than the biliproteins that have received various chromatographic manipulations.

For example, C-phycocyanin from the cyanobacterium *Spirulina platensis* has been pur-

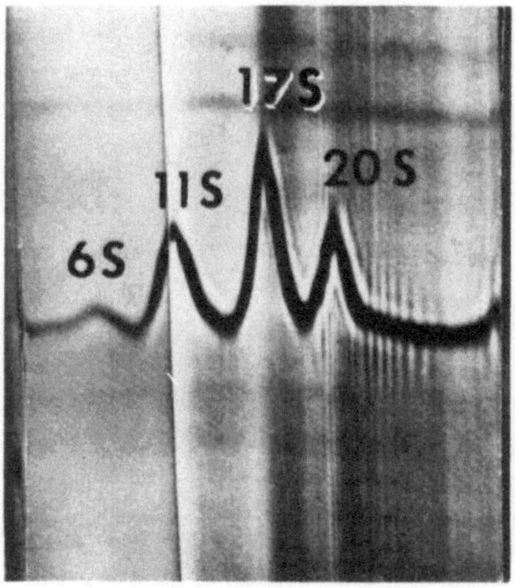

FIGURE 6. Ultracentrifuge schlieren pattern for C-phycocyanin isolated from the cyanobacterium *Spirulina platensis*. The protein was purified by ammonium sulfate fractionation and shows from left to right 6, 11, 17, and 20S aggregates. The solvent is pH 6.0 sodium phosphate buffer.

ified by two methods.[59,60] One method employed techniques including chromatography on a brushite column and resulted in protein that was mainly a trimer in pH 6.0 phosphate buffer,[60] but when only purified by ammonium sulfate fractionation a mixture of aggregates including hexamers and larger assemblies (Figure 6) was identified by the same analytic technique in the same buffer.[59] An indication that chromatography was the key factor in modifying aggregation through removal of linker polypeptides was experiments by Scott and Berns.[61] Their chromatography experiments caused modifications in the aggregation process and also showed the separation of colorless polypeptide from the biliprotein. Aggregates larger than 11S were the particular targets of these effects.

Scott and Berns,[61] Hattori et al.,[62] and Berns and Scott[63] have characterized many of the properties of this aggregation process. The 11S aggregate ($\alpha_6\beta_6$) was found to be most stable at pH 5.0 and 6.0, and the 6S aggregate ($\alpha_3\beta_3$) increased in relative concentration as the pH increased. Sodium chloride (2 *M*) increased the 11S percentage at the expense of 6S. Increasing the temperature from 5—10°C to 20—30°C likewise caused a conversion of 6 to 11S. These effects concerning 6 and 11S were reversible. A decrease in protein concentration produced a dissociation of the 11S aggregate. It was noted that the factors contributing to the 11S stabilization were those that favor hydrophobic interactions. C-Phycocyanin from *Anacystis nidulans* showed very similar behavior, but a dimer appeared to replace the trimer in the assembly scheme.[64] Osmotic pressure measurements confirmed the increased aggregation as temperature increased.[65] A study of the effect of protein concentration on the aggregation of C-phycocyanin at pH 6.0 showed that at very low protein concentrations (<0.02 mg/mℓ) the trimer and dodecamer aggregates have disappeared and only 3S ($\alpha\beta$) and 11S were left.[66] Finally, as the concentration decreased, 100% 3S was obtained. In the region where only 11 and 3S were present, a molar equilibrium constant was calculated using 35,000 for the molecular weight of monomers:

$$6\alpha\beta \rightleftharpoons \alpha_6\beta_6 \quad K = \frac{[\alpha_6\beta_6]}{[\alpha\beta]^6} = 10^{31} \tag{2}$$

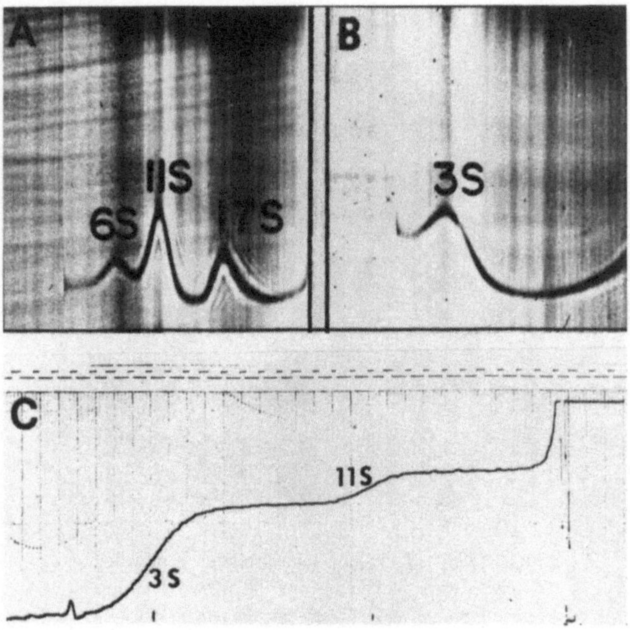

FIGURE 7. Ultracentrifugation pattern for C-phycocyanin aggregates as a function of pH. Sedimentation is from left to right. Panel A is pH 6.0 sodium phosphate buffer, panel B is pH 3.9 sodium acetate buffer, and panel C is pH 4.6 sodium acetate buffer. Panels A and B are photographs using the schlieren optics of the analytical ultracentrifuge, and panel C is measured from the absorbance at 620 nm using the photoelectric scanner of that instrument.

Also in this study, band sedimentation was performed instead of the normal boundary method to obtain sedimentation coefficients. Two aggregates were studied which gave $s_{20,w}$ of 10.7 and 16.8S. The 10.7S was the hexameric C-phycocyanin aggregate. Another band sedimentation study found 21.4S for yet another C-phycocyanin aggregate.[67] The identities of these 16.8 and 21.4S aggregates can be obtained by the following:

$$(M_1/M_2)^{2/3} = s_1/s_2 \qquad (3)$$

where M is the molecular weight and s is the sedimentation coefficient. These data showed that the 16.8S was 413,000 mol wt and 21.4S was 594,000 mol wt, using 210,000 for the molecular weight of the hexamer. Thus the 16.8 and 21.4S aggregates were stacks of two and three hexamers, respectively. Monomers of C-phycocyanin may be obtained at higher protein concentration near neutral pH through the action of salts like NaSCN and $NaClO_4$.[68] Monomers of C-phycocyanin from certain organisms were also prepared at pH 3.9 (Figures 7 and 8). When this homogeneous monomeric solution was dialyzed to pH 4.7 (the isoelectric point) an 85% conversion to hexamers occurred.[66] A detailed study of the effect of 2 *M* NaCl at several pH values on C-phycocyanin aggregates showed that the salt protected the 11S aggregate from dissociation at pH values of 7 and above.[69]

Aggregates of C-phycocyanin can be obtained from separated subunits.[70] Equimolar amounts of α and β were dissolved in 8 *M* urea at pH 8.0. The solution was dialyzed into 3 *M* urea at pH 6.7, phosphate buffer at pH 6.7, and finally into phosphate buffer at pH 7.0. The reconstituted C-phycocyanin was then purified by ion-exchange chromatography and obtained in a 40 to 60% yield. The renaturing protocol was based on experiments by Murphy

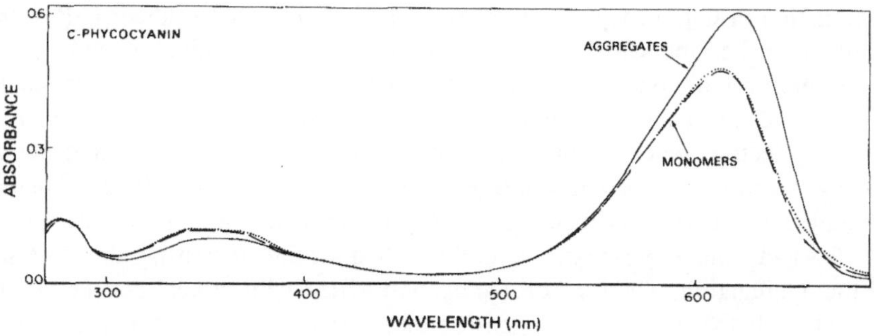

FIGURE 8. Absorption spectra of C-phycocyanin monomers. The protein is isolated from the cyanobacterium *Phormidium luridum*. The monomer spectra are in either pH 3.9 sodium acetate-acetic acid buffer or 1.0 *M* NaSCN at pH 6.0. Aggregates are in pH 6.0 sodium phosphate buffer.

Table 2
EQUILIBRIUM CONSTANTS FOR C-PHYCOCYANIN AGGREGATION AT pH 4.8 AND 6.0

	Equilibrium constants[61,66,72]	
Equilibrium	**pH 6.0**	**pH 4.8**
Hexamer-monomer	10^{-31}	9×10^{-31}
Trimer-hexamer	4×10^{4}	6×10^{5}
Trimer-monomer[a]	6×10^{-14}	7×10^{-13}

[a] The trimer-monomer equilibrium constant at pH 6.0 is not reported in the literature. It was calculated from the other equilibrium constants as follows:

$$K^2_{TM} = K_{HM} \times K_{TH}.$$

and O'Carra,[71] who started with whole C-phycocyanin in 8 *M* urea. Interestingly, the reconstitution of α and β could be accomplished even when the two subunits were from different cyanobacteria.

An ultracentrifugation study by Huang and Berns[72] of C-phycocyanin (*Phormidium luridum*) aggregation at pH 4.8, using sedimentation equilibrium and velocity methods, allowed calculation of the following molar equilibrium constants:

$$3\alpha\beta \rightleftharpoons \alpha_3\beta_3 \quad K = \frac{[\alpha_3\beta_3]}{[\alpha\beta]^3} = 1.4 \times 10^{12} \tag{4}$$

$$2\alpha_3\beta_3 \rightleftharpoons \alpha_6\beta_6 \quad K = \frac{[\alpha_6\beta_6]}{[\alpha_3\beta_3]^2} = 6.1 \times 10^{5} \tag{5}$$

$$6\alpha\beta \rightleftharpoons \alpha_6\beta_6 \quad K = \frac{[\alpha_6\beta_6]}{[\alpha\beta]^6} = 1.1 \times 10^{30} \tag{6}$$

These equilibrium constants at pH 4.8 (Table 2) can be compared with those calculated by others for C-phycocyanin at pH 6.0.[61,66,72] The literature did not contain a direct measurement of the trimer-monomer equilibrium constant at pH 6.0.

The aggregation of C-phycocyanin from red algae has also been studied. C-Phycocyanin from *C. caldarium* had the same aggregates and the same aggregation dependence on pH, temperature, and protein concentration as cyanobacterial C-phycocyanins. The stability of the hexameric and dodecameric aggregates was particularly strong.[73] Several bands were obtained on isoelectric focusing columns. Saito et al.,[74] Mizuno et al.,[75] and Iso et al.[76] studied the aggregation of C-phycocyanin (*Porphyra tenera*) by both sedimentation equilibrium and sedimentation velocity techniques. They found mainly a monomer-trimer equilibrium at pH 6.8 and a monomer-hexamer equilibrium at pH 5.4. Increasing the ionic strength increased the aggregation. The type of aggregation varied with protein concentration. In a more detailed analysis, Iso et al.[76] suggested the existence of tetramers and also, unlike other studies, found that increasing the temperature decreased the aggregation.

IV. AGGREGATION-SPECTRA RELATIONSHIP

For C-phycocyanin, monomers have a visible absorption maximum to the blue of the aggregated forms. Since the blue shift occurred for monomers produced by both low pH and sodium thiocyanate at pH 6, the shift was a property of the monomer itself and not the solvent (Figure 8). When C-phycocyanin was denatured, the visible absorption band was quite hypochromic. In addition, the near-UV band was hyperchromic, resulting in a much lower visible to near-UV absorption ratio for the denatured state (Figure 9). In Figure 9, the visible maxima were set at equal absorbances, but in actuality the denatured protein had a much lower visible absorption. The effect of SDS was nearly identical to that of detergent plus heating (Figure 9). Heating alone would have also denatured the protein so that there was a stable chromophore conformation that was characteristic of the denatured protein. Rosenberg et al.[77] have shown similar spectra for the effect of this detergent on C-phycocyanin.

Scheer and Kufer[78] have proposed that the ratio of visible to near-UV absorbance was an indicator of the conformation of the phycocyanobilins of C-phycocyanin. They suggested that an extended tetrapyrrole conformation would have a higher ratio of visible to near-UV absorbance than a cyclic conformation (Figure 9). When the protein was denatured, the tetrapyrroles were relatively free of noncovalent interactions with the apoprotein and were thus able to achieve their own stable status. Scheer and Kufer[78] offered citations of quantum chemical studies on tetrapyrroles and model compound data to validate their proposal. They noted that porphyrins which were constrained to be cyclic by covalent bonds have a very low visible to near-UV ratio. Quantum chemical studies[79-84] have determined that cyclic tetrapyrroles should indeed have a lower visible and a higher near-UV absorbance than an extended conformation. Burke et al.,[79] in studies on another tetrapyrrole-containing protein from higher plants, phytochrome, performed theoretical calculations on the chromophore that are applicable to C-phycocyanin and other biliproteins. They calculated the oscillator strengths at 380 and 700 nm and found that their values varied with assumed tetrapyrrole geometry. The value at 700 nm increased gradually from about 0.2 for a cyclic conformation to over 1.0 for an extended conformation. The 380-nm oscillator strength was also affected but in the opposite direction. It was calculated at about 1.0 for the cyclic state and about 0.1 for the extended state. Another example was the study by Chae and Song[82] on the tetrapyrrole biliverdin. The oscillator strength of a visible transition for a linear conformation was found to be 2.48 and for a near UV to be 0.06. However, when a cyclic conformation was calculated the visible transition was at 0.25 and the near UV was at 1.20. Bois-Choussy and Barbier[85] have synthesized tetrapyrroles that were constrained by covalent bonds to be in an extended conformation. As predicted, their visible absorption bands were strong absorbers, and they showed very little 340-nm absorption. An important function of the apoprotein of the biliproteins must be to maintain the bilins in a biologically useful extended conformation.

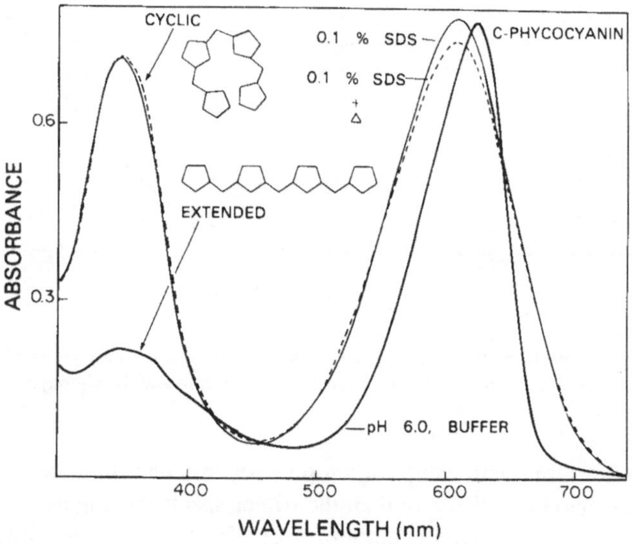

FIGURE 9. Absorption spectra of native and denatured C-phycocyanin. The protein is isolated from the cyanobacterium *Phormidium luridum*. The native protein is in pH 6.0 sodium phosphate, and the denatured protein is in 0.1% SDS. The detergent-treated protein in one case has been kept for 1 min in boiling water. The absorbances for all three samples was set at equal values for their visible absorption maxima, and the absorbance in the visible region is actually significantly greater for native than for denatured protein.

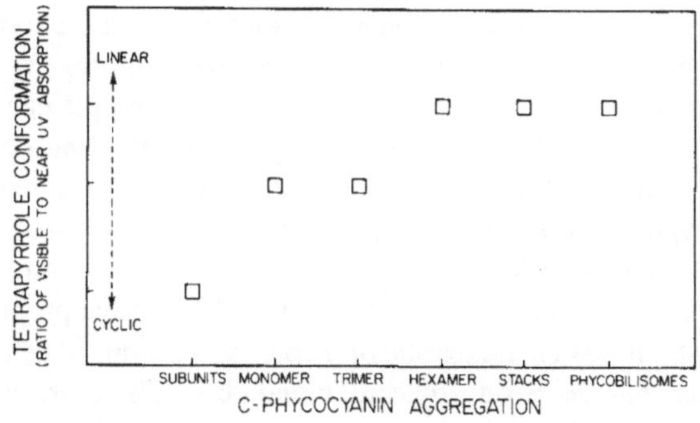

FIGURE 10. Conformation of the phycocyanobilin of C-phycocyanin as a function of aggregation. The subunit value is for subunits in totally denaturing conditions, e.g., 8.0 *M* urea at pH 3.0.

The conformation of native C-phycocyanin in the phycobilisomes was achieved in discrete steps (Figure 10).[67] When denatured C-phycocyanin was compared to monomers, the monomers showed a large increase in the visible to near-UV absorbance ratio. Monomers, therefore, have bilins which are significantly more extended than those in the denatured protein. When trimers were examined, however, their ratio was similar to that for monomers. At the hexamer level the ratio of visible to near-UV absorbance also showed a definite increase, suggesting another change in bilin conformation to a further extended state. When dodecamers and intact phycobilisomes (Figure 11) were studied, their ratios were very similar to those for hexamers. The hexamer level then determined the final conformation of the C-phycocyanin tetrapyrroles. In addition to bilin conformation, factors like the polarity of their environments will affect the spectrum of a bilin.

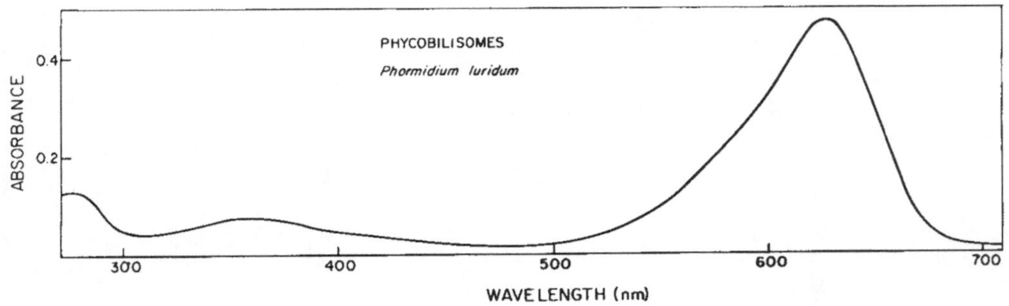

FIGURE 11. Absorption spectrum of isolated phycobilisomes. The organelles are prepared from the cyanobacterium *Phormidium luridum* and are in 0.75 *M* phosphate. C-Phycocyanin and the allophycocyanins are present in this cyanobacterium.

The aggregation-spectra relationship for allophycocyanin was quite interesting. The monomer spectrum was markedly different than the trimer spectrum (Figure 5). The trimer was characterized by a relatively sharp maximum near 650 nm and lower intensity shoulders toward the blue of the main peak. The monomer may be produced at pH 3.9 or in the presence of 1.0 *M* $NaClO_4$ at pH 7.0.[86,87] The monomer spectrum was very similar to that of monomeric C-phycocyanin, both having identical band maxima and band widths at half height. The fluorescence emission maxima of allophycocyanin monomers and trimers were also characteristic with the monomer at 642 nm and the trimer at 660 nm. Monomers can be completely reconverted to trimers by returning the solvent to pH 7.0 buffer. The fluorescence polarization spectrum of the trimer form was very flat across the entire visible band and did not prominently display the two plateau features of the phycocyanins and phycoerythrins.[86] The magnitude of the polarization was low. A low polarization for these proteins was indicative of the extent of excitation energy transfer, and the allophycocyanin trimer surprisingly had a lower polarization than did C-phycocyanin aggregates with more chromophores. The monomer polarization did have the characteristic two-plateau polarization which has been assigned by Teale and Dale[88] to sensitizing and fluorescing chromophores in the cases of phycocyanins and phycoerythrins. The absorption and fluorescence polarization spectra of allophycocyanin trimers was very unusual when compared to C-phycocyanin aggregates and monomers of both C-phycocyanin and allophycocyanin. A unique status of its chromophores was therefore anticipated. MacColl et al.[86,87] have proposed the possibility that the 650-nm band which was characteristic of trimers was produced by exciton interactions between chromophores that were on adjacent monomers. Therefore as monomers were formed by dissociation, the exciton interaction was broken and the 650-nm band was vacated. Murakami et al.[89] have likewise investigated monomer-trimer spectra of allophycocyanin but disagreed on the cause of the differences between them. They suggested the spectra were produced by the protein assuming either a tight or a relaxed state around the bilin and thereby affecting its energy levels.

It is difficult to distinguish totally between an exciton interaction[86] or a chromophore conformation[89] as the cause for the 650-nm band in allophycocyanin trimers. The tetrapyrrole chromophores are very flexible and readily able to assume a very large array of conformations that the apoprotein may impress upon them. It is quite reasonable to accept that the 650-nm band could be produced by such an event. Another major argument against exciton interaction comes from the circular dichroism (CD) spectrum of allophycocyanin (see Section V). Strongly coupled excitons which could cause the band splitting needed to produce the 650-nm band are characterized by both a negative and a positive band in their CD spectrum. Allophycocyanin has two small positive bands in the visible absorption spectra with no noticeable negative component. However, recent analysis of the CD spectrum of allophycocyanin was quite interesting. Csatorday et al.[90] showed that the allophycocyanin trimer had two such negative components, but that they were almost completely obscured by the positive bands. A small sliver of one negative band was detectable,[43] and the greater intensity of the CD signal of the monomer also supported the idea that the trimer spectrum was depleted by a negative-positive overlap.[90]

Huang[91] has studied the monomer-trimer equilibrium of allophycocyanin by stopped-flow kinetics using both light-scattering and absorbance modes. She has determined that there were two absorbance changes that occurred when trimers were converted to monomers. The faster was associated in rate with the protein dissociation, and the slower may be a conformational relaxation of bilins occurring on monomers which were newly formed from trimers:

$$\alpha_3\beta_3 \rightarrow 3(\alpha\beta) \rightarrow 3\alpha\beta \qquad (7)$$

The rate of trimer dissociation was similar to the rate of the fast absorbance change, suggesting, as was mandated by the exciton model, that the dissociation and absorbance changes are linked. These results also indicated that whether or not exciton interaction occurred, a conformational change in the bilin may accompany the monomer-trimer equilibrium.

The equilibrium between allophycocyanin monomers and trimers can be studied through their spectral differences. Beer's law equations can be derived which give monomer and trimer molar concentrations:[92]

$$[\alpha_3\beta_3] = \frac{1.76 \times 10^5\, A_{652} - 0.24 \times 10^5\, A_{620}}{11.7 \times 10^{10}} \qquad (8)$$

$$[\alpha\beta] = \frac{A_{620} - 4.55 \times 10^5\, [\alpha_3\beta_3]}{1.76 \times 10^5} \qquad (9)$$

These concentrations can be used to calculate the equilibrium constant for the assembly of monomers to trimers. The chaotropic salts, NaSCN and $NaClO_4$, were effective in dissociating allophycocyanin trimers, and equilibrium constants can be calculated at a number of different salt concentrations:

$$K = \frac{[\alpha_3\beta_3]}{[\alpha\beta]^3} \qquad (10)$$

The Gibbs free energies (ΔG) corresponding to these equilibrium constants were calculated from:

$$\Delta G = -RT \ln K \qquad (11)$$

and the free energies can be extrapolated to zero-salt concentration to give the ΔG and K for the dissociation in pH 6 and 7, 0.1-ionic strength sodium phosphate buffer. The equilibrium constant for the assembly of monomers to trimers was about 1.7×10^{15}, and the free energies were -20.4 kcal/mol at pH 7.0 and -20.8 kcal/mol at pH 6.0. This meant that each monomer-monomer contact contributed -6.9 kcal/mol at pH 6.0 to the trimer stability. For C-phycocyanin, the free energy per monomer-monomer contact was -5.6 kcal/mol in the trimer and -6.9 in the hexamer at pH 6.0 (Equations 4 and 6).

V. CD SPECTROSCOPY OF BILIPROTEINS

CD spectroscopy is measured by alternately irradiating the material being studied with

Table 3
CD BANDS OF BILIPROTEINS

	Visible CD		
Protein	**Maximum (nm)**	**Minimum (nm)**	**Ref.**
Allophycocyanin	655, 625		43
C-Phycocyanin	630, 595		93
R-Phycocyanin	627, 552		24
Phycoerythrocyanin	578		5
C-Phycoerythrin	548, 520	567	94
R-Phycoerythrin	572, 536, 496		93
B-Phycoerythrin	572, 537, 505	558	93
b-Phycoerythrin	529	560	38
CU-Phycoerythrin	555	542	30
Allophycocyanin B	680, 630	668	56
Allophycocyanin I	668	642	56
Phycocyanin 612	595, 574	627	95
Phycocyanin 645	609, 583	644	58
Phycoerythrin 545	530	562	96
Phycoerythrin 566	540, 532	574	96

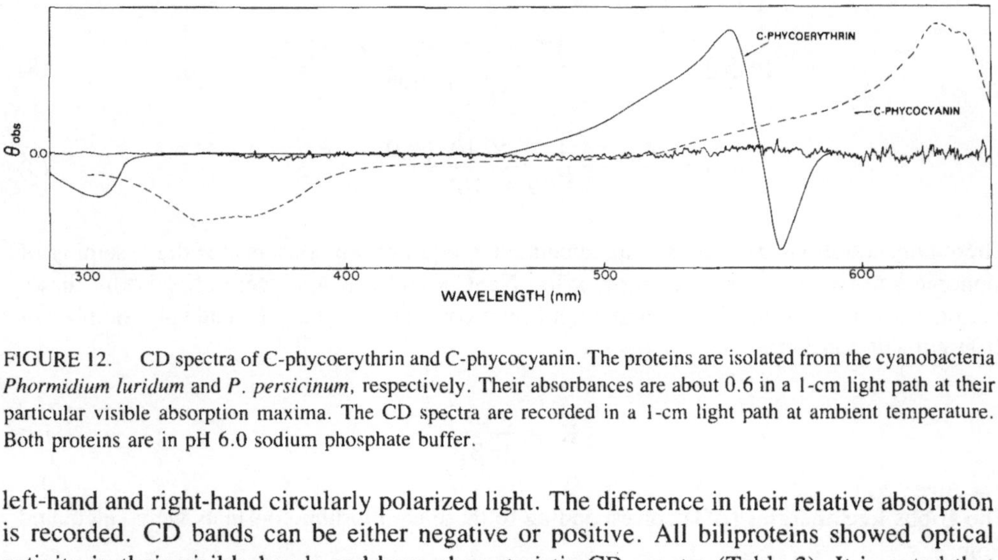

FIGURE 12. CD spectra of C-phycoerythrin and C-phycocyanin. The proteins are isolated from the cyanobacteria *Phormidium luridum* and *P. persicinum*, respectively. Their absorbances are about 0.6 in a 1-cm light path at their particular visible absorption maxima. The CD spectra are recorded in a 1-cm light path at ambient temperature. Both proteins are in pH 6.0 sodium phosphate buffer.

left-hand and right-hand circularly polarized light. The difference in their relative absorption is recorded. CD bands can be either negative or positive. All biliproteins showed optical activity in their visible bands and have characteristic CD spectra (Table 3). It is noted that some biliproteins had only positive CD bands and others had both positive and negative bands. The CD spectra of C-phycocyanin and C-phycoerythrin were representative of these two types of spectra (Figure 12).

An important use for CD spectroscopy was to determine if chromophores were interacting with each other. If, for example, a dimer of two identical chromophores occurred, their CD spectrum would be characteristic of the coupling of their oscillators. The CD for strong exciton coupling would be dependent on the distance between the chromophores and their geometry. An exciton dimer has two CD bands: one positive and one negative. The rotational strength of the exciton bands often were much larger than that of the isolated monomers. Therefore, those CD spectra showing both negative and positive CD bands were candidates for exciton interaction. However, biliproteins were complex, being composed either of more than one type of chromophore or of chemically identical chromophores in different conformations. The CD analysis was thus complicated, since these different chromophores may have either positive or negative bands, and such a CD spectrum was therefore not clear proof of exciton interaction. Two biliproteins, allophycocyanin[43,90] (see Section IV) and the cryptomonad phycocyanin 645,[97] have been studied by CD, and the possibilities of strongly coupled exciton interactions between certain pairs of chromophores have been discussed.

Canaani and Gantt[56] measured the CD spectra of allophycocyanins I, II, III, and B at room temperature and allophycocyanins I, II, and III at −196°C. At −196°C, the broad 668-nm positive CD band of allophycocyanin I from room temperature was shown to have three components at 655, 669, and 680 nm. Allophycocyanin I was not the 95,000-mol wt polypeptide alone but a complex of its 35,000-mol wt cleavage product associated with allophycocyanin.

Langer et al.[98] have studied C-phycoerythrin (*Pseudanabaena* W1173) by CD spectroscopy. The separated α and β subunits were returned to pH 7 phosphate buffer, and the summation of their CD spectra was very similar to the CD spectrum of native C-phycoerythrin. The β subunit had both negative and positive CD bands, and the positive α bands diminished the negative β band and augmented the positive. The close correspondence between the sum of the α and β subunit spectra and that of the native protein monomers suggested that very little change in bilin conformation or interactions occurred that was specific to the α-β subunit interaction of monomers. Glazer et al.[99] had previously performed the same protocols on C-phycocyanin. In this case, however, the sum of α and β subunits have a CD spectrum with much more rotational strength than native C-phycocyanin. They pointed out that self-aggregation of the α and β subunits could contribute to this difference. The possibility of aggregation is a serious factor to be prudently considered when studying separated subunits. The CD spectrum of aggregated C-phycocyanin was also significantly different than the monomeric CD spectrum. The absorption spectra, likewise, showed this difference, with the monomer having a 30% lower extinction coefficient than hexamers. Troxler et al.[100] have also studied the CD spectra of C-phycocyanin and its renatured α and β subunits, and they emphasized the difficulties in assessing the relative contributions of different aggregates vs. different chromophore environments on the same aggregate. Frackowiak et al.[101] have also used CD to monitor the properties of biliproteins.

The CD spectrum of intact phycobilisomes (*Fremyella diplosiphon*) containing C-phycoerythrin, C-phycocyanin, allophycocyanin, and allophycocyanin 680 was composed of several positive and negative bands. Rigbi et al.[102] have progressively dissociated the phycobilisomes by exposing them to low temperatures (2°C) and lowering the phosphate concentrations. Large-scale dissociation of these phycobilisomes resulted in noticeable simplication in the CD spectra. This may demonstrate that certain chromophore events were specific to phycobilisomes, but some of these changes were identified as occurring after phycobilisome dissociation and were therefore not attributed to phycobilisome-specific bands. Although less sensitive to these changes than CD, the absorption spectrum of intact phycobilisomes (*Nostoc* sp.) also showed the presence of an absorption band that disappeared upon phycobilisome dissociation.[103] Lundell and Glazer[104] have examined the CD spectra of the phycobilisome core and its dissociation complexes isolated from mutant strain AN112 of the cyanobacterium *A. nidulans*. They reported very definite evidence of chromophore changes occurring as the core was assembled. However, no CD change occurred when the C-phycocyanin rods were added to the cores.

CD spectroscopy has been used to monitor the denaturation of biliproteins. Urea denaturation of several C-phycocyanins has been monitored by both the decrease in visible absorption and the change in CD intensity of the apoprotein at 222 nm.[105,106] Both methods give similar although not identical results. Particularly at low urea concentrations the visible absorption changed more extensively. The general similarities between the two methods, however, suggested that visible absorption and probably fluorescence could be used as indirect probes of biliprotein unfolding. The visible CD spectrum also lost rotational strength upon urea denaturation.

Table 4
SPECIFIC ABSORPTIVITIES OF PURIFIED BILIPROTEINS

Protein	Solvent (pH)	Absorptivity[a]	Ref.
Allophycocyanin	7	7.3 (trimers)	43
	7, 1 *M* $NaClO_4$	5.2 (monomers)[b]	87
C-Phycocyanin	7.0	6.5 (monomers)	99
	5.5	9.5 (hexamers)[c]	99
R-Phycocyanin	7.0	7.0	24
Phycoerythrocyanin	6.8	8.5	5
C-Phycoerythrin	7.0	12.7	24
R-Phycoerythrin	6.8	8.2	34
B-Phycoerythrin	5.5	10.0	38
b-Phycoerythrin		9.7	38
Phycocyanin 645	6.0	11.4	58
Phycoerythrin 545	6.0	12.6	96

[a] Calculated for a 1 g/ℓ solution in a 1-cm light path at the visible absorption maximum.
[b] Calculated using the 7.3 value for trimers.
[c] At pH 6.0, a mixture of C-phycocyanin aggregates yields an absorptivity of 6.0.[66]

For C-phycoerythrin (*Pseudanabaena* W1173) the visible CD spectrum was also attenuated in urea.[98] At up to 6.5 *M* urea the negative and positive bands were retained, although both were diminished, but at 8 *M* urea the CD spectrum completely lost its negative band and had two positive bands instead.

The absorbance changes for C-phycocyanin[105] and C-phycoerythrin[98] with increasing urea concentration were complex and were products of both aggregate dissociation and polypeptide unfolding. At up to about 3 *M* urea the slope of the absorbance vs. urea concentration curves was smaller than above 4 *M*. Protein unfolding then may commence above 3 *M* urea.[98] The CD at 222 nm conversely showed little change up to 4 *M* urea.[98,105] The events occurring at up to 3 *M* urea which produced the absorption changes were probably disaggregation, and disaggregation then seemed to occur without much effect on the structure of the polypeptides.

VI. ABSORPTIVITIES

Frequently, visible absorption spectra are used to calculate biliprotein concentration. The specific absorptivities have been established for several biliproteins (Table 4), and there was a reasonable compliance with Beer's law over certain absorption ranges. The absorptivities of the various C-phycocyanin aggregates vary as do those for allophycocyanin monomers and trimers. Another report of the absorptivity of a mixture of B-phycoerythrin and b-phycoerythrin has shown a lower value than those in Table 4.[4] The value given have the advantage of being consistent with the magnitudes of other phycoerythrobilin-containing biliproteins, but further confirmatory work is needed.

VII. DENATURATION AND RENATURATION

Denaturation was widely used to study biliproteins. Subunits were separated after urea denaturation, and their molecular weights were determined after detergent denaturation. A comparison of the denatured to native absorption spectra was a useful indicator of chromophore conformation. Visible absorption in acidic 8 *M* urea was the method of obtaining the type and quantity of the chromophores. In this section, only certain comments on the denaturation process not covered elsewhere (see Section V) will be made.

Murphy and O'Carra[71] found 8 *M* urea to be effective in rapidly denaturing C-phycocyanin (*Nostoc punctiformis*). Removal of urea produced a reversal of the denaturation. The best results for renaturation were obtained when 8 *M* urea was diluted to 3 *M* urea and the remaining urea dialyzed away. This method restored 77% of the original absorption.

Erokhina and Krasnovskii[107] have observed that denaturing agents, including urea and pH extremes, converted the allophycocyanin spectrum into one resembling C-phycocyanin. Allophycocyanin B was studied and found to convert readily to a spectral form similar to allophycocyanin.[108] The source of biliproteins can greatly affect their denaturation proclivities. Allophycocyanin from a thermophilic cyanobacterium (*Synechococcus lividus*) will denature at much higher temperatures and much higher urea concentrations than an allophycocyanin from an organism (*Phormidium luridum*) that is normally grown at moderate temperatures (a mesophile).[109]

Allophycocyanins (*P. luridum* and *C. caldarium*) that had completely lost their 650-nm absorption through urea denaturation can regain it by returning the protein to pH 7 phosphate buffer.[43,44] Gysi and Zuber[110] recombined the α and β subunits of allophycocyanin (*M. laminosus*) in potassium phosphate buffer at pH 8.0 and found a 75% restoration of 650-nm absorbance. However, when the separated α and β subunits of allophycocyanin (several sources) were returned to undenaturing buffers without recombination of the subunits, the results were entirely different. The "renatured" α and β subunits exhibited absorption maxima between 610 and 618 nm, and there was no hint of a 650-nm band.[110] Apparently, the generation of the 650-nm band was dependent on structural factors necessitating that both α and β subunits be present, i.e., the trimer ($\alpha_3\beta_3$) structure was essential. This contrasted with results for C-phycocyanin[99] and C-phycoerythrin[98] where the separately renatured subunits had absorption spectra that when added together showed maxima identical to native protein.

Berns and Morgenstern[111] studied the effects of the protein denaturant guanidine hydrochloride on C-phycocyanin (*P. luridum*). Most surprisingly, at a concentration of 0.1 *M* guanidine hydrochloride at pH 7.0, instead of denaturing or dissociating the protein, a sharp increase in 11S (hexamer) at the expense of 6S (trimer) was observed. As the guanidine concentration was increased to 1.9 *M* it began to dissociate and denature.

The actions of several denaturants of B-phycoerythrin (*Porphyridium cruentum*) and C-phycoerythrin (*N. punctiformis*) have been investigated.[112] The freezing of solutions of B-phycoerythrin (*P. cruentum*) has been observed to produce hypochromicity in its absorption and dissociation of its aggregates.[113]

VIII. X-RAY CRYSTALLOGRAPHY AND ELECTRON MICROSCOPY

Several early researchers prepared crystals of biliproteins sometimes as a means of purification.[22,28,114,115] C-Phycocyanin (*M. laminosus*,[116] *Anabaena variabilis*,[117] and *Agmenellum quadruplicatum*[118] and B-phycoerythrin[119] (*P. cruentum*) have been studied at a preliminary level by X-ray crystallography techniques. The C-phycocyanin crystals showed structures consistent with the ability of the protein to form stacks of disk-like hexamers. Single crystals had an absorption spectrum sharply altered from solution properties.[120]

Fisher et al.[121] studied C-phycocyanin and B-phycoerythrin to a 0.5-nm resolution. They found the C-phycocyanin hexamer to be 11.0 nm in diameter and B-phycoerythrin $\alpha_6\beta_6\gamma$ was 10.7 nm in diameter. The γ polypeptide may be located in the center of the aggregate.

X-ray results compared favorably with measurements by electron microscopy. Berns and Edwards[20] found that C-phycocyanin hexamers had 13.0-nm diameter, and Kessel et al.[122] found for C-phycocyanin stacks a 12.5-nm diameter and 6.0-nm stack repeats. Each 6.0-

nm stack repeat is considered to be a hexamer. Gantt,[123] using electron micrographs of negatively stained B-phycoerythrin (*P. cruentum*), measured a 10.1-nm diameter and a 5.4-nm thickness for the disk-like particle, and Mörschel et al.[124] found allophycocyanin to be a disk of 10.0 × 3.0 nm in the trimer state. Only R-phycoerythrin showed a nondisk shape in the electron microscope and was found to be 6.3 × 5.3 nm.[125]

The structure of C-phycocyanin from *M. laminosus* has been determined at a resolution of 0.3 nm by X-ray diffraction.[126] A trimeric disk of three $\alpha\beta$ units was found to be 11.0 × 3.0 nm and had a central channel 3.5 nm in diameter. These dimensions agreed very well with results from electron microscopy. A hexamer would have dimensions of 11 × 6 nm. The phycocyanobilins were in an extended conformation as previously predicted from spectroscopic measurements (see Section IV). The pyrrole rings of each chromophore were found to be not co-planar. The shortest center-to-center distance between two chromophores in a trimer was 2.2 nm.

IX. SELF- AND DIRECTED-ASSEMBLY

Proteins may aggregate to form biologically necessary structures either through information found on their own surfaces or through help from other agents. Biliprotein assembly to phycobilisomes was apparently directed by a series of colorless linker polypeptides. However, certain stages of the assembly process occurred by a self-assembly route. Since a great deal of research was performed prior to the discovery of linker polypeptides, the exact status of certain parts of the assembly scheme have not been totally characterized.

Allophycocyanin II showed only α and β subunits on SDS gels,[9] but monomers formed specifically from this form of allophycocyanin could be totally reassembled to trimers.[87] Likewise purified α and β subunits of allophycocyanin, which were most unlikely to contain residual linker polypeptides, could be recombined to form trimers.[110] Assembly of allophycocyanin to the trimer level appeared to be a self-assembly process.

Purified α and β subunits of C-phycocyanin can also be assembled to an aggregate form which was shown by gel electrophoresis to be free of colorless polypeptides.[70] Measurements of the molecular weights of C-phycocyanin aggregates (trimer and hexamer) were consistent with summation of the molecular weights of their α and β subunits without any contribution of linker polypeptides (see Section II), but the systematic application of SDS gel electrophoresis was not used. The early steps in C-phycocyanin and allophycocyanin aggregation were therefore both driven by self-assembly.

The larger aggregates of C-phycocyanin, $(\alpha_6\beta_6)_2$ and $(\alpha_6\beta_6)_3$, and C-phycocyanin-allophycocyanin complexes showed behavior in early studies that now would be explained as being induced by linker polypeptides. Lee and Berns[127] carefully and gently purified C-phycocyanin (*Phormidium luridum*) so that very large C-phycocyanin and C-phycocyanin-allophycocyanin aggregates were retained. Since they used 0.1-ionic strength buffer, the phycobilisomes would be dissociated, but a high percentage of the stacks and stacks with some of the core was apparently retained. The role for the then undiscovered linker polypeptides was suggested from gel filtration (Sephadex G-200) experiments during which the larger aggregates were dissociated. Recombination of all the colored fractions and restoration to the original biliprotein concentration did not result in the recovery of the dissociated larger aggregates. The irreversibility of the dissociation process at this level can be explained now knowing that a function of the colorless linkers was to stabilize larger aggregate assembly.

X. ADDITIONAL STUDIES

Zeng et al.[128] have purified R-phycocyanin from the marine red alga *Polysiphonia urceolata*.

Fujiwara-Arasaki et al.[129] have studied the spectral denaturation of biliproteins from the red alga *Porphyra tenera*. They employed both urea and detergent, SDS, and studied spectral changes in the presence and absence of mercaptoethanol.

Eiserling and Glazer[130] showed that crude biliprotein preparations when allowed to remain at room temperature for 1 to 3 days formed bundles of rods. Electron micrographs indicated that the bundles were 12 nm in diameter and were made from 6-nm-thick disks. A faint 3-nm periodicity was observed in addition to the stronger one at 6 nm.

Koller and Wehrmeyer[131] and Koller et al.[132] have studied B-phycoerythrin and B-phycoerythrin-C-phycocyanin aggregates from the red alga *R. violacea*.

Ilani and Berns[133] have demonstrated an effect of ferric ion on the fluorescence intensity of C-phycocyanin (*Anabaena variabilis*). A few minutes after the addition of an excess of ferric chloride, there was a large decrease in the fluorescence of C-phycocyanin. This rapid period was followed by a long interval of gradual attenuation. Ferrous chloride produced no change in the fluorescence of the protein. The effect of ferric ion on the biliprotein may be related to the activity of biliproteins in model membrane systems (see Chapter 10).

Czeczuga has identified R-phycoerythrin, R-phycocyanin, and allophycocyanin from an Antarctic red alga *Leptosomia simplex*.[134] A discussion of the finding of biliprotein-containing organisms in various habitats is found in Chapter 7.

REFERENCES

1. **Svedberg, T. and Katsurai, T.,** The molecular weights of phycocyan and of phycoerythrin from *Porphyra tenera* and of phycocyan from *Aphanizomenon flos aquae*, *J. Am. Chem. Soc.*, 51, 3573, 1929.
2. **Airth, R. L. and Blinks, L. R.,** A new phycoerythrin from *Porphyra naiadum*, *Biol. Bull.*, 111, 321, 1956.
3. **Haxo, F., O'hEocha, C., and Norris, P.,** Comparative studies of chromatographically separated phycoerythrins and phycocyanins, *Arch. Biochem. Biophys.*, 54, 162, 1955.
4. **Gantt, E. and Lipschultz, C. A.,** Phycobilisomes of *Porphyridium cruentum:* pigment analysis, *Biochemistry*, 13, 2960, 1974.
5. **Bryant, D. A., Glazer, A. N., and Eiserling, F. A.,** Characterization and structural properties of the major biliproteins of *Anabaena* sp., *Arch. Microbiol.*, 110, 61, 1976.
6. **Fujita, Y. and Shimura, S.,** Phycoerythrin of the marine blue-green alga *Trichodesmium thiebautii*, *Plant Cell Physiol.*, 15, 939, 1974.
7. **Rippka, R., Waterbury, J., and Cohen-Bazire, G.,** A cyanobacterium which lacks thylakoids, *Arch. Microbiol.*, 100, 419, 1974.
8. **Lemberg, R. and Bader, G.,** Die Phycobiline der Rot-algen. Überfuhrung in Mesobilirubin und Dehydro-Mesobilirubin, *Liebigs Ann. Chem.*, 505, 151, 1933.
9. **Zilinskas, B. A., Zimmerman, B. K., and Gantt, E.,** Allophycocyanin forms isolated from *Nostoc* sp. phycobilisomes, *Photochem. Photobiol.*, 27, 587, 1978.
10. **Glazer, A. N. and Bryant, D. A.,** Allophycocyanin B (λ_{max}671, 618 nm). A new cyanobacterial phycobiliprotein, *Arch. Microbiol.* 104, 15, 1975.
11. **O'hEocha, C., O'Carra, P., and Mitchell, D.,** Biliproteins of cryptomonad algae, *Proc. R. Ir. Acad.*, 63B, 191, 1964.
12. **Bryant, D.,** Phycoerythrocyanin and phycoerythrin: properties and occurrence in cyanobacteria, *J. Gen. Microbiol.*, 128, 835, 1982.
13. **Honsell, E., Kosovel, V., and Talarico, L.,** Phycobiliprotein distribution in Rhodophyta: studies and interpretations on the basis of their absorption spectra, *Bot. Mar.*, 27, 1, 1984.
14. **O'Carra, P.,** Algal biliproteins, *Biochem. J.*, 119, 2P, 1970.
15. **Bennett, A. and Bogorad, L.,** Properties of subunits and aggregates of blue-green algal biliproteins, *Biochemistry*, 10, 3625, 1971.
16. **Glazer, A. N. and Cohen-Bazire, G.,** Subunit structure of the phycobiliproteins of blue-green algae, *Proc. Natl. Acad. Sci. U.S.A.*, 68, 1398, 1971.
17. **O'Carra, P. and Killilea, S. D.,** Subunit structure of C-phycocyanin and C-phycoerythrin, *Biochem. Biophys. Res. Commun.*, 45, 1192, 1971.
18. **Kobayashi, Y., Siegelman, H. W., and Hirs, C. H. W.,** C-phycocyanin from *Phormidium luridum*. Isolation of subunits, *Arch. Biochem. Biophys.*, 152, 187, 1972.

19. **Kato, M., Lee, W. I., Eichinger, B. E., and Schurr, J. M.,** Molecular weight and shape of the phycocyanin hexamer, *Biopolymers*, 13, 2293, 1974.
20. **Berns, D. S. and Edwards, M. R.,** Electron micrographic investigations of C-phycocyanin, *Arch. Biochem. Biophys.*, 110, 511, 1965.
21. **Brody, S. S. and Brody, M.,** A quantitative assay for the number of chromophores on a chromoprotein: its application to phycoerythrin and phycocyanin, *Biochim. Biophys. Acta*, 50, 348, 1961.
22. **Kylin, H.,** Uber Phykoerythrin und Phykocyan bei *Ceramium rubrum* (Huds.) Ag., *Hoppe-Seyler's Z. Physiol. Chem.*, 69, 169, 1910.
23. **Eriksson-Quensel, I.-B., LXXVI.** The molecular weights of phycoerythrin and phycocyan. I, *Biochem. J.*, 32, 585, 1938.
24. **Glazer, A. N. and Hixson, C. S.,** Characterization of R-phycocyanin. Chromophore content of R-phycocyanin and C-phycoerythrin, *J. Biol. Chem.*, 250, 5487, 1975.
25. **Tiselius, A. and Gross, D.,** Measurement of the diffusion of protein, *Kolloid Z.*, 66, 11, 1934.
26. **O'hEocha, C. and Haxo, F. T.,** Some atypical algal chromoproteins, *Biochim. Biophys. Acta*, 41, 516, 1960.
27. **Pecci, J. and Fujii. ri, E.,** Spectral change of phycoerythrin from *Hydrocoleum* species and its relationship to protein dissociation. Effect of mercurials on single- and double-peaked forms, *Biochim. Biophys. Acta*, 154, 332, 1968.
28. **Hattori, A. and Fujita, Y.,** Crystalline phycobilin chromoproteids obtained from a blue-green alga, *Tolypothrix tenuis*, *J. Biochem.*, 46, 633, 1959.
29. **Alberte, R. S., Wood, A. M., Kursar, T. A., and Guillard, R. R. L.,** Novel phycoerythrins in marine *Synechococcus* spp. Characterization and evolutionary and ecological implications, *Plant Physiol.*, 75, 732, 1984.
30. **Kursar, T. A., Swift, H., and Alberte, R. S.,** Morphology of a novel cyanobacterium and characterization of light-harvesting complexes from it: implications for phycobiliprotein evolution, *Proc. Natl. Acad. Sci. U.S.A.*, 78, 6888, 1981.
31. **Ong, L. J., Glazer, A. N., and Waterbury, J. B.,** An unusual phycoerythrin from a marine cyanobacterium, *Science*, 224, 80, 1984.
32. **Shimura, S. and Fujita, Y.,** Phycoerythrin and photosynthesis of the pelagic blue-green alga *Trichodesmium thiebautii* in the waters of Kuroshio, Japan, *Mar. Biol.*, 31, 121, 1975.
33. **Bryant, D. A., Cohen-Bazire, G., and Glazer, A. N.,** Characterization of the biliproteins of *Gloeobacter violaceus*. Chromophore content of a cyanobacterial phycoerythrin carrying phycourobilin chromophore, *Arch. Microbiol.*, 129, 190, 1981.
34. **Yu, M.-H., Glazer, A. N., Spencer, K. G., and West, J. A.,** Phycoerythrins of the red alga *Callithamnion*. Variation in phycoerythrobilin and phycourobilin content, *Plant Physiol.*, 68, 482, 1981.
35. **Stadnichuk, I. N., Odinzova, T. I., and Strongin, A. Y.,** Molecular organization and chromophore composition of R-phycoerythrin from red alga *Callithamnion rubosum*, *Mol. Biol.*, 18, 343, 1984.
36. **Van der Velde, H. H.,** The natural occurrence in red algae of two phycoerythrins with different molecular weights and spectral properties, *Biochim. Biophys. Acta*, 303, 246, 1973.
37. **Gantt, E. and Lipschultz, C. A.,** Structure and phycobiliprotein composition of phycobilisomes from *Griffithsia pacifica* (Rhodophyceae), *J. Phycol.*, 16, 394, 1980.
38. **Glazer, A. N. and Hixson, C. S.,** Subunit structure and chromophore composition of Rhodophytan phycoerythrins. *Porphyridium cruentum* B-phycoerythrin and b-phycoerythrin, *J. Biol. Chem.*, 252, 32, 1977.
39. **Koller, K. P. and Wehrmeyer, W.,** B-phycoerythrin from *Rhodella violacea*. Characterization of two isoproteins, *Arch. Microbiol.*, 104, 255, 1975.
40. **Airth, R. L. and Blinks, L. R.,** Properties of phycobilins from *Porphyra naiadum*, *J. Gen. Physiol.*, 41, 77, 1957.
41. **Almog, R. and Berns, D. S.,** Effect of acetic acid on phycocyanin-phycoerythrocyanin mixture extracted from *Anabaena variabilis*, *Arch. Biochem. Biophys.*, 228, 388, 1984.
42. **Craig, I. W. and Carr, N. G.,** C-Phycocyanin and allophycocyanin in two species of blue-green algae, *Biochem. J.*, 106, 361, 1968.
43. **Brown, A. S. Foster, J. A., Voynow, P. V., Franzblau, C., and Troxler, R. F.,** Allophycocyanin from the filamentous cyanophyte, *Phormidium luridum*, *Biochemistry*, 14, 3581, 1975.
44. **Brown, A. S. and Troxler, R. F.,** Properties and N-terminal sequence of allophycocyanin from the unicellular rhodophyte *Cyanidium caldarium*, *Biochem. J.*, 163, 571, 1977.
45. **Cohen-Bazire, G., Beguin, S., Rimon, S., Glazer, A. N., and Brown, D. M.,** Physico-chemical and immunological properties of allophycocyanins, *Arch. Microbiol.*, 111, 225, 1977.
46. **Gysi, J. and Zuber, H.,** Isolation and characterization of allophycocyanin II from the thermophilic blue-green alga *Mastigocladus laminosus* Cohn, *FEBS Lett.*, 48, 209, 1974.
47. **Gysi, J. and Zuber, H.,** Allophycocyanin I — a second cyanobacterial allophycocyanin? Isolation, characterization and comparison with allophycocyanin II from the same alga, *FEBS Lett.*, 68, 49, 1976.

48. **Troxler, R. F., Greenwald, L. S., and Zilinskas, B. A.,** Allophycocyanin from *Nostoc* sp. phycobilisomes. Properties and amino acid sequence at the NH_2 terminus of the α and β subunits of allophycocyanins I, II, and III, *J. Biol. Chem.*, 255, 9380, 1980.
49. **Ley, A. C., Butler, W. L., Bryant, D. A., and Glazer, A. N.,** Isolation and function of allophycocyanin B of *Porphyridium cruentum*, *Plant Physiol.*, 59, 974, 1977.
50. **Lundell, D. J. and Glazer, A. N.,** Allophycocyanin B. A common β subunit in *Synechococcus* allophycocyanin B (λ_{max}670 nm) and allophycocyanin (λ_{max}650 nm), *J. Biol. Chem.*, 256, 12600, 1981.
51. **Lundell, D. J., Yamanaka, G., and Glazer, A. N.,** A terminal energy acceptor of the phycobilisome: the 75,000-dalton polypeptide of *Synechococcus* 6301 phycobilisomes — a new biliprotein, *J. Cell. Biol.*, 91, 315, 1981.
52. **Lundell, D. J., Williams, R. C., and Glazer, A. N.,** Molecular architecture of a light-harveting antenna. *In vitro* assembly of the rod substructures of *Synechococcus* 6301 phycobilisomes, *J. Biol. Chem.*, 256, 3580, 1981.
53. **Redlinger, T. and Gantt, E.,** Phycobilisome structure of *Porphyridium cruentum*. Polypeptide composition, *Plant Physiol.*, 68, 1375, 1981.
54. **Rusckowski, M. and Zilinskas, B. A.,** Allophycocyanin I and the 95 kilodalton polypeptide. The bridge between phycobilisomes and membranes, *Plant Physiol.*, 70, 1055, 1982.
55. **Zilinskas, B. A.,** Isolation and characterization of the central component of the phycobilisome core of *Nostoc* sp., *Plant Physiol.*, 70, 1060, 1982.
56. **Canaani, O. D. and Gantt, E.,** Circular dichroism and polarized fluorescence characteristics of blue-green algal allophycocyanins, *Biochemistry*, 19, 2950, 1980.
57. **Guglielmi, G., Cohen-Bazire, G., and Bryant, D. A.,** The structure of *Gloeobacter violaceus* and its phycobilisomes, *Arch. Microbiol.*, 129, 181, 1981.
58. **MacColl, R., Habig, W., and Berns, D. S.,** Characterization of phycocyanin from *Chroomonas* species, *J. Biol. Chem.*, 248, 7080, 1973.
59. **MacColl, R. and Berns, D. S.,** Increased aggregation of *C*-phycocyanin produced by phenol and benezene, *Arch. Biochem. Biophys.*, 156, 161, 1973.
60. **Hefferle, P., John, W., Scheer, H., and Schneider, S.,** Thermal denaturation of monomeric and trimeric phycocyanins studied by static and polarized time-resolved fluorescence spectroscopy, *Photochem. Photobiol.*, 39, 221, 1984.
61. **Scott, E. and Berns, D. S.,** Protein-protein interaction. The phycocyanin system, *Biochemistry*, 4, 2597, 1965.
62. **Hattori, A., Crespi, H. L., and Katz, J. J.,** Association and dissociation of phycocyanin and the effects of deterium substitution on the processes, *Biochemistry*, 4, 1225, 1965.
63. **Berns, D. S. and Scott, E.,** Protein aggregation in a thermophilic protein. Phycocyanin from *Synechococcus lividus*, *Biochemistry*, 5, 1528, 1966.
64. **Neufeld, G. J. and Riggs, A. F.,** Aggregation properties of C-phycocyanin from *Anacystis nidulans*, *Biochim. Biophys. Acta*, 181, 234, 1969.
65. **Berns, D. S.,** Protein aggregation in phycocyanin — osmotic pressure studies, *Biochem. Biophys. Res. Commun.*, 38, 65, 1970.
66. **MacColl, R., Lee, J. J., and Berns, D. S.,** Protein aggregation in C-phycocyanin. Studies at very low concentrations with the photoelectric scanner of the ultracentrifuge, *Biochem. J.*, 122, 421, 1971.
67. **MacColl, R. and Berns, D. S.,** Biliproteins: some relationships among aggregation states, spectra, and excitation-energy transfer, *Isr. J. Chem.*, 21, 296, 1981.
68. **MacColl, R., Berns, D. S., and Koven, N. L.,** Effect of salts on C-phycocyanin, *Arch. Biochem. Biophys.*, 146, 477, 1971.
69. **Kao, O. H. W., Berns, D. S., and Town, W. R.,** The characterization of C-phycocyanin from an extremely halo-tolerant blue-green alga, *Coccochloris elabens*, *Biochem. J.*, 131, 39, 1973.
70. **Glazer, A. N. and Fang, S.,** Formation of hybrid proteins from the α and β subunits of phycocyanins of unicellular and filamentous blue-green algae, *J. Biol. Chem.*, 248, 663, 1973.
71. **Murphy, R. F. and O'Carra, P.,** Reversible denaturation of C-phycocyanin, *Biochim. Biophys. Acta*, 214, 371, 1970.
72. **Huang, C. and Berns, D. S.,** An ultracentrifuge study of C-phycocyanin aggregation, *Biochemistry*, 20, 7016, 1981.
73. **Kao, O. H. W., Edwards, M. R., and Berns, D. S.,** Physical-chemical properties of C-phycocyanin isolated from an acido-thermophilic eukaryote, *Cyanidium caldarium*, *Biochem. J.*, 147, 63, 1975.
74. **Saito, T., Iso, N., and Mizuno, H.,** Solution properties of phycocyanin. I. Studies of dissociation-association by sedimentation measurement, *Bull. Chem. Soc. Jpn.*, 47, 1375, 1974.
75. **Mizuno, H., Saito, T., and Iso, N.,** Solution properties of phycocyanin. III. Studies of the sedimentation equilibrium of phycocyanin, *Bull. Chem. Soc. Jpn.*, 48, 3496, 1975.

76. **Iso, N., Mizuno, H., Saito, T., Nitta, N., and Yoshizaki, K.**, Solution properties of phycocyanin. IV. Studies of the self-association equilibrium of phycocyanin in a pH 6.8 solution, *Bull. Chem. Soc. Jpn.*, 50, 2892, 1977.
77. **Rosenberg, R. M., Crespi, H. L., and Katz, J. J.**, Nuclear magnetic resonance study of the interaction of sodium dodecyl sulfate with phycocyanin, *Biochim. Biophys. Acta*, 175, 31, 1969.
78. **Scheer, H. and Kufer, W.**, Studies on plant bile pigments. IV. Conformational studies on C-phycocyanin from *Spirulina platensis*, *Z. Naturforsch.*, 32c, 513, 1977.
79. **Burke, M. J., Pratt, D. C., and Moscowitz, A.**, Low-temperature absorption and circular dichroism studies of phytochrome, *Biochemistry*, 11, 4025, 1972.
80. **Blauer, G. and Wagnière, G.**, Conformation of bilirubin and biliverdin in their complexes with serum albumin, *J. Am. Chem. Soc.*, 97, 1949, 1975.
81. **Suzuki, H., Sugimoto, T., and Ishikawa, K.**, On the models for phytochrome chromophore. II, *J. Phys. Soc. Jpn.*, 38, 1110, 1975.
82. **Chae, Q. and Song, P.-S.**, Linear dichroic spectra and fluorescence polarization of biliverdin, *J. Am. Chem. Soc.*, 97, 4176, 1975.
83. **Wagniere, G. and Blauer, G.**, Calculations of optical properties of biliverdin in various conformations, *J. Am. Chem. Soc.*, 98, 7806, 1976.
84. **Scheer, H., Formanek, H., and Schneider, S.**, Theoretical studies of biliprotein chromophores and related bile pigments by molecular orbital and Ramachandran type calculations, *Photochem. Photobiol.*, 36, 259, 1982.
85. **Bois-Choussy, M. and Barbier, M.**, Isomerisations and cyclisations in bile pigments, *Heterocycles*, 9, 677, 1978.
86. **MacColl, R., Csatorday, K., Berns, D. S., and Traeger, E.**, Chromophore interactions in allophycocyanin, *Biochemistry*, 19, 2817, 1980.
87. **MacColl, R., Csatorday, K., Berns, D. S., and Traeger, E.**, The relationship of the quaternary structure of allophycocyanins to its spectrum, *Arch. Biochem. Biophys.*, 208, 42, 1981.
88. **Teale, F. W. J. and Dale, R. E.**, Isolation and spectral characterization of phycobiliproteins, *Biochem. J.*, 116, 161, 1970.
89. **Murakami, A., Mimuro, M., Ohki, K., and Fujita, Y.**, Absorption spectrum of allophycocyanin isolated from *Anabaena cylindrica*: variation of the absorption spectrum induced by changes of the physico-chemical environment, *J. Biochem. (Tokyo)*, 89, 79, 1981.
90. **Csatorday, K., MacColl, R., Csizmadia, V., Grabowski, J., and Bagyinka, C.**, Exciton interaction in allophycocyanin, *Biochemistry*, 23, 6466, 1984.
91. **Huang, C.**, Private communication.
92. **MacColl, R.**, Stability of allophycocyanin's quaternary structure, *Arch. Biochem. Biophys.*, 223, 24, 1983.
93. **Pecci, J. and Fujimori, E.**, Mercurial-induced circular dichroism changes of phycoerythrin and phycocyanin, *Biochim. Biophys. Acta*, 188, 230, 1969.
94. **Fujimori, E. and Pecci, J.**, Circular dichroism of single- and double-peaked phycoerythrin: mercurial-induced changes, *Biochim. Biophys. Acta*, 221, 132, 1970.
95. **MacColl, R. and Guard-Friar, D.**, Phycocyanin 612: a biochemical and photophysical study, *Biochemistry*, 22, 5568, 1983.
96. **MacColl, R., Berns, D. S., and Gibbons, O.**, Characterization of cryptomonad phycoerythrin and phycocyanin, *Arch. Biochem. Biophys.*, 177, 265, 1976.
97. **Jung, J., Song, P.-S., Paxton, R. J., Edelstein, M. S., Swanson, R., and Hazen, E. E., Jr.**, Molecular topography of the phycocyanin photoreceptor from *Chroomonas* species, *Biochemistry*, 19, 24, 1980.
98. **Langer, E., Lehner, H., Rüdiger, W., and Zickendraht-Wendelstadt, B.**, Circular dichroism of C-phycoerythrin: a conformational analysis, *Z. Naturforsch.*, 35c, 367, 1980.
99. **Glazer, A. N., Fang, S., and Brown, D. M.**, Spectroscopic properties of C-phycocyanin and of its α and β subunits, *J. Biol. Chem.*, 248, 5679, 1973.
100. **Troxler, R. F., Foster, J. A., Brown, A. S., and Franzblau, C.**, The α and β subunits of *Cyanidium caldarium* phycocyanin: properties and amino acid sequences at the amino terminus, *Biochemistry*, 14, 268, 1975.
101. **Frackowiak, D., Grabowski, J., and Manikowski, H.**, Circular dichroism spectra of biliproteins, *Photosynthetica*, 10, 204, 1976.
102. **Rigbi, M., Rosinski, J., Siegelman, H. W., and Sutherland, J. C.**, Cyanobacterial phycobilisomes: selective dissociation monitored by fluorescence and circular dichroism, *Proc. Natl. Acad. Sci. U.S.A.*, 77, 1961, 1980.
103. **Gray, B. H. and Gantt, E.**, Spectral properties of phycobilisomes and phycobiliproteins from the blue-green alga-*Nostoc* sp., *Photochem. Photobiol.*, 21, 121, 1975.
104. **Lundell, D. J. and Glazer, A. N.**, Molecular architecture of a light-harvesting antenna. Quaternary interactions in the *Synechococcus* 6301 phycobilisome core as revealed by partial tryptic digestion and circular dichroism studies, *J. Biol. Chem.*, 258, 8708, 1983.

105. **Chen, C.-H., Kao, O. H. W., and Berns, D. S.,** Denaturation of phycocyanin by urea and determination of the enthalpy of denaturation by microcalorimetry, *Biophys. Chem.*, 7, 81, 1977.
106. **Chen, C.-H. and Berns, D. S.,** Comparison of the stability of phycocyanins from the thermophilic, mesophilic, psychrophilic and halophilic algae, *Biophys. Chem.*, 8, 203, 1978.
107. **Erokhina, L. G. and Krasnovskii, A. A.,** Investigation of the spectral and photochemical effects of denaturation of allophycocyanin of blue-green algae, *Mol. Biol.*, 8, 651, 1974.
108. **Erokhina, L. G., Shubin, L. M., and Krasnovskii, A. A.,** Spectral effects of allophycocyanin B denaturing, *Fiziol. Rast.*, 27, 483, 1980.
109. **MacColl, R., Edwards, M. R., and Haaksma, C.,** Some properties of allophycocyanin from a thermophilic blue-green alga, *Biophys. Chem.*, 8, 369, 1978.
110. **Gysi, J. R. and Zuber, H.,** Properties of allophycocyanin II and its α- and β-subunits from the thermophilic blue-green alga *Mastigocladus laminosus*, *Biochem. J.*, 181, 577, 1979.
111. **Berns, D. S. and Morgenstern, A.,** Two denaturant effects of guanindine salts on the protein C-phycocyanin, *Arch. Biochem. Biophys.*, 123, 640, 1968.
112. **Erokhina, L. G. and Krasnovskii, A. A.,** Spectral effects of denaturation of B- and C-phycoerythrins, *Biokhimiya*, 41, 1594, 1976.
113. **Leibo, S. P. and Jones, R. F.,** Freezing of the chromoprotein phycoerythrin from the red algae *Porphyridium cruentum*, *Arch. Biochem. Biophys.*, 106, 78, 1964.
114. **Molisch, H.,** Das Phycocyan, ein krystallisirbarer Eiweisskörper, *Bot. Ztg.*, 53, 131, 1895.
115. **Kitasato, Z.,** Biochemische Studien über Phycoerythrin und Phycocyan, *Acta Phytochim.*, 2, 75, 1925.
116. **Dobler, M., Dover, S. D., Laves, K., Binder, A., and Zuber, H.,** Crystallization and preliminary crystal data of C-phycocyanin, *J. Mol. Biol.*, 71, 785, 1972.
117. **Sweet, R. M., Fuchs, H. E., Fisher, R. G., and Glazer, A. N.,** Preliminary crystallographic investigations of two phycobiliproteins, *J. Biol. Chem.*, 252, 8258, 1977.
118. **Hackert, M. L., Abad-Zaptero, C., Stevens, S. E., Jr., and Fox, J. L.,** Crystallization of C-phycocyanin from the marine blue-green alga *Agmenellum quadruplicatum*, *J. Mol. Biol.*, 111, 365, 1977.
119. **Abad-Zapatero, C., Fox, J. L., and Hackert, M. L.,** The quaternary structure of a unique phycobiliprotein: B-phycoerythrin from *Porphyridium cruentum*, *Biochem. Biophys. Res. Commun.*, 78, 266, 1977.
120. **Priestle, J. P., Jr., Rhyne, R. H., Jr., Salmon, J. B., and Hackert, M. L.,** Phycobiliproteins: comparison of solution and single crystal fluorescence for C-phycocyanin and B-phycoerythrin, *Photochem. Photobiol.*, 35, 827, 1982.
121. **Fisher, R. G., Woods, N. E., Fuchs, H. E., and Sweet, R. M.,** Three-dimensional structures of C-phycocyanin and B-phycoerythrin at 5-Å resolution, *J. Biol. Chem.*, 255, 5082, 1980.
122. **Kessel, M., MacColl, R., Berns, D. S., and Edwards, M. R.,** Electron microscope and physical chemical characterization of C-phycocyanin from fresh extracts of two blue-green algae, *Can. J. Microbiol.*, 19, 831, 1973.
123. **Gantt, E.,** Properties and ultrastructure of phycoerythrin from *Porphyridium cruentum*, *Plant Physiol.*, 44, 1629, 1969.
124. **Mörschel, E., Koller, K.-P., and Wehrmeyer, W.,** Biliprotein assembly in the disc-shaped phycobilisomes of *Rhodella violacea*. Electron microscopical and biochemical analyses of C-phycocyanin and allophycocyanin aggregates, *Arch. Microbiol.*, 125, 43, 1980.
125. **Talarico, L. and Kosovel, V.,** Properties and ultrastructure of R-phycoerythrin from *Gracilaria verrucosa* (*Gigartinales, Florideae*) (Huds.) Papenfuss, *Photosynthetica*, 12, 369, 1978.
126. **Schirmer, T., Bode, W., Huber, R., Sidler, W., Zuber, H.,** X-ray crystallographic structure of the light-harvesting biliprotein C-phycocyanin from the thermophilic cyanobacterium *Mastigocladus laminosus* and its resemblance to globulin structures, *J. Mol. Biol.*, 184, 257, 1985.
127. **Lee, J. J. and Berns, D. S.,** Protein aggregation. Studies of larger aggregates of C-phycocyanin, *Biochem. J.*, 110, 457, 1968.
128. **Zeng, F.-J., Yang, Z.-X., and Jiang, L.-J.,** Isolation and characterization of R-phycocyanin from *Polysiphonia urceolata*, *Hydrobiologia*, 116/117, 594, 1984.
129. **Fujiwara-Arasaki, T. Yamamoto, M., and Kakiuchi, K.,** Spectroscopic behavior of biliprotein from *Porphyra tenera* upon denaturation with urea and sodium dodecyl sulfate (SDS), *Hydrobiologia*, 116/117, 588, 1984.
130. **Eiserling, F. A. and Glazer, A. N.,** Blue-green algal proteins: assembly forms of C-phycocyanin from *Synechococcus* sp., *J. Ultrastruc. Res.*, 47, 16, 1974.
131. **Koller, K. P. and Wehrmeyer, W.,** Isolierung und Charakterisierung der Biliproteide von *Rhodella violacea* (Bangiophycidae), *Arch. Microbiol.*, 100, 253, 1974.
132. **Koller, K. P., Wehrmeyer, W., and Schneider, H.,** Isolation and characterization of disc-shaped phycobilisomes from the red alga *Rhodella violacea*, *Arch. Microbiol.*, 112, 61, 1977.
133. **Ilani, A. and Berns, D. S.,** The effect of ferric ion on phycocyanin fluorescence, *Biochem. Biophys. Res. Commun.*, 45, 1423, 1971.
134. **Czeczuga, B.,** Studies on phycobiliproteins in algae. VII. Light-harvesting phycobiliprotein pigments of the red alga *Leptosomia simplex* from the Antarctic, *Polar Biol.*, 4, 179, 1985.

Chapter 5

PHYCOBILISOME: LINKERS AND CORE STRUCTURES

I. DISCOVERY OF THE LINKERS

The phycobilisomes of the cyanobacteria and red algae are the sites of biliproteins. Tandeau de Marsac and Cohen-Bazire[1] discovered several bands in sodium dodecyl sulfate (SDS) gel electrophoresis experiments of phycobilisomes that do not correspond to the molecular weights of the biliprotein subunits. These bands appeared to be colorless and were observed for eight species of cyanobacteria.

To establish these "colorless" polypeptides as integral parts of the phycobilisomes, Tandeau de Marsac and Cohen-Bazire centrifuged the extracted phycobilisome preparations on a sucrose density gradient. They obtained a series of fractions from the gradient and measured the quantities of the "colorless" polypeptides and the biliprotein subunits by scanning Coomassie® blue stained gels at 640 ± 10 nm. The quantities of both protein groups showed a beautiful correspondence having a constant ratio of the total amounts of "colorless" polypeptides to biliprotein subunits for all fractions. The individual "colorless" polypeptides also had a fairly uniform distribution in all phycobilisome-containing fractions.

The next step in ascertaining the functions of these polypeptides was quite ingenious and has been used extensively by subsequent investigators. Tandeau de Marsac and Cohen-Bazire[1] made use of a very interesting property of certain C-phycoerythrin-containing cyanobacteria. These cyanobacteria had the property of complementary chromatic adaption. Their biliprotein contents depended on the spectral distribution of the growth light. When grown in red light which was mainly absorbed by C-phycocyanin the C-phycoerythrin content of these cyanobacteria was diminished and in some circumstances virtually eliminated. In green light which was mainly absorbed by the red-colored C-phycoerythrins the C-phycoerythrin content increased. White light may give an intermediate mixture of C-phycocyanin and C-phycoerythrin. Phycobilisomes from *Lyngbya-Plectonema-Phormidium* group 7409 and *Fremyella* 7601 grown in red, green, and white light were isolated. Certain of the "colorless" polypeptides were also affected. From the green light-grown phycobilisomes, a new "colorless" polypeptide was found, while from the red light-grown phycobilisomes, two polypeptides only very weakly observed in the green became quite prominent. Apparently, certain "colorless" polypeptides were associated with particular biliproteins.

Two functions were proposed for these polypeptides: attachment to thylakoid membrane and positioning and assembly of the biliproteins. The largest molecular weight polypeptide (70,000) was observed both in phycobilisomes and in phycobilisome-free thylakoid membranes and was a candidate for attachment of the organelles to the membranes.

Subsequent experiments from other laboratories have significantly modified only one aspect of this initial discovery and rather extensive investigation. It is now known that the largest molecular weight polypeptide isolated from phycobilisomes was colored. The other polypeptides, which apparently lack chromophores, are now known as linker polypeptides, owing to their role in determining biliprotein assembly.

II. LINKERS IN RED ALGAE

Phycobilisomes from the red alga *Porphyridium cruentum* have been shown to have nine colorless linker polypeptides (Figure 1).[2] In Figure 1, the bands marked by an asterisk were pigmented. As was typical in all cases, the highest-molecular weight polypeptide had a chromophore and was the long-wavelength emitter of allophycocyanin I. Of the remaining

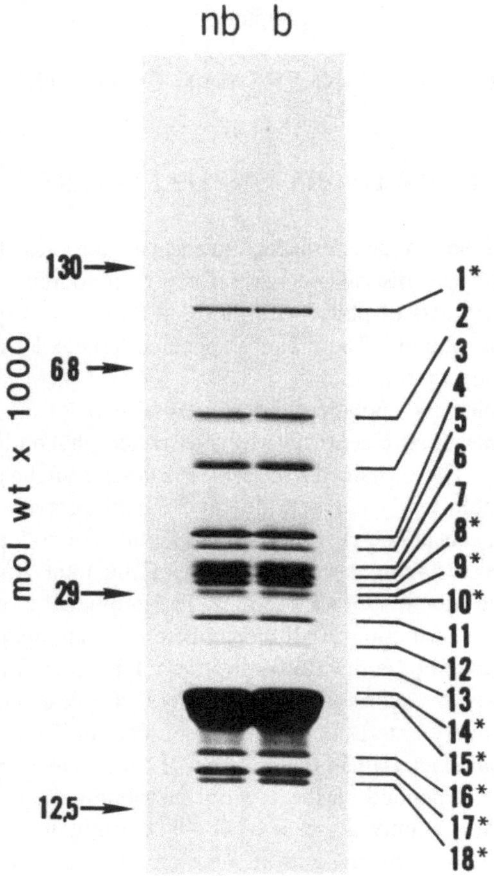

FIGURE 1. SDS gels of phycobilisomes from the red alga *Porphyridium cruentum*. The phycobilisomes were either boiled (b) or not boiled (nb) prior to electrophoresis. The asterisks indicate pigmented subunits. Photograph courtesy of E. Gantt. (From Redlinger, T. and Gantt, E., *Plant Physiol.*, 68, 1375, 1981. With permission.)

linkers, bands 12 and 13 (Figure 1) were present in very small amounts, leaving seven potentially useful linker polypeptides (bands 2 to 7 and 11). Bands 8 to 10 were the γ subunits of B-phycoerythrin, and bands 14 to 18 were the α and β subunits of allophycocyanin, R-phycocyanin, B-phycoerythrin, and b-phycoerythrin. As with cyanobacteria, about 15% of this red algal phycobilisome was linker. In a second study on red algal phycobilisomes, four organisms were studied by SDS gel electrophoresis and linkers were noted in all cases.[3]

III. FUNCTIONS OF LINKERS

A. Linkers in *Anacystis nidulans*

Phycobilisomes have been isolated from the cyanobacterium *A. nidulans*[4] and the linkers studied in detail.[5-7] These phycobilisomes did not contain phycoerythrin and were composed of four major linkers at 75,000, 33,000, 30,000, and 27,000 mol wt; they are compared in Table 1 with linkers from other cyanobacteria.[8-11]

Yamanaka et al.[6] have produced a nitrosoguanidine-induced mutant of *A. nidulans* that

Table 1
MOLECULAR WEIGHTS OF PRINCIPAL[a] LINKERS VIA SDS POLYACRYLAMIDE GEL ELECTROPHORESIS ($\times 10^{-3}$)

Anabaena variabilis[8]	*Nostoc sp.*[9]	*Synechocystis* 6701[10]	*Anacystis nidulans*[4]	*Pseudanabaena*[11]
120	95	99	75	94
32.5	34.5	33.5	33	34.2
30.5	34	31.5	30	32.6
29	32	30.5		31.3
27	29	27	27	27.4

[a] In several cases, 45K (45,000 mol wt) and approximately 9 to 10K "colorless" polypeptides are also detected. The 45K occurs in very small amounts. The linker polypeptides are named after their molecular weights from SDS gel electrophoresis. To make this system less cumbersome, the convention found in the literature of mol wt being 1000 times (K) the number given is used. The 75 to 120K polypeptide is the 680-nm emitter of allophycocyanin I.

has a molar ratio of C-phycocyanin to allophycocyanin of 1.4:1 compared to a ratio of 3.8:1 for the wild type. This mutant was a good system for linker study because it also was devoid of the 30 and 33K linkers. The 27 and 75K linkers were retained. The phycobilisome core was unaltered and was composed of two cylinders, each cylinder consisting of four trimer-like disks. Instead of the usual rods projecting from the core as in the wild type, the mutant (AN112) had only two to six single disks attached to each core. Yamanaka et al.[6] proposed that the 30 and 33K linkers were involved in the rod structure and the 27 and 75K polypeptides were associated with the core.

In a major contribution to the study of phycobilisomes, Lundell et al.[5] purified four linkers. The 30 and 33K linkers were isolated by treating phycobilisomes with 8 *M* urea at pH 5.0, heating to 45°C, and applying the cooled solution to an SP-Sephadex C-25 column. The biliproteins eluted at a lower ammonium chloride concentration than the linkers. SP-Sephadex C-25 chromatography was repeated twice, resulting in purified 30 and 33K polypeptides. In order to purify the 27 and 75K linkers, the AN112 mutant which was devoid of the 30 and 33K was selected. In this case, after SP-Sephadex chromatography the 27 and 75K linkers were dissolved in 6 *M* guanidine hydrochloride at pH 7.0 and chromatographed on a Sephadex G-200 column. The 75K linker was obtained quite pure at this point, and the 27K linker was finally purified by its almost quantitative precipitation in dilute acetic acid.

These purified linkers were then used with highly purified biliproteins to determine their function.[5] In general, C-phycocyanin, which was hexameric or smaller, was mixed with the selected linker in 8.5 *M* urea at pH 3, dialyzed to 3 *M* urea at pH 8.0, then dialyzed to 0.6 *M* sodium phosphate at pH 8.0, and centrifuged in a sucrose density gradient. The urea or guanidine were needed to solubilize the pure linkers. The resulting materials were analyzed by absorption and fluorescence spectroscopy, electron microscopy, and SDS gel electrophoresis. The 27K linker caused the C-phycocyanin to form disks in high yield but no rods. Note that disks in relation to rod structures are always hexamers in these discussions. The 30 and 33K linkers produced extensive aggregation to disks and rods (stacks of disks). The 75K polypeptide did not complex with the biliprotein. The 33 and 30K linkers together may result in more and longer rods than either alone. When the 27K linker was used in combination with either the 30K, the 33K, or the 30 + 33K linker, much shorter rods resulted. Two-disk stacks were common results from these types of mixed reactions. The stoichiometry of linker per hexamer was roughly one linker per C-phycocyanin hexamer when the 27, 30, and 33K linkers were used alone, but when the 30 or 33K linkers were used in combination with the 27K linker the ratio was always greater than 1. Lundell et al.[5] suggested from these

Table 2
EFFECTS OF MUTATIONS ON CYANOBACTERIAL LINKERS

Organism	Mutagen	Mutation	Effect on linkers	Ref.
Anacystis nidulans	Nitrosoguanidine	AN112, no rods, single disks attached to core; core appears normal	No 33 or 30K linkers	6
Synechocystis sp.	UV light	UV16, little C-phycocyanin made, rods and cores form but are not attached	Possible defect in 33.5K linker	13
Synechocystis sp.	Nitrosoguanidine	6701-NTG31, rod attachment incomplete; a typical finding is a core with only two rods attached	No 33.5K linker	10
Synechocystis sp.	Nitrosoguanidine	12 mutants all low in C-Phycoerythrin content	Amounts of 30.5 and 31.5K linkers decrease proportionally to C-phycoerythrin content; 27 and 33.5K linkers unchanged	14,15

data that the 30 and 33K forms were responsible for rod assembly. The 27K linker was proposed as possibly connecting the first disk of a rod and the core, since it reacted with rod-proteins but did not produce rod elongation.

Very shortly thereafter Lundell et al.[7] developed a new purification scheme for the 75K linker. This more rapid method relied on the precipitation of the linkers with 0.5 *M* ammonium thiocyanate. Using mutant AN112 phycobilisomes, the 0.5 *M* NH_4SCN precipitate was dissolved in 6 *M* guanidine hydrochloride and chromatographed on a Sephacryl® S-200 column. The 75K polypeptide was now clearly observed to have a chromophore attached. Its absorption spectrum closely resembled that of phycocyanobilin. To avoid precipitation, the 75K polypeptide was only partially renatured into either 3 *M* urea or 0.5% Triton® X-100. The fluorescence emission maximum of the partially renatured 75K linker was at 676 nm. It has now been clarified that this 75K linker was in an enzymatically cleaved form the actual emitter for the allophycocyanin I aggregate isolated by Zilinskas et al.[12] The initial isolation of this 75K linker had bleached the chromophores. Exposure of denatured biliproteins to neutral pH rapidly caused loss of color.

B. Linkers in *Synechocystis* 6701

Synechocystis 6701 is a cyanobacterium which has C-phycoerythrin in addition to C-phycocyanin and the allophycocyanins. Its linkers have been examined by employing both mutations[10,13-15] and chromatic adaptation[14] (Table 2 and 3). Its polypeptide composition showed one more linker than *A. nidulans* in the 30,000-mol wt range (Table 1). Its phycobilisome rods had two disks each of C-phycoerythrin and C-phycocyanin.

Using nitrosoguanidine, two mutants of *Synechocystis* 6701 were prepared which lacked the 33.5K linker (Table 2).[10] The C-phycoerythrin and C-phycocyanin contents of these mutants were reduced, but the core structures appeared normal. The wild type phycobilisomes of this cyanobacterium had the very common structure consisting of a three-cylinder core and had six rods projecting from the core. The mutants had incomplete structures showing variable numbers of rods. They speculated that perhaps the absent linker controlled C-phycocyanin stability, particularly in disk-to-disk assembly of this biliprotein.

Gingrich et al.[14] have isolated 12 nitrosoguanidine mutants, and all have phycobilisomes

Table 3
EFFECTS OF COMPLEMENTARY CHROMATIC ADAPTATION ON LINKERS

Organism	Growth light	Effects on biliproteins	Effect on linkers	Ref.
Synechocystis sp.	Red	C-Phycoerythrin reduced, in one case to four hexamers per core	30.5K linker is absent; 31.5K linker is reduced	14
Nostoc sp.	White	One type of rod with 2:1 molar ratio of phycoerythrin to phycocyanin	Rods have 29, 32, and 34K linkers	9
	Red	One type of rod, almost all C-phycocyanin, only small amount of C-phycoerythrin	Rods have 34.5 and 29K linkers	
	Pink	Appeared to be a mix of the two rod types	Rods have 29, 32, 34, and 34.5K linkers	

deficient in C-phycoerythrin content. The 30.5 and 31.5K linkers also were reduced in the mutants while the 27K linker was unaffected. The amounts of these 30.5 and 31.5K linkers were directly proportional to the amounts of C-phycoerythrin that were retained. Moreover, the 30.5K linker vanished first, and the 31.5K linker was reduced more gradually until both the 31.5K linker and the C-phycoerythrin content were entirely depleted. This suggested that the 31.5K linker stabilized the phycoerythrin hexamer nearest the C-phycocyanin, while the 30.5K linker was associated with the C-phycoerythrin hexamer on the periphery. The universality of the 27K linker supported its suggested function as a link between the core and rods.

An UV mutant of this cyanobacterium was also defective in phycobilisome structure.[13] The core and C-phycoerythrin rods appeared normal, but the rods did not attach to the cores. The rods contained the 30.5 and 31.5K linkers, and the core had the 27 and 99K linkers. Perhaps there was insufficient C-phycocyanin or a defect in the 33.5K linker (Table 2).

This phycobilisome has also been studied via chromatic adaptation by growing the organism in red, green, and white light.[14] The amounts of C-phycoerythrin retained by an organism grown in red light depended on light intensity, temperature, and the age of the culture. In a particular set of growth conditions — 23°C, low light intensity, and red light — phycobilisomes were produced in which there were only four C-phycoerythrin disks per phycobilisome (Table 3). There was no 30.5K linker in the red light-grown organelles. This strongly supported the idea that the function of the 30.5K linker was the addition of the outer C-phycoerythrin hexamer. The C-phycocyanin content was unchanged in the red light chromatic adaptation of this organism.

The functions of the 31.5 and 30.5K linker were further characterized by complementation experiments.[15] A mutant phycobilisome was used that was completely devoid of C-phycoerythrin, 31.5, and 30.5K linkers. A complex of C-phycoerythrin and the 31.5K linker added onto the mutant phycobilisome. However, when a C-phycoerythrin and 30.5K linker complex was used no addition was observed. This supported the previously mentioned data for these linkers. The 31.5K linker stabilized the first C-phycoerythrin hexamer and may help connect this disk to the C-phycocyanin. The 30.5K linker added the second C-phycoerythrin disk and perhaps signaled rod termination.

C. Linkers in *Nostoc* sp.

This was a second example of studying linkers in a C-phycoerythrin-containing organism, but for *Nostoc* sp. there was only one C-phycocyanin disk per rod instead of the two found in *Synechocystis* 6701. Both *Nostoc* sp. and *Synechocystis* 6701 had two C-phycoerythrin disks per rod under "normal" growth conditions. The linkers were observed at 34.5, 34, 32, and 29K (Table 1).

Glick and Zilinskas[16] have obtained phycobilisomes from *Nostoc* sp. which have the 2:1 molar ratio of C-phycoerythrin to C-phycocyanin. These phycobilismes were dissociated in low concentrations of potassium phosphate in the presence of a proteolytic inhibitor. The partially dissociated phycobilisomes were applied to a calcium phosphate column. A mixture of linkers, C-phycoerythrin, and C-phycocyanin was obtained free of allophycocyanin. This mix was centrifuged through a linear sucrose density gradient in the presence of 0.75 *M* potassium phosphate, and two bands were isolated: the first had a 4:1 molar ratio of C-phycocyanin to C-phycoerythrin, and the second a 2:1 molar ratio of C-phycoerythrin to C-phycocyanin. The allophycocyanin left behind on the original elution from the calcium phosphate column was also removed for reassociation experiments.

Reassociation experiments were carried out for the various dissociated rod materials together with the allophycocyanin in 0.75 *M* potassium phosphate and examined after centrifugation in a sucrose density gradient.[16] These various C-phycoerythrin-C-phycocyanin solutions reassembled to phycobilisomes with the allophycocyanin preparation. The linker polypeptide constitution of these fractions was shown by SDS gel electrophoresis to be for the 4:1 molar phycocyanin-phycoerythrin complex, 34.5 and 29K linkers, and for the 2:1 molar phycoerythrin-phycocyanin complex, 32, 34, and 29K linkers. Apparently, the 32 and 34K linkers were involved in C-phycoerythrin disk formation and assembly, and the 34.5K linker stabilized the C-phycocyanin disk assembly. These authors argued that the 29K linker serves the same function of connecting rods to core in *Nostoc* sp. phycobilisomes as the 27K linker did in *A. nidulans* and *Synechocystis* 6701. The evidence for this latter assignment was that C-phycocyanin-C-phycoerythrin mixtures which lacked the 29K linker did not form phycobilisomes with the allophycocyanin pool. In addition, rods can be reconstituted in the presence of other linkers but in the absence of the 29K linker.

SDS gels showed the various patterns for *Nostoc* sp. phycobilisomes and some dissociation products (Figure 2).[17] It was clear that the 34.5, 34, 32, and 29K linkers were absent from the allophycocyanin fractions but present in the phycoerythrin-phycocyanin mixture. The 80K linker was a partial breakdown product of the 95K form. In the absence of inhibitors, the 95K linker went to 80K then to 43K and 39K forms. These gels (Figure 2) also showed that allophycocyanin III differed from allophycocyanin II in that the former had, in addition to the α and β subunits, a 10K linker.

These phycobilisomes from *Nostoc* sp. have also been studied by chromatic adaptation in different growth lights.[9] When grown in white light each rod had one C-phycocyanin disk and two C-phycoerythrin disks, whereas in red light the C-phycoerythrin was greatly reduced and there were then two C-phycocyanin disks on each rod. The white light phycobilisomes had linkers at 34, 32, and 29K, while the red light phycobilisomes had 29 and 34.5K rod-associated linkers. When grown in pink light, the 34.5, 34, 32, and 29K linkers were all found. As noted above, the phycobilisomes can be dissociated in low phosphate, and the C-phycoerythrin and C-phycocyanin can be separated from the core components by calcium phosphate chromotography. When this technique was applied to phycobilisomes grown in white light, the phycoerythrin-phycocyanin elutant was reassociated to rods composed of two phycoerythrin disks and one disk of phycocyanin. The pink light phycobilisomes, however, reassembled into two rod types: one identical to the white light rods and the second, in a greater proportion, a rod of two C-phycocyanin disks with only a trace of C-phycoerythrin. From the red light phycobilisomes, a rod was composed of two C-phycocyanin disks with a small amount of C-phycoerythrin. When white light-grown phycobilisomes were dissociated, SDS gel electrophoresis experiments demonstrated that the phycocyanin fraction contained the 34 and 29K linkers and the phycoerythrin contained either 32K alone

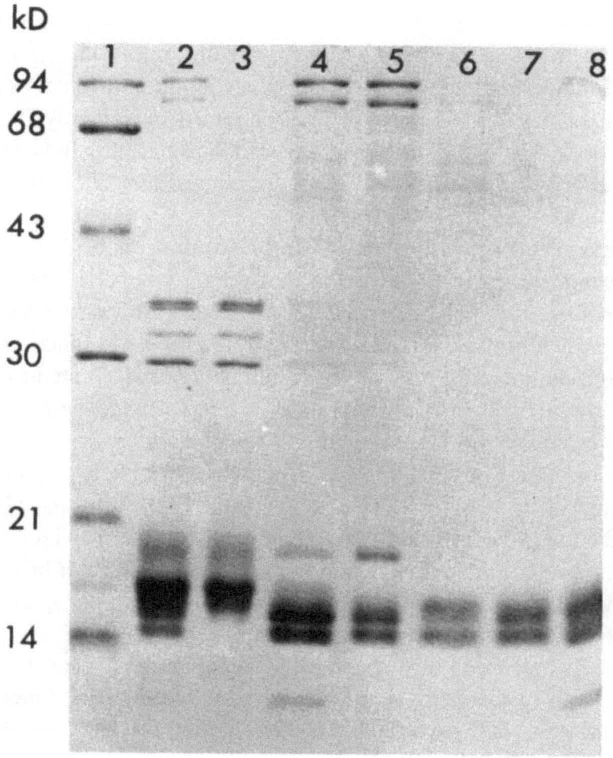

FIGURE 2. SDS gels of phycobilisomes and biliproteins from *Nostoc* sp. Lane 1, molecular weight markers; lane 2, phycobilisomes; lane 3, phycoerythrin-phycocyanin mixture; lane 4, all allophycocyanins; lane 5, allophycocyanin I; lane 6, allophycocyanin B; lane 7, allophycocyanin II; and lane 8, allophycocyanin III. Photograph courtesy of B. Zilinskas. (From Rusckowski, M. and Zilinskas, B. A., *Plant Physiol.*, 70, 1055, 1982. With permission.)

or 34K in combination with 32K linker. This result suggested that the 32 and 34K linkers were arranged in the rods such that the 34K linker attached a phycoerythrin disk to the phycocyanin and the 32K connected two phycoerythrin disks. The 34.5K linker may serve to join the two C-phycocyanin disks in the red light-grown phycobilisomes. The red light-grown phycobilisomes had only one linker that was controlling rod development (34.5K). This number was even less than that in *A. nidulans* (Table 1) where two linkers controlled C-phycocyanin assembly to rods.

Elegant experiments on *Synechocystis* 6701 have clearly demonstrated that the two linkers responsible for C-phycoerythrin rod assembly were ordered so that one (30.5K) was at the periphery of the rod and the other (31.5K) was nearer the C-phycocyanin disk. For *A. nidulans*, there was no definite experimental evidence for the type of linker specialization within the rods. It was noted, however, that phycocyanin complexes with the 30K linker fluoresce at slightly higher energies than complexes with the 33K linker. Since energy transfer proceeds best from high to low energy, the 33K linker may be located nearer the core than the 30K. For *Nostoc* sp. the dissociation pattern of the white light-grown phycobilisomes provided support for a linker distribution pattern in which the 32K linker was farther from the core than the 34K linker.[9]

Canaani and Gantt[18] have also studied the phycobilisomes of *Nostoc* sp., and their SDS gels showed linkers at 38, 34, 31, and 95K. These first three were slightly higher in molecular weight than the 34, 32, and 29K reported above but were obviously the same polypeptides.

It is not unusual to obtain slightly different molecular weights for identical polypeptides from SDS gels run in different laboratories. Slight differences in technique or the use of different proteins for a standardization curve can easily produce differences of this magnitude. When the 29K linker (their 31K) was partially degraded by proteolysis, recombination of rods and core did not occur. This further supported the role of this linker in connecting rods and core. Rod stability was unaffected by the cleavage of the 29K linker.

D. Linkers in Phycoerythrocyanin-Containing Cyanobacteria

Phycoerythrocyanin was present in two cyanobacteria whose phycobilisomes have been studied, *Anabaena variabilis* and *Mastigocladus laminosus*. Phycoerythrocyanin has a higher energy absorption than C-phycocyanin and was therefore likely to function like phycoerythin by transferring excitation energy to C-phycocyanin. Fluorescence studies of phycobilisome dissociation into low-phosphate concentrations[19] provided evidence for a phycoerythrocyanin-phycocyanin complex from *A. variabilis*. Phycobilisomes in 0.75 *M* phosphate were diluted to a final concentration of 0.02 *M* phosphate. Immediately upon dilution to 0.02 *M* phosphate the emission maximum shifted to 642 nm, indicating that C-phycocyanin had uncoupled from the allophycocyanins. Excitation was at 530 nm, which will mainly excite phycoerythrocyanin. At a much slower rate the emission maximum broadened and shifted to still lower wavelengths (near the phycoerythrocyanin maximum), indicating the dissociation of a phycoerythrocyanin-phycocyanin complex.

For *M. laminosus*, a phycoerythrocyanin-phycocyanin complex has been isolated.[20] Nies and Wehrmeyer[20] isolated these phycobilisomes, partially dissociated them by lowering the phosphate concentration, and separated the dissociation products by centrifugation in a sucrose density gradient. Of the several bands obtained, one was a stable phycoerythrocyanin-phycocyanin complex. Unlike phycoerythrin-phycocyanin complexes which were aggregates of hexameric disks, this complex consisted of a trimer of C-phycocyanin tightly bound to a trimer of phycoerythrocyanin. Its fluorescence emission maximum was at 645 nm, indicating energy transfer from phycoerythrocyanin to phycocyanin.

Lundell et al.[5] used purified C-phycocyanin from *A. variabilis* together with linkers from *Anacystis nidulans* and found that the heterologous biliprotein was affected by the linker in ways similar to the C-phycocyanin from *A. nidulans*. Yu et al.[8] and Yu and Glazer[21] isolated complexes of C-phycocyanin and the 27 and 32.5K linkers from dissociation of phycobilisomes of *Anabaena variabilis*. The phycoerythrocyanin was associated with the 30.5K linker. The effect on linker function of trypsin cleavage of the 27 and 32.5K linkers was investigated.

Isono and Katoh[22,23] have also studied phycobilisomes from *A. variabilis*. The two strains of *Anabaena*, one studied here and the other mentioned above,[21] were quite different in their phycobilisome structure. Both the *M. laminosus*[20] and *A. variabilis*,[21] mentioned above, had the typical hemi-discoidal structure with a three-cyclinder core and six rods radiating from the core. The strain studied by Isono and Katoh[22,23] was quite distinctive. It appeared in electron micrographs as a cylinder 52.5-nm long. The core did not exhibit the usually characteristic circular face view, and only two rods radiated from the ends of this core. A possible explanation for this type of phycobilisome might be a mutation in linker structure. The SDS gels of these phycobilisomes had major linker bands at 30, 32, 34, 35, 105, and 115K. Table 1 shows the other strain to also have five linkers. The higher molecular weights in the Isono and Katoh[22,23] experiments could easily be caused by experimental differences in electrophoresis. In any case, their strain apparently had a full group of linkers, and the mutation concept was not supported. Both strains had in common a higher molecular weight than the other cyanobacteria (Table 1) for the allophycocyanin I polypeptide (120 and 115K). They isolated a 23S allophycocyanin component after partial phycobilisome dissociation. It also did not have the appearance of other allophycocyanin cores in the electron microscope.

The 23S complex can be further dissociated into 6.1 and 14.5S allophycocyanin units. Both the 680-nm emission and the 115 or 105K linker were in the 14.5S portion.

E. Linkers in Red Algae

A phycoerythrin-phycocyanin complex from the red alga *Porphyridium sordidum* was shown to require a 30K linker for stability.[24] Kursar et al.[25] examined phycobilisomes from the wild type and mutants of the red alga *Gracilaria tikvahiae*. One of these mutants, NMY-1, was very unusual in that its phycobilisomes had 71% phycoerythrin, 27% allophycocyanin, and insufficient phycocyanin to fulfill its normal role as energy transfer conduit from phycoerythrin to allophycocyanin. Nonetheless, excitation of these mutant phycobilisomes at 498 nm resulted in emission at 670 to 674 nm. Undoubtedly, energy was being transferred from phycoerythrin to allophycocyanin. Kursar et al.[25] suggested that direct contact and transfer between phycoerythrin and the core was occurring. Wild-type phycobilisomes contained linkers of 29, 34, and 89K, but NMY-1 had only the 89K. They noted that others have found a 27 to 29K linker that connected phycocyanin to the core in the cyanobacterial proteins.

F. Linkers?

Siegelman and Kycia[26] have brought forth a provocative set of data on the linkers. They have obtained phycobilisomes and produced SDS gel electrophoresis patterns that showed several bands having molecular weights in the low 30,000 region exactly as anticipated for the linkers. However, in their experiments, these bands were colored and were considered to be aggregates of the α and β biliprotein subunits. Molecular weights in the low 30,000 range were of course good candidates for dimerization of biliprotein subunits which are in the range of 15,000 to 18,000. They studied red and white light-grown *Fremyella diplosiphon* and obtained SDS gels of distinctively colored banding for the chromatically adapted structures. Again, in their view, these bands were artifactual. They maintained that the colorless bands reported by other researchers could be readily produced by bleaching of the chromophores. This latter argument was easily documented. Lundell et al.[5] first isolated the 75K linker from *Anacystis nidulans* in a colorless state and later,[7] using a modified protocol, in the true pigmented condition. The bleaching of denatured biliproteins occurred readily at pH values >4 or in the presence of reducing agents such as mercaptoethanol (Figure 3).[27] The native biliproteins have chromophores that are protected from these effects. To further evaluate this possibility, the amino acid compositions of Lundell et al.[5] were recalculated, arbitrarily setting the aspartic acid results at the same value for all the linkers as well as α and β subunits of both C-phycocyanin and allophycocyanin all from the same cyanobacterium, *A. nidulans* (Table 4). If the linkers were aggregates of α and β, then their normalized amino acid compositions must be averages of α and β subunit combinations that formulated them. All the linkers and biliprotein subunits did have amino acid compositions that were fairly similar. Nonetheless, all the linkers had certain amino acid contents that were not possible combinations of the α and β subunits. Phenylalanine was present in greater amounts in all the linkers than in any biliprotein subunit. Histidine was more prevalent in two of the linkers, and proline occurred in greater concentrations in three linkers than in any biliprotein subunit. There were also other examples that basically cast doubt on the linkers being aggregates of biliprotein subunits.

A summary is given in Table 5 of the proposed linker functions.

IV. PHYCOBILISOMES AND COMPLEMENTARY CHROMATIC ADAPTATION

Tandeau de Marsac[28] has examined 44 strains of cyanobacteria for the occurrence of

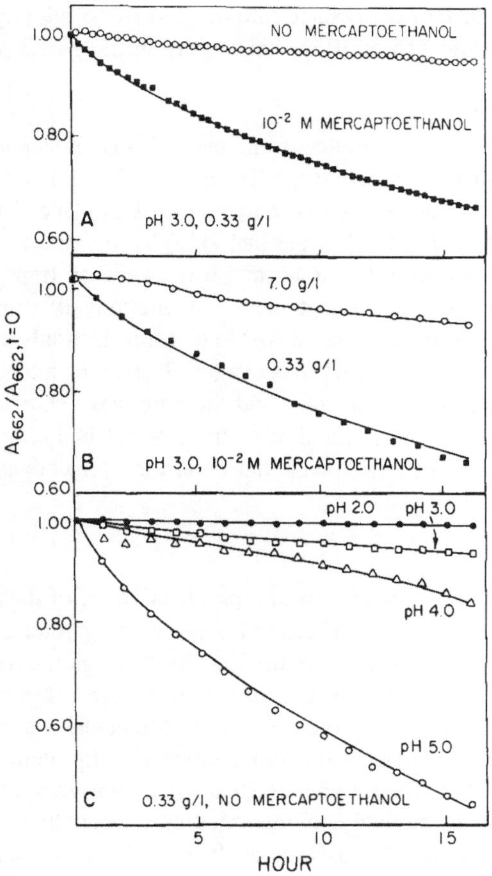

FIGURE 3. Factors causing chromophore bleaching of denatured C-phycocyanin. A, effect of mercaptoethanol at pH 3.0; B, effect of protein concentration (7.0 and 0.33 g/ℓ); and C, effect of pH.

chromatic adaptation (also see Chapter 7). In 12 strains, no adaptation was noted. The rest changed their biliprotein contents by one of two methods. More phycoerythrins in green light and less in red light were produced in 7 strains, while leaving the amount of C-phycoerythrin unchanged; 25 strains have altered both the amounts of phycoerythrin and phycocyanin. In this latter case the rate of phycoerythrin synthesis was greater in green than in red light, and the rate for C-phycocyanin was lower in green light than in red. No cases of the complete cessation of C-phycocyanin synthesis were uncovered.

Chromatic adaptation has been used as mentioned above to study linker function in several cyanobacteria.[9,14] Bryant et al.[29] have used this technique to show that the variations in phycobilisome size and pigment composition were produced by variations in rod length and composition. Rosinski et al.[30] also showed that the phycobilisomes from red light and white light-grown *F. diplosiphon,* while differing greatly in pigment compositions, have very similar phycobilisome structures except for rod length.

Siegelman and Kycia[26] have examined the phycobilisomes of chromatically adapted cyanobacteria. For *Tolypothrix tenuis,* the white light-grown phycobilisomes have two phycoerythrin disks and two phycocyanin disks per rod (Figure 4). When grown in red light, each rod had five phycocyanin disks and no phycoerythrin. The red light-grown rods were then longer than the white. For *Nostoc* sp. and *Synechocystis* 6701, chromatic adaptation in red and white lights yielded different results. The *Nostoc* sp. phycobilisomes[9] responded

Table 4
AMINO ACID COMPOSITIONS OF LINKERS AND BILIPROTEIN SUBUNITS FROM *ANACYSTIS NIDULANS*[5]

Amino acid	Linkers				C-Phycocyanin		Allophycocyanin	
	75K	33K	30K	27K	α	β	α	β
Asp	25	25	25	25	25	25	25	25
Thr	17	13	15	11	13	11	19	10
Ser	21	29	26	15	18	13	19	25
Glu	45	33	32	26	13	13	35	20
Pro	12	14	4	11	8	4	8	5
Gly	32	19	19	21	17	15	31	20
Ala	34	28	34	29	36	37	33	35
Val	24	27	19	17	9	15	27	18
Met	1	2	1	0	0	4	4	5
Ile	17	12	13	11	9	13	21	17
leu	28	25	20	22	21	17	25	28
Tyr	14	15	14	15	13	6	15	18
Phe	20	11	13	13	8	7	6	2
Lys	15	8	5	10	11	6	23	10
His	1	4	6	1	1	0	0	0
Arg	28	22	27	20	8	14	42	15

Note: Number of residues calculated based on 25 aspartic acids in each polypeptide.

Table 5
PROPOSED LINKER FUNCTIONS

Linker (K)	Organism	Function	Comments
27	*Anacystis nidulans*, *Nostoc* sp., *Synechocystis* sp.	Joins rods to core	Always present in normal phycobilisomes; found in cores that lack rods; forms complexes with C-phycocyanin that are disks; rods and cores do not join in its absence
30	*A. nidulans*	C-Phycocyanin rod assembly	Pure 30K assembles C-phycocyanin to rods; absent in rodless mutant
33	*A. nidulans*	C-Phycocyanin rod assembly	Pure 33K assembles C-phycocyanin to rods; absent in rodless mutant
30.5	*Synechocystis* sp.	C-Phycoerythrin rod assembly associated with outer C-phycoerythrin	Remove some C-phycoerythrin and lose 30.5K linker; 30.5K-phycoerythrin complex does not add to C-phycocyanin core structures
31.5	*Synechocystis* sp.	C-Phycoerythrin rod assembly associated with inner C-phycoerythrin	Removal of all C-phycoerythrin removes 31.5K; 31.5K-phycoerythrin complex adds onto C-phycocyanin core structures
33.5	*Synechocystis* sp.	Stabilizes C-phycocyanin assembly	Mutants lacking 33.5K add rods to core poorly; present in phycobilisomes lacking C-phycoerythrin
34.5	*Nostoc* sp.	Joins disks of C-phycocyanin	Only found in phycobilisomes with two C-phycocyanin disks
34	*Nostoc* sp.	C-Phycoerythrin rod assembly connects C-phycoerythrin disk to C-phycocyanin disk	Absent in phycoerythrinless phycobilisomes; some dissociates with C-phycocyanin from rods, rest with C-phycoerythrin
32	*Nostoc* sp.	C-Phycoerythrin rod assembly connects two C-phycoerythrin disks	Absent in phycoerythrinless phycobilisomes; dissociates with C-phycoerythrin

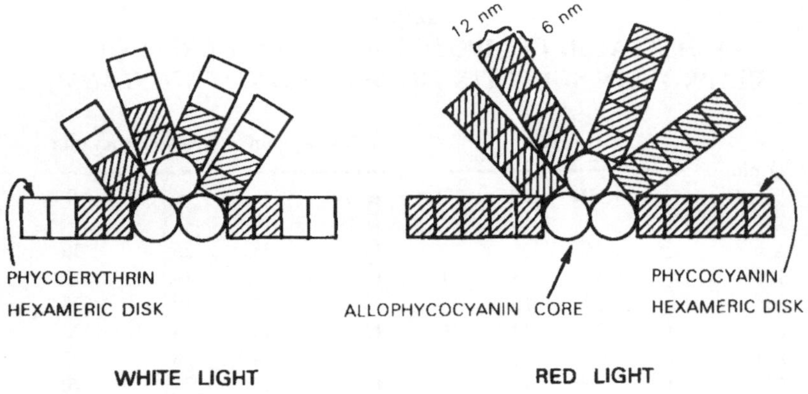

FIGURE 4. Schematic diagram of phycobilisomes from *Tolypothrix tenuis* grown in red or white light. Disks in the rods are hexameric aggregates of C-phycocyanin or C-phycoerythrin. Phycobilisomes are shown in face view. Circular core structures are cylinders of four trimeric disks.

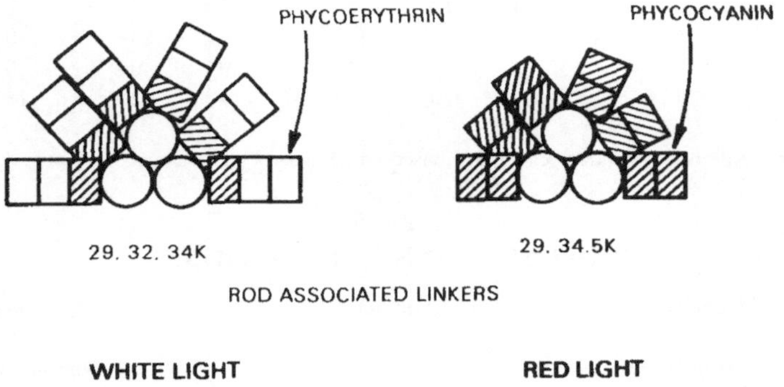

FIGURE 5. Schematic diagram of phycobilisomes from *Nostoc* sp. grown in red or white light.

in a slightly different manner than *T. tenuis*[26] upon chromatic adaptation. In this case the amount of phycocyanin increased in red light, but the total rod length diminished (Figure 5). *Synechocystis* 6701 showed totally different behavior.[14] Here while the phycoerythrin diminished under red light growth, the phycocyanin remained constant (Figure 6). It was noted that very different linker changes resulted in chromatic adaptation of *Nostoc* vs. *Synechocystis*. For *Nostoc* sp., red light produced an additional C-phycocyanin disk plus a new linker (34.5K). This linker was apparently needed to connect the new disk in the rod structure. For *Synechocystis* the C-phycocyanin portion was unaltered and thus no new linker was required.

To put these structures in better perspective, their chromophore composition can be calculated. The phycobilisome structures can be characterized in terms of numbers of disks. A C-phycoerythrin hexameric disk has 36 or 30 phycoerythrobilin chromophores depending on the organism, each C-phycocyanin hexamer has 18 phycocyanobilins, and each allophycocyanin trimer has 6 phycocyanobilins. Using *T. tenuis*, both white and red light-grown organelles have six rods. In the white light-grown organelles, each rod has two phycoerythrin disks and two phycocyanin disks: $12 \times 36 + 12 \times 18 = 648$ total rod chromophores. The red light grown phycobilisomes have five phycocyanin disks per rod: $30 \times 18 = 540$ total rod chromophores. In addition, there are approximately 72 phycocyanobilins in each three-cylinder core. The AN112 mutant of *A. nidulans* has in comparison 108 phycocyanobilins in its truncated rods and about 48 in its bicylindrical core.

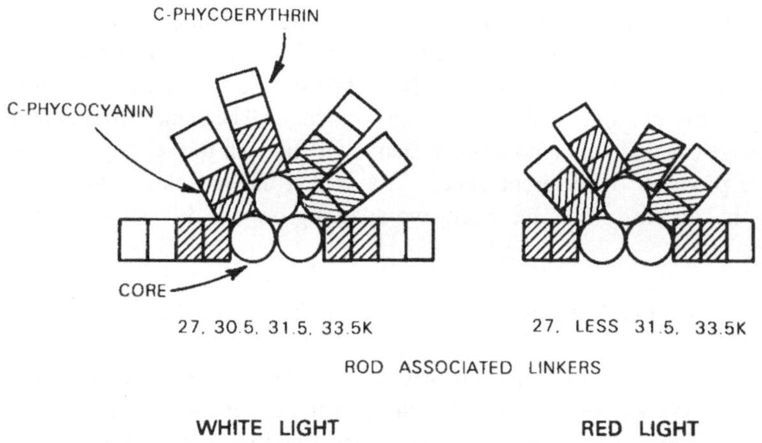

FIGURE 6. Schematic diagram of phycobilisomes from *Synechocystis* 6701 grown in red or white light.

Table 6
MOLECULAR WEIGHTS OF SOME CYANOBACTERIAL PHYCOBILISOMES[26]

Cyanobacteria	Growth light	Mol wt ($\times 10^{-6}$)
Fremyella diplosiphon	White	6.38
	Red	4.46
Tolypothrix tenuis	White	6.28
	Red	6.84
Lyngbya-Plectonema-Phormidium sp.	White	4.64
	Red	7.68
Phormidium persicinum	White	16.6
P. luridum	White	4.46
Anabaena flos-aquae	White	4.17
Mastigocladus laminosus	White	2.97

Some very interesting phycobilisomes which did not chromatically adapt have also been studied by Siegelman and Kycia[26] in terms of biliprotein content and size. *Phormidium persicinum* had an extremely low content of C-phycocyanin and allophycocyanin. The percentages of biliproteins content were 86% for C-phycoerythrin, 9% for C-phycocyanin, and 5% for allophycocyanin. The very high C-phycoerythrin content and the large size of its phycobilisomes (Table 5) suggested that the usual phycobilisome structure probably was somewhat modified for this species. A situation of high allophycocyanin content was found in phycobilisomes from *Microcystis aeruginosa,* where the biliprotein percentages were 0% for C-phycoerythrin, 80.3% for C-phycocyanin, and 19.7% for allophycocyanin. Siegelman and Kycia[26] have calculated molecular weights for a variety of phycobilisomes (Table 6). The results showed that the basic phycobilisome structure can be stretched to accomodate a wide variety of sizes.

Another level of phycobilisome complexity has been revealed by Bryant and Cohen-Bazire[11] and Bryant.[31] They examined phycobilisomes from several cyanobacteria, which chromatically adapted by changing the contents of both C-phycocyanin and C-phycoerythrin, by SDS gel electrophoresis. Of the 31 species examined, 24 exhibited two α and two β C-phycocyanin subunits instead of the normal content of one each. One α and one β subunit were the usual number of subunits found in nonadapting cells, while the second pair was specifically synthesized in red light-grown cells. The amino acid compositions of all the α and β subunits from red light-grown *Pseudanabaena* 7409 showed that the two α and the two β subunits have similar compositions. Nonetheless, the two α subunits differed from each other especially in valine and phenylalanine contents. The same two amino acids were also present in clearly different amounts for each β subunit. All four C-phycocyanin subunits were therefore products of different genes. Strains that did not chromatically adapt or in which the C-phycoerythrin was the exclusive adapting protein did not have the second α and β subunits.

V. THE 18.3K BILIPROTEIN CORE SUBUNIT

In a study of the allophycocyanin core structure of the AN112 mutation of *A. nidulans*, an 18,300-mol wt polypeptide was isolated bearing a single phycocyanobilin chromophore.[32] An 18S core complex was treated with trypsin, and fractions were separated by hydroxylapatite chromatography. One of the four major fractions when further studied by isoelectric focusing in the presence of 9.2 *M* urea, dissociated into three major components: the α and β subunits of allophycocyanin and a pigmented polypeptide of 18.3K. The partial N terminal amino acid sequence of the 18.3K polypeptide differed from those of the α and β subunits of either allophycocyanin or C-phycocyanin. The bilin content of this new biliprotein subunit was calculated indirectly from the properties of its complex.

A 16.2K polypeptide has been isolated from *Mastigocladus laminosus*.[33] Its partial N terminal amino acid sequence closely resembled the 18.3K biliprotein subunit. Its absorption maximum was 630 nm, and its fluorescence emission maximum was 648 nm.

VI. THE CORE STRUCTURE

The AN112 mutant strain of *A. nidulans* has been a major source of our increased understanding of the core structure of the phycobilisome. Like all hemi-discoidal phycobilisomes, the core of AN112 has two cylindrical structures contiguous to the thylakoid membrane. The length of each cylinder lies flat on the membrane surface, and each cylinder is composed of four trimer-like disks (Figure 7). Unlike most other phycobilisomes it lacked the third cylinder which was usually found on top of the first two, and unlike other phycobilisomes its mutation has eliminated the six rods which were replaced by six single hexameric C-phycocyanin disks (Figure 7). Yamanaka et al.[34] have partially dissociated these phycobilisomes and have separated three complexes in a sucrose density gradient. One of these complexes sedimented at 18S. This complex had all the 75K polypeptide, most of the 27K linker, and a portion of the C-phycocyanin and allophycocyanin subunits. Additional speculations in this report on the organization of the core have been modified and refined in the later work of this same group. They noted that one danger in studying the core was the possibility of subunit exchanging among different biliproteins during isolation.[34,35]

Lundell and Glazer[32,36,37] have extensively studied this system and have developed a model for positioning all the polypeptide components within the core. These components were the α and β subunits of allophycocyanin, the 18.3K biliprotein subunit, the 75K polypeptide, and α^B subunit which is the 680-nm emitter of allophycocyanin B, and the 10.5K linker. The 18S core component was treated with trypsin and separated into three complexes. Two of these complexes were core related: an $\alpha_3\beta_3$ trimer of allophycocyanin and an $\alpha_2\beta_2$-18.3K-40K-11K complex. The 40 and 11K polypeptides were believed to be proteolytic cleavage products of the 75K polypeptide.[32]

Lundell and Glazer[36] then isolated and purified two more allophycocyanin-containing complexes: $\alpha_3\beta_3$-10.5K and $\alpha^B\alpha_2\beta_3$-10.5K, where again α and β were the normal allophycocyanin subunits. The portions of the dissociated phycobilisome not comprising the 18S

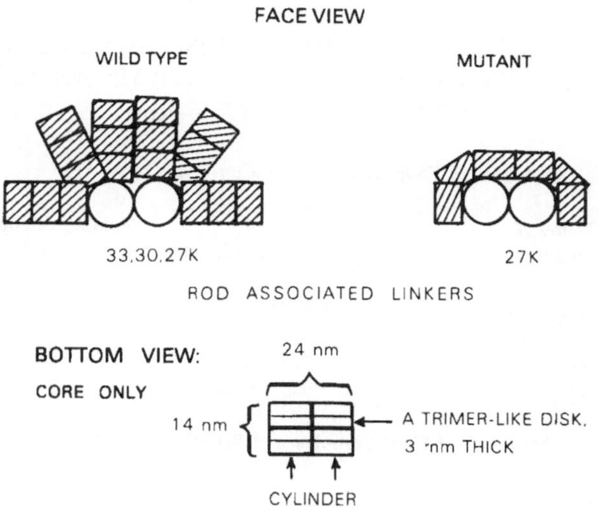

FIGURE 7. Schematic diagram of phycobilisomes from wild type *Anacystis nidulans* and its mutant AN112. A face view is given of both phycobilisomes, and a bottom view is shown of the core. The cores of both are composed of two cylinders lying on the thylakoid membrane. For the mutant, each cylinder has been resolved into four individual trimer-like disks. The face-to-face assembly of the trimeric disks forms the cylinders.

complex were the sources of these complexes. From the 6S fraction an $\alpha_3\beta_3$-10.5K complex was purified by elution through a hydroxylapatite column and the second, the $\alpha^B\alpha_2\beta_3$-10.5K complex, was purified by a combination hydroxylapatite and DEAE-cellulose chromatography. Unlike prior studies in which allophycocyanin and allophycocyanin B were shown to exchange subunits,[35] the $\alpha_3\beta_3$-10.5K complex did not dissociate very rapidly.[36] Each of these four core complexes was generally similar in size to a trimeric disk of allophycocyanin subunits. Remembering that there were eight such disks in the core of *A. nidulans* suggested that each of these complexes occurred once in each cylinder. Lundell and Glazer[36] used a variety of calculation methods to verify that two of each was consistent with the properties of these phycobilisomes (Table 7). The 45K polypepetide observed in these phycobilisome preparations was neither present in sufficient amounts for one copy per organelle nor a cleavage product of the 75K linker. One possibility is offered that it could be a membrane protein carried along in the preparations.

Lundell and Glazer[37] have isolated a complex containing three of the four core complexes: $\alpha_3\beta_3$,$\alpha_2\beta_2$-18.3K-75K, and $\alpha^B\alpha_2\beta_3$-10.5K. Therefore, the fourth complex, $\alpha_3\beta_3$-10.5K, was on the terminal of each core cylinder. The two components of the 18S complex would be adjacent to each other in the core. A proposal (Table 7), although certainly not a rigorously proven one, was made for the arrangement of the four complexes within the core of AN112.[37] In the presentation of this data, the Zilinskas et al.[12] nomenclature of allophycocyanins I, II, III, and B is used. This structure put the two 680-nm emitters, 75K linker and α^B, on different but adjacent complexes in the middle of the core. The reason for two different emitters on each phycobilisome was unclear, but it may be expected that each emitter transfers excitation energy to different receptors. A scheme for excitation energy transfer of the phycobilisome reflected this duality of emission (Figure 8).

Another cyanobacterium from which details of the allophycocyanin core were obtained was *Nostoc.* sp. Zilinskas et al.[12] isolated four complexes containing allophycocyanin. Although these complexes were not characterized in as much detail as those from *A. nidulans* mutant AN112, there seemed to be a general correspondence[4] of the four AN112 complexes

Table 7
STRUCTURE OF THE PHYCOBILISOME CORE OF *ANACYSTIS NIDULANS* MUTANT STRAIN AN112[37] AND *SYNECHOCYSTIS* 6701[39]

Structure of Four Isolated Core Complexes of AN112

Name of complex	Structure	Number of copies per phycobilisome
Allophycocyanin I	$\alpha_2\beta_2$-18.3K-75K	2
Allophycocyanin II	$\alpha_3\beta_3$	2
Allophycocyanin III	$\alpha_3\beta_3$-10.5K	2
Allophycocyanin B	$\alpha^B\alpha_2\beta_3$-10.5K	2
Structure of Core Complexes of *Synechocystis* 6701		
Allophycocyanin I	$\alpha_2\beta_2$-18.5K-99K	2
Allophycocyanin II	$\alpha_3\beta_3$	4
Allophycocyanin III	$\alpha_3\beta_3$-10K	4
Allophycocyanin B	$\alpha^B\alpha_2\beta_3$-10K	2
Proposed Arrangement of Complexes Within the Core of AN112		
Cylinder 1	II/I/B/III	
Cylinder 2	III/B/I/II	

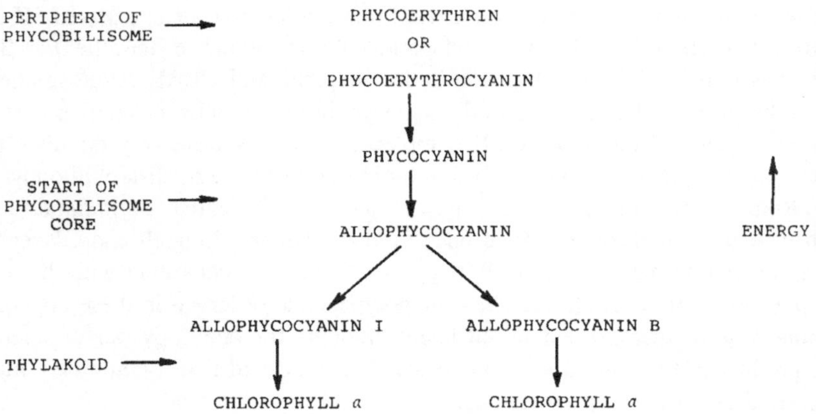

FIGURE 8. Excitation energy transfer pathway of a typical phycobilisome.

to the *Nostoc* sp. complexes of allophycocyanin I, II, III, and B. This may indicate an analogous core structure for both organisms. *Nostoc* sp. had three cylinders in its core.

Zilinskas[38] has further characterized the core of *Nostoc* sp. phycobilisomes. The core was separated into three complexes: 19, 10.3, and 5.5S. The SDS gels of these complexes (Figure 9) showed that the 19S complex had the 95K polypeptide and the α and β allophycocyanin subunits. The 10.3S complex had the α and β subunits of allophycocyanin plus two fairly faint bands at 46 and 44K, and the 5.5S complex had the α and β subunits plus a 10K linker.

Zilinskas[38] purified the 95K polypeptide from the α and β allophycocyanin subunits by gel filtration in the presence of 8 *M* urea at pH 8.0. SDS gels demonstrated the purity of the 95K polypeptide (Figure 10). Attempts to remove fully the urea from the solvent resulted in precipitation of the 95K polypeptide when the α and β subunits were absent.

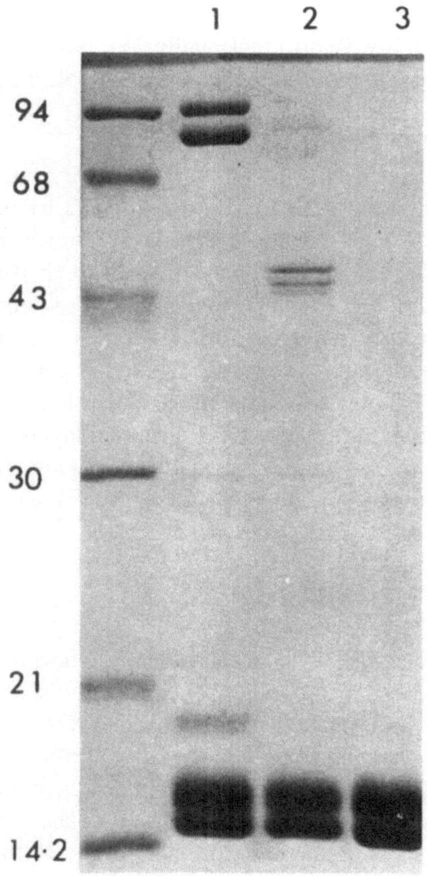

FIGURE 9. SDS gel of three core components of the phycobilisomes from *Nostoc* sp. Lane on left, standard molecular weight markers; lane 1, 19S complex of allophycocyanin; lane 2, 10.3S complex; and lane 3, 5.5S complex. Photograph courtesy of B. Zilinskas. (From Zilinskas, B. A., *Plant Physiol.*, 70, 1060, 1982. With permission.)

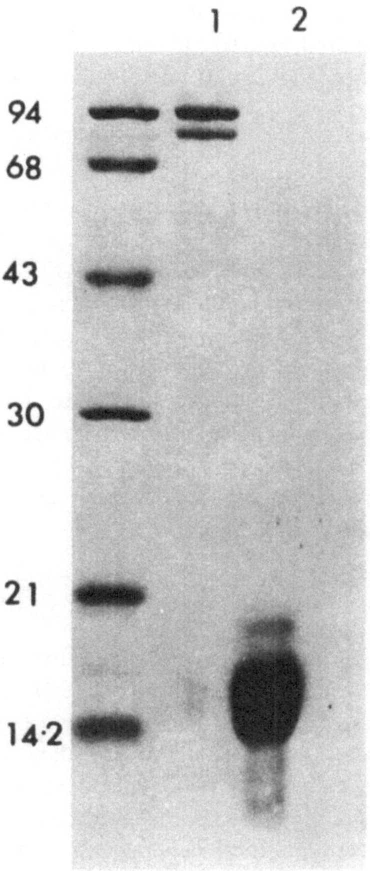

FIGURE 10. SDS gel electrophoresis of the *Nostoc* sp. 95K polypeptide (lane 1). Lane on left shows standard molecular weight markers, and lane 2 shows α and β subunits of allophycocyanin. Photograph courtesy of B. Zilinskas. (From Zilinskas, B. A., *Plant Physiol.*, 70, 1060, 1982. With permission.)

Gingrich et al.[39] have analyzed the core structure of *Synechocystis* 6701 phycobilisomes (Table 7). They found that the two cylinders adjacent to the thylakoid membrane were identical to those of *A. nidulans* AN112. The cylinder not on the membrane surface was found to be composed of two trimers of allophycocyanin II and two trimers of allophycocyanin III. They also cautioned that some of these complexes may dissociate upon extraction and then reassociate to form new complexes not found in the phycobilisomes.

VII. AMINO ACID SEQUENCE OF A LINKER

An 8.9K polypeptide has been isolated from the phycobilisomes of the cyanobacterium *M. laminosus*.[33] This colorless material has been completely sequenced and was shown to be a product of two genes. The polypeptides were each 67-amino acid residues long and differed only at residue 9 where the amino acid was either cysteine or serine. It was found to be associated with allophycocyanin and was therefore probably a core component.

Füglistaller et al.[33] proposed that this polypeptide might be identical in function to the

10.5K linker of *A. nidulans*. They noted that the core structure of Lundell and Glazer[37] required two 10.5K linkers in each cylinder and each with different binding locales. They proposed that the two 8.9K linkers might differ in order to bind to these different complexes.[33]

VIII. DETERGENT-FREE PHYCOBILISOMES

Hiller et al.[40] have isolated phycobilisomes from the red alga *Griffithsia monilis* by use of the enzyme trypsin. Normally, detergent, e.g., Triton® X-100, was used to remove the organelles from the thylakoids. The trypsin-released phycobilisomes band in a sucrose density gradient at the same position as detergent-treated samples. The absorbance and fluorescence properties of both preparations were identical. The 94K polypeptide of this organism was comparable to other high-molecular weight polypeptides that have been discussed as possible anchors of the phycobilisomes to the thylakoid.[1] This idea was supported by its role as the final excitation energy transfer agent to the chlorophyll *a* and its presence in washed thylakoid membranes. However, the trypsin-treated phycobilisomes still had a small residual amount of the 94K polypeptide which raised questions about how these phycobilisomes were released from the membrane.

IX. REASSOCIATION OF PHYCOBILISOMES

In order to isolate stable and intact phycobilisomes, high concentrations of phosphate were employed (typically 0.75 *M*). Kume et al.[41] have studied the stability of phycobilisomes from *Anabaena variabilis* in the absence of high phosphate. Using viscosity, refractive index, and light scattering, they found that high phycobilisome concentrations promoted organelle stabilization at low phosphate concentrations. The high phycobilisome concentrations served to promote reassembly in the association-dissociation equilibrium. They estimated that the local concentration of phycobilisomes in a cell was extremely high and that this concentration effect produced the in vivo stabilization.

A few selected studies on phycobilisome reassociation that have not already been mentioned will be discussed here. Phycobilisomes from *Nostoc* sp. can be dissociated into a phycoerythrin-phycocyanin complex and an allophycocyanin fraction in low-phosphate concentrations.[42] Significant reassociation occurred when these two fractions were mixed and returned to 0.75 *M* potassium phosphate. Sucrose (2 *M*) improved the reassociation.

This type of reassociation has also been performed on mixed systems of species.[43] Allophycocyanin from *Nostoc* sp. reassembled intact phycobilisomes with a phycoerythrin-phycocyanin complex from *F. diplosiphon* and vice versa. Not all cyanobacterial phycobilisomes were competent to enter into such mixed systems, however. The ability of any of these phycobilisomes to reassemble of course indicated that the two mixtures have the required linkers. The mixed reassociation was suggestive of the finding that linkers from *Anacystis nidulans* will aggregate purified C-phycocyanin from other cyanobacteria.[5]

It has been demonstrated that not only can dissociated phycobilisomes be reformed but also that dissociated phycobilisomes can be reunited with thylakoid membranes to form a functional photosynthetic system.[44] Phycobilisomes still attached to thylakoid membranes have been isolated from *F. diplosiphon*.[44] When the ionic strength was lowered, dissociation of the phycobilisome membrane structure occurred. When the ionic strength was then increased, the phycobilisome membrane structure was renewed. In adition, there was a return of energy transfer from C-phycoerythrin to chlorophyll *a* as measured by fluorescence emission and photosystem II activity.

X. EFFECTS OF LINKERS ON SPECTRA

Yu et al.[8] isolated two C-phycocyanin trimers, one having a 27K linker and the other a

32.5K linker. Both the absorption and fluorescence spectra of these complexes were different. This suggested that linkers in addition to stabilizing aggregates may modulate their spectral properties, perhaps to foster increased energy transfer efficiencies. The linker complexes from *A. nidulans* also showed these spectral properties.[5] Bekasova et al.[45] have discussed the occurrence of various spectral forms of biliproteins in phycobilisomes.

XI. KINETICS OF LINKER APPEARANCE

Lönneborg et al.[46] have studied the rate of appearance of 30 and 33K linkers after shifting *A. nidulans* from growth in white to red light. There was an immediate increase in the ratios of both these linkers to the 27K linker. At the same time C-phycocyanin synthesis also increased sharply. Later the rate of C-phycocyanin synthesis slowed, and at that time the ratio of the 30 and 33K linkers to the 27K linker became constant. The authors noted that the 30 and 33K linkers were involved in rod elongation. They proposed that initially the phycobilisomes remained constant in number but increased in rod length, causing an increase in 30K and 33K relative to 27K. The slower C-phycocyanin synthesis period would correspond to the formation of new phycobilisomes which would also require more 27K as well as 30 and 33K.

XII. ENERGY TRANSFER WITHIN THE CORE

The isolated core from a mutant of *Synechocystis* 6701 can be readily obtained.[13] These cores have been studied by ultrafast kinetics.[47] The rise time of fluorescence was faster than the resolution of their instrument. Energy transfer within the cores, therefore, must be very rapid (see Chapter 5 for other kinetics studies).

REFERENCES

1. **Tandeau de Marsac, N. and Cohen-Bazire, G.,** Molecular composition of cyanobacterial phycobilisomes, *Proc. Natl. Acad. Sci. U.S.A.*, 74, 1635, 1977.
2. **Redlinger, T. and Gantt, E.,** Phycobilisome structure of *Porphyridium cruentum*. Polypeptide composition, *Plant Physiol.*, 68, 1375, 1981.
3. **Mörschel, E.,** Accessory polypeptides in phycobilisomes of red algae and cyanobacteria, *Planta*, 154, 251, 1982.
4. **Yamanaka, G., Glazer, A. N., and Williams, R. C.,** Cyanobacterial phycobilisomes. Characterization of the phycobilisomes of *Synechococcus* sp. 6301, *J. Biol. Chem.*, 253, 8303, 1978.
5. **Lundell, D. J., Williams, R. C., and Glazer, A. N.,** Molecular architecture of a light-harvesting antenna. *In vitro* assembly of the rod substructures of *Synechococcus* 6301 phycobilisomes, *J. Biol. Chem.*, 256, 3580, 1981.
6. **Yamanaka, G., Glazer, A. N., and Williams, R. C.,** Molecular architecture of a light-harvesting antenna. Comparison of wild type and mutant *Synechococcus* 6301 phycobilisomes, *J. Biol. Chem.*, 255, 11004, 1980.
7. **Lundell, D. J., Yamanaka, G., and Glazer, A. N.,** A terminal energy acceptor of the phycobilisome: the 75,000-dalton polypeptide of *Synechococcus* 6301 phycobilisomes — a new biliprotein, *J. Cell Biol.*, 91, 315, 1981.
8. **Yu, M.-H., Glazer, A. N., and Williams, R. C.,** Cyanobacterial phycobilisomes. Phycocyanin assembly in the rod substructures of *Anabaena variabilis*, *J. Biol. Chem.*, 256, 13,130, 1981.
9. **Zilinskas, B. A. and Howell, D. A.,** Role of the colorless polypeptides in phycobilisome assembly in *Nostoc* sp., *Plant Physiol.*, 71, 379, 1983.
10. **Williams, R. C., Gingrich, J. C., and Glazer, A. N.,** Cyanobacterial phycobilisomes. Particles from *Synechocystis* 6701 and two pigment mutants, *J. Cell Biol.*, 85, 558, 1980.

11. **Bryant, D. A. and Cohen-Bazire, G.,** Effects of chromatic illumination on cyanobacterial phycobilisomes. Evidence for the specific induction of a second pair of phycocyanin subunits in *Pseudanabaena* 7409 grown in red light, *Eur. J. Biochem.*, 119, 415, 1981.
12. **Zilinskas, B. A., Zimmerman, B. K., and Gantt, E.,** Allophycocyanin forms isolated from *Nostoc* sp. phycobilisomes, *Photochem. Photobiol.*, 27, 587, 1978.
13. **Anderson, L. K., Rayner, M. C., and Eiserling, F. A.,** Ultra-violet mutagenesis of *Synechocystis* sp. 6701: mutations in chromatic adaptation and phycobilisome assembly, *Arch. Microbiol.*, 138, 237, 1984.
14. **Gingrich, J. C., Blaha, L. K., and Glazer, A. N.,** Rod substructure in cyanobacterial phycobilisomes: analysis of *Synechocystis* 6701 mutants low in phycoerythrin, *J. Cell Biol.*, 92, 261, 1982.
15. **Gingrich, J. C., Williams, R. C., and Glazer, A. N.,** Rod substructure in cyanobacterial phycobilisomes: phycoerythrin assembly in *Synechocystis* 6701 phycobilisomes, *J. Cell Biol.*, 95, 170, 1982.
16. **Glick, R. E. and Zilinskas, B. A.,** Role of the colorless polypeptides in phycobilisome reconstitution from separated phycobiliproteins, *Plant Physiol.*, 69, 991, 1982.
17. **Rusckowski, M. and Zilinskas, B. A.,** Allophycocyanin I and the 95 kilodalton polypeptide. The bridge between phycobilisomes and membranes, *Plant Physiol.*, 70, 1055, 1982.
18. **Canaani, O. and Gantt, E.,** Native and in vitro-associated phycobilisomes of *Nostoc* sp. composition, energy transfer, and effect of antibodies, *Biochim. Biophys. Acta*, 723, 340, 1983.
19. **MacColl, R., O'Connor, G., Crofton, G., and Csatorday, K.,** Phycoerythrocyanin: its spectroscopic behavior and properties, *Photochem. Photobiol.*, 34, 719, 1981.
20. **Nies, M. and Wehrmeyer, W.,** Biliprotein assembly in the hemidiscoidal phycobilisomes of the thermophilic cyanobacterium *Mastigocladus laminosus* Cohn. Characterization of dissociation products with special reference to the peripheral phycoerythrocyanin-phycocyanin complexes, *Arch. Microbiol.*, 129, 374, 1981.
21. **Yu, M.-H. and Glazer, A. N.,** Cyanobacterial phycobilisomes. Role of the linker polypeptides in the assembly of phycocyanin, *J. Biol. Chem.*, 257, 3429, 1982.
22. **Isono, T. and Katoh, T.,** Cylindrical phycobilisomes from a blue-green alga, *Anabaena variabilis*, *Plant Cell Physiol.*, 23, 1347, 1982.
23. **Isono, T. and Katoh, T.,** Subparticles of *Anabaena* phycobilisomes. I. Reconstitution of allophycocyanin cores and entire phycobilisomes, *Plant Cell Physiol.*, 24, 357, 1983.
24. **Lipschultz, C. A. and Gantt, E.,** Association of phycoerythrin and phycocyanin: in vitro formation of a functional energy transferring phycobilisome complex of *Porphyridium sordidum*, *Biochemistry*, 20, 3371, 1981.
25. **Kursar, T. A., van der Meer, J., and Alberte, R. S.,** Light-harvesting system of the red alga *Gracilaria tikvahiae*. II. Phycobilisome characteristics of pigment mutants, *Plant Physiol.*, 73, 361, 1983.
26. **Siegelman, H. W. and Kycia, J. H.,** Molecular morphology of cyanobacterial phycobilisomes, *Plant Physiol.*, 70, 887, 1982.
27. **Guard-Friar, D. and MacColl, R.,** Spectroscopic properties of tetrapyrroles on denatured biliproteins, *Arch. Biochem. Biophys.*, 230, 300, 1984.
28. **Tandeau de Marsac, N.,** Occurence and nature of chromatic adaptation in cyanobacteria, *J. Bacteriol.*, 130, 82, 1977.
29. **Bryant, D. A., Guglielmi, G., Tandeau de Marsac, N., Castets, A.-M., and Cohen-Bazire, G.,** The structure of cyanobacterial phycobilisomes: a model, *Arch. Microbiol.*, 123, 113, 1979.
30. **Rosinski, J., Hainfeld, J. F., Rigbi, M., and Siegelman, H. W.,** Phycobilisome ultrastructure and chromatic adaptation in *Fremyella diplosiphon*, *Ann. Bot.*, 47, 1, 1981.
31. **Bryant, D. A.,** The photoregulated expression of multiple phycocyanin species. A general mechanism for the control of phycocyanin synthesis in chromatically adapting cyanobacteria, *Eur. J. Biochem.*, 119, 425, 1981.
32. **Lundell, D. J. and Glazer, A. N.,** Molecular architecture of a light-harvesting antenna. Structure of the 18S core-rod subassembly of the *Synechococcus* 6301 phycobilisome, *J. Biol. Chem.*, 258, 894, 1983.
33. **Füglistaller, P., Rümbeli, R., Suter, F., and Zuber, H.,** Minor polypeptides from the phycobilisome of the cyanobacterium *Mastigocladus laminosus*. Isolation, characterization and amino-acid sequences of a colorless 8.9-kDa polypeptide and of a 16.2-kDa phycobiliprotein, *Hoppe-Seyler's Z. Physiol. Chem.*, 365, 1085, 1984.
34. **Yamanaka, G., Lundell, D. J., and Glazer, A. N.,** Molecular architecture of a light-harvesting antenna. Isolation and characterization of phycobilisome subassembly particles, *J. Biol. Chem.*, 257, 4077, 1982.
35. **Lundell, D. J. and Glazer, A. N.,** Allophycocyanin B. A common β subunit in *Synechococcus* allophycocyanin B (λ_{max} 670 nm) and allophycocyanin (λ_{max} 650 nm), *J. Biol. Chem.*, 256, 12,600, 1981.
36. **Lundell, D. J. and Glazer, A. N.,** Molecular architecture of a light-harvesting antenna. Core substructure in *Synechococcus* 6301 phycobilisomes: two new allophycocyanin and allophycocyanin B complexes, *J. Biol. Chem.*, 258, 902, 1983.

37. **Lundell, D. J. and Glazer, A. N.,** Molecular architecture of a light-harvesting antenna. Quaternary interactions in the *Synechococcus* 6301 phycobilisome core as revealed by partial tryptic digestion and circular dichroism studies, *J. Biol. Chem.*, 258, 8708, 1983.
38. **Zilinskas, B. A.,** Isolation and characterization of the central component of the phycobilisome core of *Nostoc* sp., *Plant Physiol.*, 70, 1060, 1982.
39. **Gingrich, J. C., Lundell, D. J., and Glazer, A. N.,** Core substructure in cyanobacterial phycobilisomes, *J. Cell. Biochem.*, 22, 1, 1983.
40. **Hiller, R. G., Post, A., and Stewart, A. C.,** Isolation of intact detergent-free phycobilisomes by trypsin, *FEBS Lett.*, 156, 180, 1983.
41. **Kume, N., Isono, T., and Katoh, T.,** Stability of cyanobacterial phycobilisomes in reference to their concentration, *Photobiochem. Photobiophys.*, 4, 25, 1982.
42. **Canaani, O., Lipschultz, C. A., and Gantt, E.,** Reassembly of phycobilisomes from allophycocyanin and a phycocyanin-phycoerythrin complex, *FEBS Lett.*, 115, 225, 1980.
43. **Canaani, O. and Gantt, E.,** Formation of hybrid phycobilisomes by association of phycobiliproteins from *Nostoc* and *Fremyella*, *Proc. Natl. Acad. Sci. U.S.A.*, 79, 5277, 1982.
44. **Kirilovsky, D., Kessel, M., and Ohad, I.,** In vitro reassociation of phycobiliproteins and membranes to form functional membrane bound phycobilisomes, *Biochim. Biophys. Acta*, 724, 416, 1983.
45. **Bekasova, O. D., Muslimov, I. A., and Krasnovski, A. A.,** Fractionation of phycobilisomes from the blue-green alga *Nostoc muscorum*, *Mol. Biol.*, 18, 262, 1984.
46. **Lönneborg, A., Lind, L. K., Kalla, S. R., Gustafsson, P., and Öquist, G.,** Acclimation processes in the light-harvesting system of the cyanobacterium *Anacystis nidulans* following a light shift from white to red light, *Plant Physiol.*, 78, 110, 1985.
47. **Glazer, A. N., Chan, C., Williams, R. C., Yeh, S. W., and Clark, J. H.,** Kinetics of energy flow in the phycobilisome core, *Science*, 230, 1051, 1985.

Chapter 6

EXCITATION ENERGY TRANSFER

I. EXCITONS

Excitation energy can be transferred from chromophore to chromophore by a mechanism described in general by Förster.[1-3] The transfer occurs by dipole-dipole interaction between donor and acceptor molecules (Figure 1). The spectra of the donor and acceptor are assumed to be unaffected by interaction. The mechanism is a radiationless, resonance interaction between the two dipoles. The rate of energy transfer depends on a coupling of energy levels between donor and acceptor, i.e., the fluorescence emission of the donor should overlap the absorption spectrum of the acceptor. The rate also depends on the distance between donor and acceptor dipoles and the orientation of the donor and acceptor transition dipoles.

Upon irradiation, the potential donor is excited to one of the vibrational levels of its first excited state (Figure 1). Then the excitation drops toward the lower vibrational levels of the first excited state through nonradiative thermal equilibriation with the environment. The donor can return to the ground state from these vibrational levels by a variety of processes. If a suitable acceptor is at the proper distance and orientation, the excitation energy can be passed directly from donor to acceptor. These transitions are coupled and do not entail radiation. When the interaction is small, the spectra of the donor and acceptor are unaffected.

Since the interaction is a dipole-dipole type, it is inversely proportional to the distance to the third power. The probability of transfer is proportional to the interaction matrix element squared and is therefore inversely related to the sixth power of the distance,

$$k_T = \frac{9(\ln 10)\, \kappa^2\, \phi_D J}{128\, \pi^5 n^4 N R^6 \tau_D} \quad (1)$$

where k_T is the rate, R is the distance between donor (D) and acceptor (A), n is the refractive index, N is Avogadro's number, ϕ_D is quantum yield of the donor, κ^2 is an orientation factor for the dipole, τ_D is the lifetime of the donor in the absence of the acceptor, and J is the modified spectral overlap. The overlap integral is usually calculated when expressed as a function of wavelength (λ) as

$$J = \int F_D(\lambda)\, a_A(\lambda)\, \lambda^4\, d\lambda \quad (2)$$

when a_A is the molar absorptivity of the acceptor in $M^{-1}cm^{-1}$, J is $M^{-1}cm^3$, and $F_D(\lambda)$ is the fluorescence emission intensity normalized to unity. The rate of energy transfer (k_T) can be written

$$k_T = \frac{(R_0/R)^6}{\tau_D} \quad (3)$$

where R_0 is distance where the efficiency of energy transfer is 50% and

$$R_0^6 = 8.8 \times 10^{-25} (\phi_D J \kappa^2/n^4)\text{cm} \quad (4)$$

The efficiency of energy transfer (E) is

$$E = \frac{R_0^6}{R_0^6 + R^6} \quad (5)$$

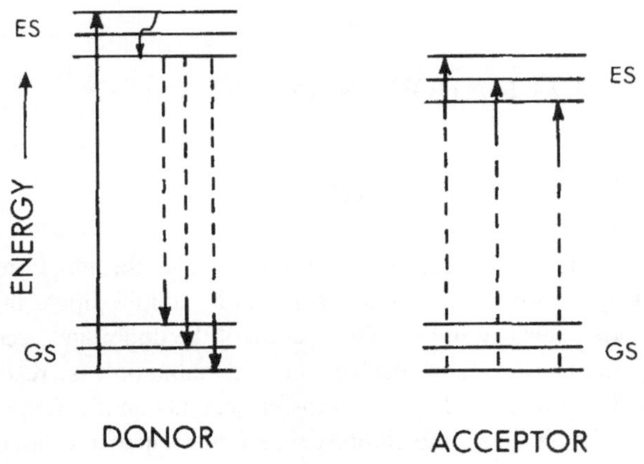

FIGURE 1. Schematic diagram of the excitation energy transfer mechanism by very weak coupling of dipoles. Transitions between donor and acceptor are coupled so that the transfers are radiationless. Solid lines are radiative transitions; wavy line is nonradiative decays; and dashed lines are nonradiative transfer of excitation energy from donor to acceptor.

The orientation factor (κ^2) is given by

$$\kappa^2 = (\cos \theta_{DA} - 3 \cos \theta_D \cos \theta_A)^2 \qquad (6)$$

where θ_{DA} is the angle between the transition moment vectors of A and D and the other two are the angles between these dipoles and the vector joining A and D. κ^2 may vary from 0 to 4, and a random direction distribution gives it a value of 2/3. It is difficult to assess the value of κ^2, although some aspects of the problem have been clarified.[4] Very often the 2/3 value is assumed in the calculations. Since the sixth root is taken to obtain distance, if κ^2 is reasonably close to 2/3, little error is introduced.

Although it appears that this type of mechanism may be appropriate in most cases for describing the excitation energy transfer properties of biliproteins, there are a few situations where other types of interactions have been discussed. In such cases the chromophores are coupled more strongly to each other, the excitation is delocalized, and the original spectra are altered.[3,5,6] This type of interaction can occur for biliproteins when chromophores are close to each other (<2 nm). Circular dichroism (CD) spectroscopy is an excellent tool to evaluate the presence of these strong exciton interactions.[7] Since it can be shown that the rotational strength of the exciton-coupled bands should total zero, the CD spectra of an interacting dimer will characteristically have both positive and negative components. The CD spectra of chlorophylls have been interpreted in terms of coupled excitons.[8,9] This type of coupling has been called "strong" coupling[10] of dipoles, and the mechanism in which there is no change in spectra is called "very weak" coupling.[3] It is chiefly a convenience to consider only the strong or very weak coupling cases, but what is so far known about the electronic properties of biliproteins seems adequately treated by these two limits.

Arnold and Oppenheimer[11] have discussed these energy transfer mechanisms specifically for biliproteins.

II. ENERGY TRANSFER IN INTACT ORGANISMS

The measurement of fluorescence after excitation of phycoerythrin in cyanobacteria and red algae revealed in addition to chlorophyll *a* emission some emission from phycocyanin.[12,13]

These were the first demonstrations of phycocyanin serving as an excitation energy transfer intermediary between phycoerythrin and chlorophyll. Subsequent studies on isolated phycobilisomes and their constituents revealed that after leaving phycocyanin, energy migrated to allophycocyanin and then to allophycocyanin 680 before reaching the chlorophyll receptors in the thylakoid membrane.

Tomita and Rabinowitch[14] studied the rates and efficiencies of energy transfer of phycoerythrin to phycocyanin and phycocyanin to chlorophyll *a* in the red alga *Porphyridium cruentum* and phycocyanin to chlorophyll *a* in the cyanobacterium *Anacystis nidulans*. The efficiencies of energy transfer from pigment to pigment may depend on the growth conditions and relative proportions of the pigments.[15-17] Wang and Myers[18] have proposed that reverse energy transfer from chlorophyll *a* to allophycocyanin can occur in the cyanobacterium *A. nidulans*.

Ley[19] measured the effective absorption cross-sections for O_2 production in *P. cruentum*. The technique employed consisted of illuminating cells on a platinum surface with various types of light. The effective absorption cross-sections were measured as the flash energy dependence of O_2 flash yields. The experimental results revealed that apparently all the reaction centers of photosystem II received excitation energy from phycobilisomes. In addition, several reaction centers shared the excitation energy from a single phycobilisome. An estimate was made that three to four reaction centers of photosystem II jointly used the light absorbed by a single phycobilisome. A different result was obtained by Diner[20] in studies of energy transfer from the phycobilisomes in the red alga *Cyanidium caldarium*. Light at either 600 or 436 nm was used to excite C-phycocyanin or chlorophyll *a*, respectively, and the yields of O_2 were recorded. The results obtained were interpreted to signify that only half of the reaction centers of photosystem II were attached to the phycobilisomes.

The physical proximity of a phycobilisome chromophore to the appropriate chlorophyll-protein receptor in the thylakoid is essential for energy transfer to occur. Studies have appeared which probe the factors that can disrupt this attachment.[21-24] Hydrostatic pressure at 1200 atm produced a large increase in the fluorescence of phycoerythrin and phycocyanin when applied to the red alga *Porphyra perforata*.[21] The uncoupling of the excitation energy transfer system suggested a change in the thylakoid membrane. The effect was fully reversible. Another factor was that a short period of preillumination prior to freezing *A. nidulans* affected the fluorescence intensity.[22] Again a proposal was made that the critical factor was nearness between the phycobilisome and the thylakoid membrane. In Chapter 8, preillumination of cells is discussed in terms of placing them in either state 1 or state 2. These states differ in the amount of energy transfer from photosystem II to photosystem I. Finally, cold pretreatment ($<5°C$) produced a decrease in energy transfer efficiency among the biliproteins,[23] and various ions and pH values affected the cold-induced uncoupling.[24]

III. ENERGY TRANSFER IN ISOLATED PHYCOBILISOMES

Gantt and Lipschultz[25] performed fluorescence experiments on isolated phycobilisomes from *Porphyridium cruentum*. Excitation of B- and b-phycoerythrin at 545 nm resulted in an emission maximum at 675 to 680 nm. This indicated that R-phycocyanin was not the final emitter, since its normal emission maximum was at 636 nm. Allophycocyanin was then implicated as a next step from phycocyanin in the excitation energy transfer scheme. Discovery of the two forms of allophycocyanin 680, allophycocyanins I and B,[26-28] in the phycobilisomes then completed the scheme of biliprotein to biliprotein energy transfer (Figure 2).

Energy from phycoerythrin (B, b, R, C, and CU) in cyanobacteria and red algae is always channeled through phycocyanin (R and C) prior to reaching the allophycocyanin in the core of the phycobilisome. This route is not required by the concepts of excitation energy transfer

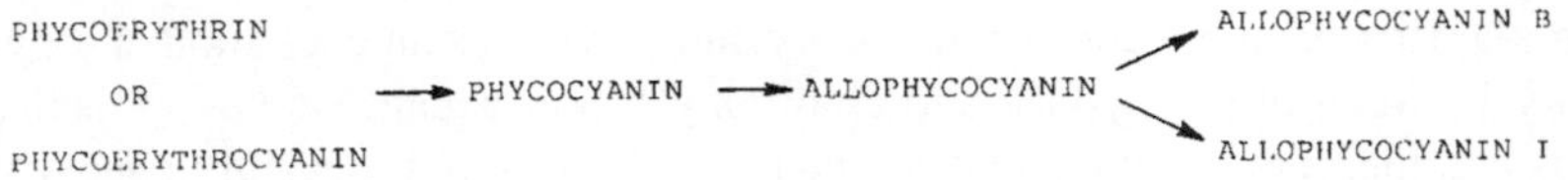

FIGURE 2. Route of energy transfer within a typical phycobilisome.

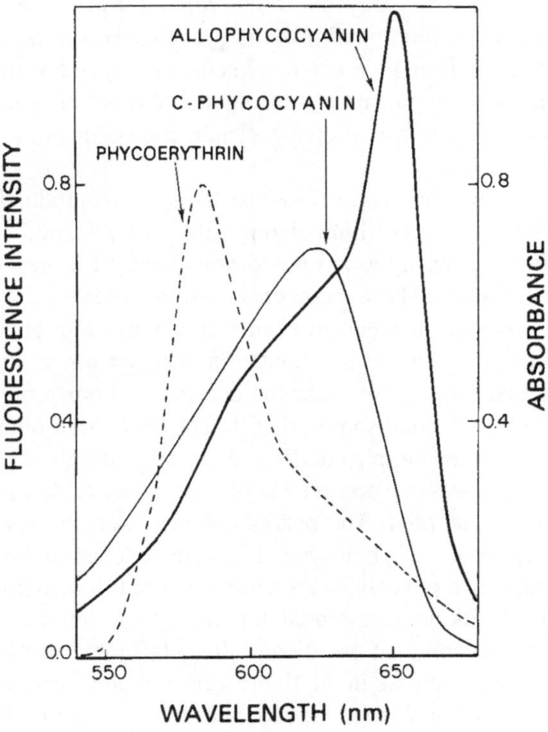

FIGURE 3. Fluorescence emission spectrum of phycoerythrin and absorption spectra of C-phycocyanin and allophycocyanin.

(Equation 1). Emission from the phycoerythrins fulfills the criterion of spectral overlap with the absorption of allophycocyanin which is laid down by the Förster mechanism of energy transfer (Figure 3). The overlap J between C-phycoerythrin (*Calothrix membranacea*) and C-phycocyanin is calculated from Equation 2 to be 8.6 × 10^{-10} cm^6/mol, and from C-phycoerythrin to allophycocyanin, it is 6.2 × 10^{-10} cm^6/mol. Grabowski and Gantt[29] showed that in some cases, phycoerythrins overlapped better with allophycocyanin than with C-phycocyanins (Table 1). Therefore phycoerythrins were almost as spectrally well suited for direct transfer to an allophycocyanin as to a C-phycocyanin receptor if both receptors were the same distance and orientation from the donor (which, of course, they are not). C-Phycoerythrin could even transfer energy to chlorophyll *a*, since the J was 1.4 × $10^{-10}$$cm^6$/mol. The reason for the total specificity of phycocyanin as the intermediary is the structural design of the phycobilisome which always physically places phycocyanin between phycoerythrin and allophycocyanin.[30] The importance of the stepwise transfer from phycoerythrin or phycoerythrocyanin (Figure 2) is then twofold: first, in most cases, it improves the extent of spectral overlap for the next transfer; and second, perhaps more critically, it provides a more unidirectional flow by setting up controlled steps which would tend to stop energy from migrating back toward the periphery of the phycobilisome. For example, the overlap between a C-phycocyanin donor and a C-phycoerythrin receptor was only 0.001 to 0.004

Table 1
SPECTRAL OVERLAP INTEGRALS (J) AND R_o VALUES FOR DONORS AND ACCEPTORS FROM THE CYANOBACTERIUM *NOSTOC* SP.[29]

Donor	Acceptor	J(cm^6/mol) × 10^{10}	R_o (nm)[a]
C-Phycoerythrin II[b]	C-Phycoerythrin II	1.8	4.8
	C-Phycoerythrin I	2.9	5.2
	C-Phycocyanin	6.8	5.9
	Allophycocyanin	7.6	6.1
C-Phycocyanin[c]	C-Phycoerythrin I	0.0004	1.7
	Allophycocyanin	11.7	6.5
Allophycocyanin[c]	C-Phycocyanin	1.1	4.4
	Allophycocyanin	8.5	6.1

[a] R_o calculated at $\kappa^2 = 2/3$ and $n = 1.576$.

[b] Phycoerythrin II differs from phycoerythrin I principally in that the former has a shoulder at 535 to 545 nm.

[c] Although the data were not presented here, both C-phycocyanin and allophycocyanin would on spectral evidence alone be able to transfer energy to chlorophyll *a*. The basis for normally neglecting this route is the maximum emission of isolated phycobilisomes at 670 to 680 nm which is a longer maximum wavelength than either phycocyanin or allophycocyanin can normally emit.

$\times\ 10^{-10}$ cm^6/mol.[29] Back transfer from a lower- to a higher-energy biliprotein was therefore much less probable than C-phycoerythrin transferring energy to another C-phycoerythrin, since the J values for self-transfer were between 1 and 3 $\times\ 10^{-10}$ cm^6/mol.[29]

The efficiencies of energy transfer among the biliproteins was quite high. For phycobilisomes from the red alga *Gloiopeltis furcata,* under optimal conditions the efficiencies were for R-phycoerythrin to phycocyanin, 93%; for phycocyanin to allophycocyanin, 98%; and for allophycocyanin to allophycocyanin 680, 96%.[31] For phycobilisomes from *P. cruentum, Nostoc* sp., and *Fremyella diplosiphon* the phycoerythrin to phycocyanin energy transfer efficiencies were in a narrow range of 94 to 98%.[32] For whole organisms, the transfer efficiencies were found to be similar, and for *P. cruentum,* phycoerythrin to phycocyanin transfer was 96%.[14] The relative dipole transition moments have been obtained for the biliproteins in a phycobilisome.[33]

Quite often experiments were performed on phycobilisomes by measuring the uncoupling of the energy transfer between biliproteins by fluorescence spectroscopy. These increases in fluorescence from phycoerythrin, phycocyanin, and allophycocyanin were universally viewed as meaning the biliproteins were dissociating. Kume and Katoh[34] noted that a shift in donor-acceptor distance and orientation (Equation 1) could possibly contribute to these observations prior to the actual dissociation of the organelles. In order to investigate this possibility, they studied the dissociation of phycobilisomes from *Anabaena variabilis* by both light scattering and fluorescence. The kinetics of the light-scattering changes which arose solely from molecular weight changes were similar to the kinetics of energy transfer uncoupling from fluorescence. Therefore it was concluded that the fluorescence changes from energy transfer uncoupling were indeed totally produced by protein dissociation.

IV. ENERGY TRANSFER IN PURIFIED BILIPROTEINS

A. Fluorescence Polarization

Goedheer and Birnie[35] found that pH affected the value of the fluorescence polarization for C-phycocyanin solutions. The polarization was found to be highest at either high- or low-pH values. They noted that these solvent conditions favored dissociation of the C-phycocyanin aggregates. Their interpretation was that energy transfer was more extensive on the larger aggregates which caused depolarization.

Fluorescence polarization is measured by use of polarizing filters. The fluorescence polarization is calculated by

$$p = \frac{I_{vv} - G\,I_{vh}}{I_{vv} + G\,I_{vh}} \tag{7}$$

where I_{vv} is the intensity of fluorescence excitation measured with both polarizers in the vertical position, I_{vh} is the intensity measured with the excitation polarizer vertical and the emission polarizer horizontal, and G is a grating correction factor. G is given by I_{hv}/I_{hh} and is independent of sample but dependent on emission wavelength. Polarization results are also sometimes reported as anisotropy (r), which can be obtained from the polarization by

$$r = \frac{2p}{3 - p} \tag{8}$$

The values of p are related to the angle between absorption and emission oscillators,[36]

$$p = \frac{3\cos^2\theta - 1}{\cos^2\theta + 3} \tag{9}$$

The maximum and minimum values of p occur when the absorption and emission oscillator are either perpendicular or parallel and give a range of p values from $-1/3$ to $+1/2$. Factors like energy transfer and motion of the chromophores will readily change the measured value of p from the extreme. In the absence of these factors, a p less than $+0.5$ is a measure of the angle between the absorption and fluorescence oscillators.

Vernotte,[37] Teale and Dale,[38] and MacColl et al.[39] used fluorescence polarization to study C-phycocyanin dissociation. For monomers[37,39] the highest polarization values obtained were at $+0.36$ to $+0.41$. For allophycocyanin monomers[40] the polarization reached a high of $+0.37$. For the separated subunits of allophycocyanin[41] the polarizations were similar to the monomers and were both about $+0.4$. For C-phycocyanin[42] the α subunit had a p of $+0.40$, and β was $+0.35$. Using a value of $+0.4$ in Equation 9, the angle was found to be 23° between absorbing and emitting oscillators in the absence of energy transfer. The lack of energy transfer for these highest polarization values was particularly well documented for the separated α subunits of allophycocyanin and C-phycocyanin because they contained only a single chromophore. The correspondence of the other values confirmed that they too were produced by single nontransferring chromophores. Motion was apparently not a factor in these measurements. The chromophores are attached to the apoprotein and will move slowly compared to the lifetime of the excited state. The chromophores themselves probably are held rigidly by the apoprotein as was suggested by the high quantum yields of the biliproteins.

Fluorescence polarization spectra across the lowest energy-excited state should be constant when the bands are composed of a single transition.[43] For phycoerythrin and phycocyanin from cyanobacteria and red algae, the fluorescence polarization spectra across the visible absorption bands were not constant.[38,44]

B. s and f Chromophores

The fluorescence properties of a biliprotein with a complex absorption spectrum, R-phycoerythrin (*Ceramium rubrum*), enabled the suggestion to be offered that excitation energy is passed from the higher-energy absorbers to the lowest-energy absorber.[45,46] The lowest-energy absorber then was responsible for fluorescence emission. This biliprotein is comprised of three absorption bands at 565, 545, and 500 nm. The 565-nm band was hypochromic in the presence of the sulfhydryl reagent, *p*-chloromercuribenzoate. The other absorption bands were unaffected, but the entire fluorescence excitation spectrum also decayed with the 565-nm absorption. It was suggested from this result that the 565-nm form received energy by a nonradiative process from both the 500- and 545-nm chromophores. The 565-nm chromophore was then the sole fluorescing chromophore in the native protein.

Additional study on R-phycoerythrin (*Antithamnion* sp.) showed that in fully denatured protein the chromophores were no longer coupled. The fluorescence emission spectra were variable depending on the excitation wavelengths,[47] whereas emission was mainly independent of excitation wavelength for intact biliproteins.

The concept of energy transfer on individual biliprotein aggregates was characterized in a seminal contribution by Teal and Dale.[38] They examined several biliproteins from cyanobacteria and red algae: R-phycoerythrin (*Rhodymenia* sp.); C-phycoerythrin (*Schizothrix calcicola*); B-phycoerythrin (*P. cruentum*); and C-phycocyanin (*S. calcicola* and *Anacystis nidulans*). These proteins were purified and studied by absorption, fluorescence, and fluorescence polarization spectroscopy. The fluorescence polarization spectra were complex in all cases, showing a low polarization in the higher-energy portion of the visible absorption band and then rising sharply in the red end of the band. Teale and Dale[38] compared this behavior to the spectrum of the zinc complex of phycoerythrobilin, which in a high-viscosity solvent, glycerol, had a constant and high polarization across the entire absorption band. The spectral complexity of the isolated protein aggregates was assigned to the existence of two types of chromophores. These two types were the s chromophore which absorbed light at higher energies and then transferred (sensitized) the energy to the lower-energy type and the f chromophore which either received energy from the s chromophore or directly absorbed the lower-energy light and then fluoresced. Of course, when these aggregates were in the phycobilisomes the f chromophores did not fluoresce but were the agents of transfer to the next protein in the energy migration chain. Phycoerythrocyanin had not yet been discovered when Teale and Dale were carrying out their studies, but it also has the same type of fluorescence polarization spectrum (Figure 4). The shapes of these polarization spectra were determined by the orientation and absorption of the s and f chromophores. It must be noted that some emission leakage occurs even from s chromophores, although the energy transfer between s and f chromophores was quite efficient.[38] The fluorescence excitation spectra of all the biliproteins examined showed excellent correspondence to their absorption spectra.

These studies were particularly revealing for C-phycoerythrin and C-phycocyanin.[38] These proteins were each composed of but one chromophore type, and their absorption spectra showed a single maximum. The fluorescence polarization spectra of both clearly suggested at least two chromophore types. It is apparent that the same chromophores may be influenced by the protein to have different spectral characteristics.

Dale and Teale[44] have developed the concept of s and f chromophores in more detail. Spectra corresponding to the s and f chromophores were calculated assuming complete energy transfer from s to f occurs. The suggestion of very efficient s to f transfer was endorsed by the correspondence of the absorption and fluorescence excitation spectra. The anisotropy (r) in this case is

$$r = \frac{r^f F + r^s S}{F + S} \tag{10}$$

where r^f is the anisotropy of the f chromophore, F is the absorbance (A) of the f chromophore, and r^s and S are the corresponding s chromophore characteristics.

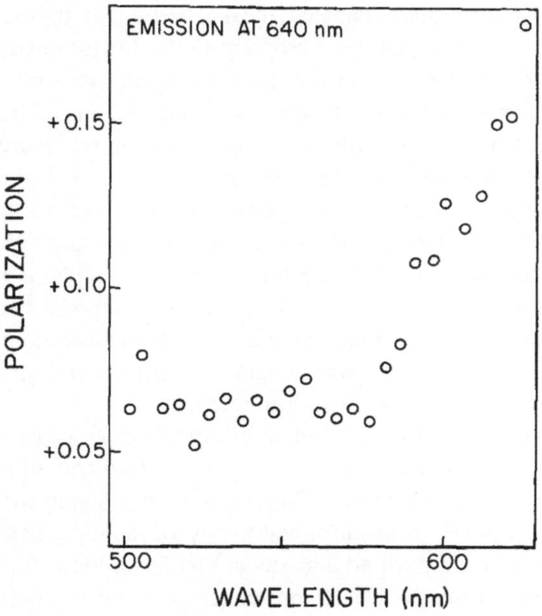

FIGURE 4. Fluorescence polarization spectrum of phycoerythrocyanin. The solvent is pH 6.0, 0.1-ionic strength sodium phosphate buffer. The spectrum was measured at room temperature. The protein had an absorbance of 0.05 at the visible maximum in a 1-cm light path.

$$F + S = A \tag{11}$$

then

$$rA = r^f F + r^s(A - F) \tag{12}$$

giving

$$F = A \frac{[r - r^s]}{[r^f - r^s]} \tag{13}$$

changing to polarization (Equation 8),

$$F = \frac{A\,[3 - p^f]\,[p - p^s]}{[p^f - p^s]\,[3 - p]} \tag{14}$$

where p^f and p^s are the polarizations of the f and s chromophores. Application of Equation 14 indicated that s chromophores of C-phycocyanin had a maximum at about 600 to 615 nm and f chromophores at 630 to 635 nm. Using Förster's[1-3] very weak coupling of dipoles, the distance between s and f chromophores in C-phycocyanin was found to be 3.3 to 3.5 nm.

Vernotte[37] followed up on the s and f chromophore development by performing fluorescence polarization measurements on the individual monomers, trimers, and hexamers of C-phycocyanin purified from the cyanobacterium *Phormidium luridum*. The s and f chromophore spectra were obtained by Equation 14 and showed s chromophores with a maximum at 590 nm and f at 630 nm. The fluorescence polarization spectrum of monomers had an s plateau below 600 nm with a polarization of about +0.3 and an f plateau above 610 nm at about +0.4. The corresponding regions of trimers were lower in polarization, and those of hexamers were lower still. A subsequent study on C-phycocyanin monomers determined the s and f polarizations to be lower.[42]

Zickendraht-Wendelstadt et al.[48] have measured the fluorescence polarization spectra of C-phycoerythrin from *Pseudanabaena* W 1173 and its α and β subunits. In all three cases, the high-energy portion of the visible absorption spectra had a low-polarization plateau which steeply rose near the red portion of the band. For the α subunit which has two chromophores, the fluorescence polarization spectrum of the individual α subunit showed that one of the chromophores is an s type and the second an f. The β subunit has three chromophores, and the deconvolution of its absorption spectrum showed that two were probably sensitizing. For the native C-phycoerythrin, there was a proposal of two f and three s chromophores. A model was presented in which energy transfer occurred both within and between α and β subunits.

The evidence presented for s and f chromophores was threefold: a lack of coincidence between mirror image of the fluorescence emission spectrum and the absorption spectrum, the near identity of the fluorescence excitation and the absorption spectra, and the shape of the fluorescence polarization spectra.[48] The mirror image of the emission of many fluorescing compounds is an almost exact duplicate of their absorption spectra, but for C-phycoerythrin, its α and β subunits, and in general many biliproteins, this was not the case. The absorption spectrum was always wider than the emission. The lack of coincidence indicated the presence of negligibly fluorescing chromophores. The coincidence of the excitation and absorption spectra for many biliproteins, however, showed that the low-fluorescing chromophores contributed to the emission, and since they did not fluoresce appreciably, their contribution was by energy transfer to the f chromophores. The polarization spectrum was not constant and thereby further fostered an acceptance of multichromophores in the phycocyanins, phycoerythrins, and phycoerythrocyanin.

Zickendraht-Wendelstadt et al.[48] have measured the fluorescence polarization spectrum of frozen solutions of C-phycoerythrin. The polarization values did not change on freezing, which further argued against rotations contributing to the lowering of p. Again, the low p values must be produced by energy transfer.

C. Fluorescence Polarization of Allophycocyanin

Allophycocyanin is a trimer when isolated near neutral pH. Its fluorescence polarization spectrum was very low[29,40,41,49] and constant[29,40,41] across its visible absorption band. The clearly defined s and f plateaus of other biliproteins were not observed.

When allophycocyanins dissociated to monomers,[40] however, the s and f regions appeared (Figure 5). The f chromophore polarization region of allophycocyanin monomers was identical to that of C-phycocyanin. In Figure 5, the pH 3.9 results were for homogenous monomers, pH 7.0 for trimers, and pH 4.7 for a mixture of trimers and monomers. The pH 4.7 polarization spectrum was influenced by the greater absorption of trimers over monomers in the 630 to 660-nm region.

Canaani and Gantt[50] have recorded the fluorescence polarization spectra of allophycocyanins I and B. The typical nonconstant spectra of the phycoerythrins and phycocyanins were observed.

D. Energy Transfer in Cryptomonad Biliproteins

Several cryptomonad biliproteins — phycocyanins 612 and 645 and phycoerythrins 545 and 566 — have fluorescence polarization spectra resembling the s and f chromophore situation for cyanobacterial and red algal biliproteins.[51-54] For phycocyanin 645 from *Chroo-*

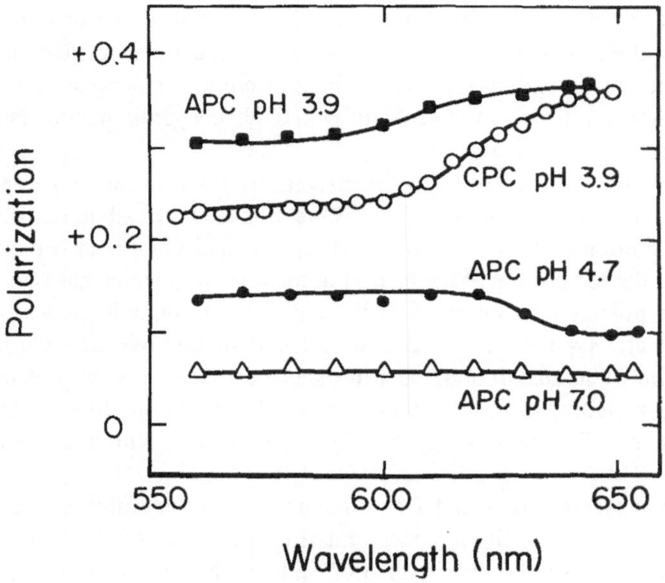

FIGURE 5. Fluorescence polarization spectra of allophycocyanin (APC) at pH 3.9, 4.7, and 7.0 and C-phycocyanin (CPC) at pH 3.9. The aggregation states of allophycocyanin are pH 7.0 for the trimers; pH 3.9 for the monomers; and pH 4.7 for a mixture of trimers and monomers. At pH 3.9, C-phycocyanin is a monomer. (Reprinted with permission from *Biochemistry*, 19, 2817, 1980. Copyright 1980 American Chemical Society.)

monas sp. the polarization showed three clearly defined plateaus.[52,54] MacColl and Guard-Friar[51] have shown that the fluorescence polarization spectrum of phycocyanin 612 from *Hemiselmis virescens* had at least two plateaus. The transition region from the lower polarization to the higher was in the spectral region, suggesting energy transfer from the cryptoviolin to phycocyanobilin chromophores. The CD spectrum of this biliprotein further suggested that there were at least two types of phycocyanobilins also present in the energy transfer system.

The fluorescence excitation spectrum of phycoerythrin 545 from *Rhodomonas lens* agreed very well with its absorption spectrum (Figure 6). This agreement suggested efficient energy from the s to f chromophores. Likewise with phycocyanin 612, the emission spectrum was independent of excitation wavelengths.[51] This result also demonstrated a high degree of efficiency for energy transfer from s to f chromophores. Also in line with results from cyanobacterial and red algal biliproteins, cryptomonad biliproteins have emission spectra which were not mirror images of their absorption spectra. The cryptomonad biliproteins were then readily characterized in terms of s and f chromophores. For phycocyanins 612 and 645, the s chromophores included cryptoviolins which absorbed at higher energies than phycocyanobilins. Results for phycocyanin 612 also suggested diversity among the phycocyanobilins, and they too may be segregated into s and f states. This agreed with the C-phycocyanin case where only phycocyanobilins occurred and s and f chromophores were observed. With phycoerythrin 545 from the cryptomonad *R. lens*, cryptoviolins also occurred,[55] but since they were at a lower energy than phycoerythrobilins, they probably served as f chromophores.

Jung et al.[54] have proposed a detailed topography of the chromophore arrangement for phycocyanin 645. An important detail of this model was the spectral splitting of a strongly coupled exciton pair produced by the close proximity of two phycocyanobilins. These interacting chromophores were located on adjacent β subunits and were calculated to be 0.9

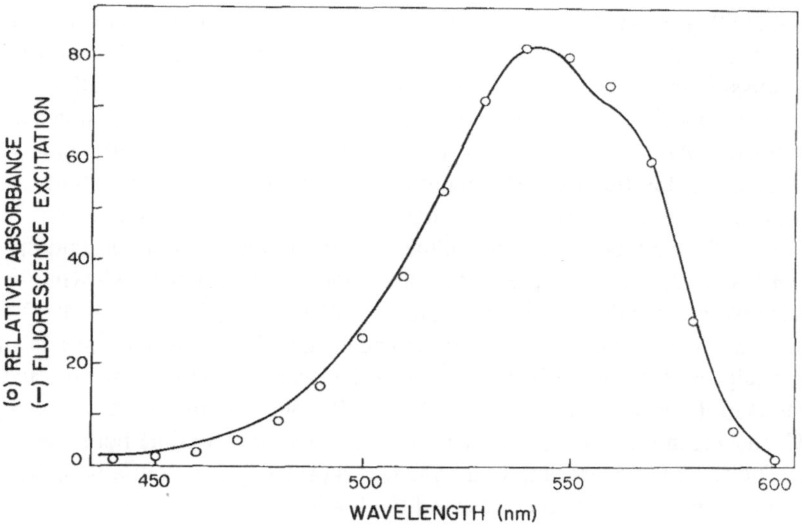

FIGURE 6. Fluorescence excitation and absorption spectrum of phycoerythrin 545 (*Rhodomonas lens*). Excitation spectrum was fully corrected for instrumentation factors.

nm apart. Although this interaction proposal[54] was very viable, other aspects of the chromophore arrangement must be modified because the number of chromophores on phycocyanin 645 has since been reevaluated.[56] MacColl and Guard-Friar[56] have shown that there are two phycocyanobilins and one cryptoviolin on each β subunit, and the α subunits have one 697-nm bilin each. Chapter 9 has a detailed account of the properties of cryptomonad biliproteins.

V. PICOSECOND TIME-RELATED STUDIES ON INTACT CYANOBACTERIA AND RED ALGAE

The preceding studies on excitation energy transfer have been separated from the time-resolved studies on the picosecond level mainly for chronological reasons. Picosecond studies generally became accessible to biliprotein research after various steady-state and kinetics methods had already led to the development of certain hypotheses on the energy transfer scheme (Figure 2).

Porter et al.[57] have performed a very detailed examination of the energy scheme in the red algae *Porphyridium cruentum*. The organism was illuminated with 530-nm light in the form of 6-psec pulses. The 530-nm light was obtained by frequency doubling the output of a mode-locked neodymium-glass laser. A Pockels cell was used to isolated a single pulse. The intensity of these pulses on the sample was kept below 10^{14} photons/cm^2. The fluorescence of these samples was recorded using a streak camera. The emissions of the biliproteins and chlorophyll *a* of these algae were discriminated among by several interference filters: B-phycoerythrin by 576 nm; R-phycocyanin, 640 nm; allophycocyanin, 661 nm; and chlorophyll *a*, 685 nm. Note that allophycocyanin 680 emission would also be included with the 685-nm filter which has an 11-nm bandwidth, and b-phycoerythrin would be grouped with B-phycoerythrin. Steady-state measurements had proven that excitation energy was passed very efficiently between biliproteins. The picosecond experiments were sufficiently sensitive to measure the small amounts of energy not transferred but given off radiatively.

The 530-nm pulse excited primarily B- and b-phycoerythrin.[57] The fluorescence of B-phycoerythrin decayed as exp-At$^{1/2}$, but the fluorescence of R-phycocyanin and allophycocyanin decayed by exp-kt. The fluorescence decay times were found to be 70 psec for B-phycoerythrin, 90 psec for R-phycocyanin, and 110 psec for allophycocyanin. The chlorophyll *a* from dark-adapted algae decayed at 175 psec. The fluorescence rise times to the maximum emission intensity were 12 psec for R-phycocyanin, 24 psec for allophycocyanin, and 50 psec for chlorophyll *a*. A 99% efficiency was estimated for energy transfer between any two pigments. An argument was presented that exciton-exciton annihilation was not a significant factor at this fluence (10^{14} photon/cm^2). These results were entirely consistent with the previously devised scheme of excitation energy migration (Figure 2).

Brody et al.[58,59] have reported results on the cyanobacterium *A. nidulans* and the red alga *P. cruentum* that were more complex. Here the fluorescence kinetics following the 6-psec 530-nm pulse were dependent on the temperature. When the red algae was at 37°C a 715-nm emission rose very rapidly and clearly preceded both 666- and 690-nm rise times. At 0°C the same algae were quite different in their fluorescence kinetics behavior, and the rise time increased in the following order: 576, 640, 690, and 715 nm. For *A. nidulans* at both 0 and 37°C, the 715-nm rise time was faster than the 690-nm rise time but slower than that at 666 nm. The rise of 715-nm emission, which was probably not from a biliprotein, ahead of 666-nm emission which would correspond to allophycocyanin, was difficult to interpret. If the 715-nm component was indeed chlorophyll *a* emission, it was a result that seemed to violate the basic migration scheme proposed for the phycobilisome (Figure 2).

Yamazaki et al.[60,61] have studied the picosecond fluorescence of seven intact organisms: the red algae *P. aerugineum* and *P. cruentum* and the cyanobacteria *A. nidulans, Anabaena cylindrica, A. variabilis, Nostoc* sp., and *F. diplosiphon*. Their system was capable of measuring entire emission spectra at a series of picosecond times. After B- and b-phycoerythrins (*P. cruentum*) were excited at 540 nm, bands representing R-phycocyanin, allophycocyanin, and chlorophyll *a* were measured as a function of time. The rise times were 30 psec for R-phycocyanin, 57 psec for allophycocyanin, and 150 psec for chlorophyll *a*. For *Anacystis nidulans* after excitation of C-phycocyanin at 580 nm, the rise times were 60 psec for allophycocyanin and 120 psec for chlorophyll *a*.

The kinetics properties of allophycocyanin 680 were also measured,[60] and a question was raised concerning its function. Since the two components of allophycocyanin 680 — allophycocyanins I and B — emitted at wavelengths similar to those for intact phycobilisomes, they have been assigned the role of energy transmitters from the phycobilisomes to chlorophyll *a* (Figure 2). Yamazaki et al.,[60] however, contended that their data did not support this model. The decay constant of allophycocyanin 680 was found to be slower than that of allophycocyanin, and they therefore doubted its proposed function as a link from the other biliproteins to chlorophyll *a*.

VI. PICOSECOND STUDIES ON PHYCOBILISOMES

Searle et al.[62] have performed the picosecond time-resolved fluorescence studies on *P. cruentum* using isolated phycobilisomes. Experimental techniques similar to those used for the intact organism were used.[57] The rise time for allophycocyanin emission to reach its maximum was found to be 120 psec. This was the time for energy transfer from B-phycoerythrin to allophycocyanin and allophycocyanin 680 and was somewhat faster than the times for transfer found earlier[14] for *P. cruentum* of 300 psec. As fluences were increased exciton-exciton annihilation seemed to occur for allophycocyanin in the phycobilisomes but not for B-phycoerythrin or R-phycocyanin.[62]

Phycobilisomes from this alga have also been investigated by Wendler et al.[63] using picosecond techniques. The fluorescence decays of B-phycoerythrin and R-phycocyanin were not exponential. The fluorescence emission from the phycobilisomes had two components at 1.0 and 1.8 nsec. Searle et al.[62] found a decay of the long-wavelength emission of 4 nsec. An explanation was proferred that the preparations of Searle et al.[62] were contaminated with

Table 2
ENERGY TRANSFER TIMES WITHIN PHYCOBILISOMES

Transfer	Organism	Type of phycobilisome	Time (psec)[a]	Ref.
B-Phycoerythrin to R-phycocyanin	*Porphyridium cruentum*	Hemi-ellipsoidal	300 ± 200	14
	P. cruentum	Hemi-ellipsoidal	280 ± 40	32
C-Phycoerythrin to C-phycocyanin and allophycocyanin	*Nostoc* sp.	Hemi-discoidal	34	64
B-Phycoerythrin to allophycocyanin	*Rhodella violacea*	Hemi-discoidal	55	65
B-Phycoerythrin to allophycocyanin 680	*P. cruentum*	Hemi-ellipsoidal	70	63
B-Phycoerythrin to allophycocyanin	*P. cruentum*	Hemi-ellipsoidal	24	57
C-Phycocyanin to allophycocyanin	*Anacystis nidulans*	Hemi-discoidal	120	67
B-Phycoerythrin to allophycocyanin	*P. cruentum*	Hemi-ellipsoidal	57	60
C-Phycocyanin to allophycocyanin	*A. nidulans*	Hemi-discoidal	60	60

[a] The last seven lines were data from picosecond time-resolved studies. They all showed significantly faster energy transfer times than the first two methods.

chlorophyll *a*. This would also explain the differences in allophycocyanin rise times in which Wendler et al.[63] found that excitation of B-phycoerythrin caused an allophycocyanin fluorescence rise time of 70 psec compared to the Searle et al.[62] value of 120 psec. Their longer rise time in this analysis resulted from the transfer step to chlorophyll *a*.

Using a picosecond laser technique which generated 6-psec pulses at 530 nm, Pellegrino et al.[64] found a 34-psec rise time of allophycocyanin emission (to the 90% mark) and hence a 34-psec transfer time of energy from C-phycoerythrin to allophycocyanin for hemi-discoidal phycobilisomes from the cyanobacterium *Nostoc* sp. The energy transfer time was then only slightly faster than the time found for the hemi-ellipsoidal phycobilisomes from the red alga *P. cruentum* (Table 2). Increased fluences above 10^{14} photons/cm^2 caused a decline in the quantum yields for all three biliproteins: C-phycoerythrin, C-phycocyanin, and allophycocyanin. The interpretation for the fluence dependence of the fluorescence was exciton-exciton annihilation. These results differed greatly from those for phycobilisomes from *P. cruentum* where the B-phycoerythrin and R-phycocyanin were fluence independent in the same range.[62]

Phycobilisomes from the red alga *Rhodella violacea* have been studied by picosecond techniques which now include wavelength-tunable picosecond pulses.[65] The tunable source allowed direct excitation of B-phycoerythrin, C-phycocyanin, and allophycocyanin of these hemi-discoidal phycobilisomes. Direct excitation of B-phycoerythrin produced a 55-psec rise time for allophycocyanin fluorescence (Table 2), and good agreement was noted for cyanobacterial, red algal, hemi-discoidal, and hemi-ellipsoidal phycobilisomes. Although exciton annihilation will not occur at the low fluences used, the decays were found to be nonexponential.

Phycobilisomes from the cyanobacterium *Mastigocladus laminosus* contained phycoerythrocyanin, C-phycocyanin, and allophycocyanin, which were present in hemi-discoidal phycobilisomes. These isolated phycobilisomes were studied by a picosecond time-resolved fluorescence system after excitation with 600-nm light at 10^{13} photons/cm^2.[66] C-Phycocyanin was primarily excited at this wavelength. An interference filter at 614 nm gave a polarized, 70-psec decay time, and the remaining filters at 630, 654, and 668 nm gave depolarized emission and two decays. The faster of the two decays was at 150 psec and may be related to energy transfer and the slower was radiation from allophycocyanin. The depolarization of the emission was produced by the energy transfer among chromophores.

Phycobilisomes from the cyanobacterium *A. nidulans* have C-phycocyanin and the allophycocyanins. Its unusual allophycocyanin cores have only two instead of the normal three cylindrical elements. Suter et al.[67] using fluorescence picosecond techniques have measured an 120-psec transfer time from C-phycocyanin to the allophycocyanins. The decay time of the final emission at 680 nm was 1.8 to 1.9 nsec in agreement with other phycobilisomes results. They noted that the 120-psec time was longer than those found in some other phycobilisomes (Table 2) and suggested that since phycobilisome rods vary in length and pigment content, these variations were understandable. Gillbro et al.[68] have examined the same phycobilisomes by polarized absorption picosecond techniques. Their results showed two transfer times from C-phycocyanin to allophycocyanin of 12 and 84 psec. They suggested that the two decays might represent C-phycocyanin chromophores transferring to the allophycocyanins from different distances.

The concept of the relationships being phycobilisome rod lengths and rate of energy transfer from the rods to the core[67] was established for wild types and mutants of *Synechocystis* 6701. The wild type had rods of two C-phycocyanin disks and one or two C-phycoerythrin disks, and mutant CM25 had no C-phycoerythrin and just the two C-phycocyanin disks in each rod of their phycobilisomes. These phycobilisomes were excited with 570-nm light, and the rise time at 680 nm was 56 psec for the wild type and 25 psec for CM25.[69] The difference in rise time of 31 psec showed the time lag for energy absorbed by C-phycoerythrin to reach the core. Yamazaki et al.[61] examined red and green light-grown *F. diplosiphon* by picosecond techniques and found that the phycocyanin transferred its energy more rapidly for phycoerythrin-rich green light-grown cells. Since the red light-grown cells had more phycocyanin and thus probably longer phycocyanin rods than the green light-grown cells, the longer average distance to the core explained the kinetics results.

VII. PICOSECOND STUDIES ON INDIVIDUAL BILIPROTEINS

A. Ultrafast Studies of the s and f Chromophores

The model of individual biliproteins being characterized by s and f chromophores can be tested by ultrafast kinetics. The techniques now being used can measure decays that are no faster than several picoseconds. Energy transfer by very weakly coupled dipoles may well occur in this measurable range. Faster decays will require applications of laser techniques on the femtosecond level. The confirmation and quantitation of s and f chromophores (see Section IV.B) were prime goals of the initial application of picosecond experiments to individual biliproteins. All biliproteins, when isolated, will have a fluorescence lifetime greater than 1 nsec, and fluorescence measurements will be analyzed to distinguish this long-lived decay from the faster decays which were candidates for s and f transfer.

Steady-state measurements have demonstrated that excitation energy transfer from s to f chromophores was very efficient. The detection of the s to f transfer will require instrumentation sensitive enough to detect the small amount of light emitted from some s chromophores. Picosecond absorption studies were also important for these systems. As already mentioned the possibility of exciton-exciton annihilation was a factor in these laser studies.

B. Ultrafast Studies on C-Phycocyanin

C-Phycocyanin was isolated from the cyanobacterium *Phormidium luridum* as hexamers, trimers, and monomers having 18, 9, and 3 phycocyanobilin chromophores, respectively. The conditions for obtaining these aggregates after ammonium sulfate purification were hexamers at pH 5.5 (88% yield), trimers at pH 8.0 (85% yield), and monomers at pH 3.9 (100% monomers). The aggregates should not transfer energy beetween each other so that the transfers are among chromophores on the same aggregates. After excitation at 530 nm, which primarily excited s chromophores, each aggregate exhibited two decays. The slower time constant decay was greater than 1 nsec and corresponded to a monomolecular decay rate of the excited state. The faster decays were characteristic of the different aggregates:

hexamers at 32 psec, trimers at 56 psec, and monomers at 85 psec. The decays were found to be depolarized. It was suggested that these fast decays might possibly be s to f transfer times, and the depolarization results suggested that the oscillators in this energy transfer process randomized the excitation. The finding of different s to f times in different aggregates was not surprising. The monomers and trimers had phycocyanobilins with different conformations than the hexamers,[71] and all three had distinctive absorption maxima and absorptivities.

These picosecond absorption experiments by Kobayashi et al.[70] were performed at somewhat higher fluences (about 10^{15} photons/cm^2) than experiments already discussed on the phycobilisomes. Wong et al.[72] and Doukas et al.[73] showed that fluences greater than 10^{14} photons/cm^2 can result in exciton annihilation effects. Although annihilation will occur at high fluences, subsequent studies have confirmed the validity of the s to f transfer. The fluorescence and time-dependent fluorescence polarization results of Hefferle et al.[74] completely established that even at fluences considered safe from annihilation effects, a relatively fast component was found for several different biliprotein aggregates. For C-phycocyanin trimers, two isotropic decays were detected at 120 and 1750 psec. A 50-psec lifetime decayed anisotropically.

Yet another time-resolved fluorescence study on isolated biliproteins detected fast components. Hefferle et al.[75] argued that annihilation cannot be responsible for their fast components, since the fluences were low (10^{13} photons/cm^2), and the number of chromophores on these trimers was small (nine). Their fast decays varied between 70 and 400 psec depending on experimental factors. The fast decays were believed to be related to the s and f and f to f energy transfers.

To evaluate better whether the s to f energy transfer times were actually being observed in these picosecond time-resolved experiments on isolated biliproteins, results from the phycobilisomes were of interest. In some cases quite fast components in the decays of phycobilisomes have been detected. For phycobilisomes from the red alga *Porphyridium cruentum* a 12-psec decay was ascribed to energy transfer within B-phycoerythrin monomers.[63] For phycobilisomes from the cyanobacterium *A. nidulans* a fast fluorescence decay of 20 psec was considered to be energy transfer from s to f chromophores within C-phycocyanin.[67] For phycobilisomes also from *A. nidulans* an anisotropic decay of 10 psec was observed by picosecond absorption.[68] The various studies on both isolated biliproteins[70,74,75] and analysis of the fast components of phycobilisome decay[63,67,68] conclusively established that s to f energy transfer was measurable on the picosecond time scale. These observations then provided a direct validation and quantitation of the concept of energy transfer between nonidentical chromophores on isolated biliproteins as described first from steady-state fluorescence[38,44,45] and fluorescence polarization[38,44] measurements.

C. Ultrafast Studies on Cryptomonad Biliproteins

Substantial picosecond data have been obtained for the biliproteins from cryptomonads. Phycobilisomes do not occur in these organisms. The individual biliproteins, phycocyanins, or phycoerythrins are isolated as dimeric ($\alpha_2\beta_2$) aggregates. They do not show the proclivity to aggregate that is characteristic of the cyanobacterial and red algal counterparts (see Chapter 9 for data on cryptomonad biliproteins).

Kobayashi et al.[70] used picosecond absorption methods to investigate phycocyanin 645 from the cryptomonad *Chroomonas* sp. Like the C-phycocyanins,[70] phycocyanin 645 has a time-resolved decay that included a fast (about 8 psec) and a slow (greater than 1 nsec) component. They assigned this fast component to s to f energy transfer. Again the question of whether this assignment was valid can be raised because of the possibility of exciton-exciton annihilation at high fluences. There were two reports[76,77] that clearly established the validity of the s to f transfer on the picosecond level. Holzwarth et al.[76] have used picosecond fluorescence kinetics to study phycocyanin 645. They found two major decays, after excitation with a low fluence, at 15 psec and 1.44 nsec. A small-amplitude component was also observed at 360 to 680 psec. They concluded that their 15-psec component was related to energy transfer from s to f chromophores. Noting that Kobayashi et al.[70] used absorption kinetics and Holzwarth et al.[76] used emission, the agreement between 8 and 15 psec was satisfactory.

A final proof that the fast component in phycocyanin 645 was energy transfer and not annihilation related came from a study by Guard-Friar et al.[77] on fluence effects on this biliprotein. Unlike biliproteins from cyanobacteria and red algae, this cryptomonad biliprotein had a decay that was fluence independent up to 10^{17} photons/cm^2. Guard-Friar et al.[77] have studied three cryptomonad biliproteins — phycocyanins 645 and 612 and phycoerythrin 545 — by picosecond fluorescence methods and compared these results with those from C-phycocyanin. The emission decay time from the C-phycocyanin dropped as fluences increased (Figure 7). In this experimental system, however, the cryptomonad phycocyanins 612 and 645 showed no change in fluorescence emission decay (Figure 7) as fluence was increased.

Phycoerythrin 545 isolated from the cryptomonad *Rhodomonas lens* was studied on the same system.[77] The decay as a function of fluence behaved very differently from both the cryptomonad phycocyanin and the cyanobacterial C-phycocyanin. For this biliprotein up to almost 10^{15} photons/cm^2 the picosecond decay was unaffected by fluence, but above 10^{15} photons/cm^2 the decay became biphasic (Figure 8). The rapid component had a decay time of about 100 psec.

A very influential factor in possibly explaining the variations in fluence behavior of these biliproteins was the number of chromophores per aggregate. The C-phycocyanin used in this study was primarily a hexameric protein — 18 chromophores; the phycocyanins 612 and 645 were dimeric proteins — 8 chromophores; and phycoerythrin 545 was also a dimer protein with 11 to 12 chromophores. The low number of chromophores on the cryptomonad phycocyanins was viewed as the critical factor in avoiding exciton-exciton annihilation effects.[77]

This factor of the number of chromophores can be used now to analyze further results taken at higher fluences.[70] The C-phycocyanin monomers used in this study had a total of three chromophores and were therefore unlikely to show any annihilation. It is important to note that a fast component is observed even for the monomers, and therefore while the 32-psec decay of the hexamers may be slightly faster due to annihilation, it is clearly not caused by it; assignment of these decays to the s to f transfer is further documented.

Hanzlik et al.[78] have studied the cryptomonad biliprotein phycocyanin 612 by picosecond fluorescence spectroscopy with excitation at 532 nm in which the spectral properties of the fast component were obtained. In one type of experiment, 10-nm band pass interference filters below 590 nm detected emission, having a single fast exponential decay of 10 to 12 psec (Figure 9). The whole unfiltered fluorescence (Figure 7B) showed a much slower decay of 2.3 nsec. When only emission longer than 650 nm was monitored, an entirely different result was obtained, and only a slow decay of about 1 nsec was obtained. This slow decay which was caused by radiative emission from f chromophores had a 9- to 11-psec rise time.

Other experiments were performed in which the entire fluorescence emission spectra of the decay was simultaneously recorded at various times: at $t = -15$ psec the maximum of the decay was at 593 nm, and at $t = +15$ psec the decay showed a maximum at 637 nm.[78] These times were relative to the center of a 30-psec excitation. The 593-nm band was indicative of sensitizing chromophores, and the 637-nm band was related to the fluorescence spectrum of the f chromophores. The kinetics and the spectra of both the spectral resolution and filtering experiments were nicely coordinated to determine this assignment. Phycocyanin 612 has an absorption spectrum with two visible bands at 585 and 612 nm. The absorption

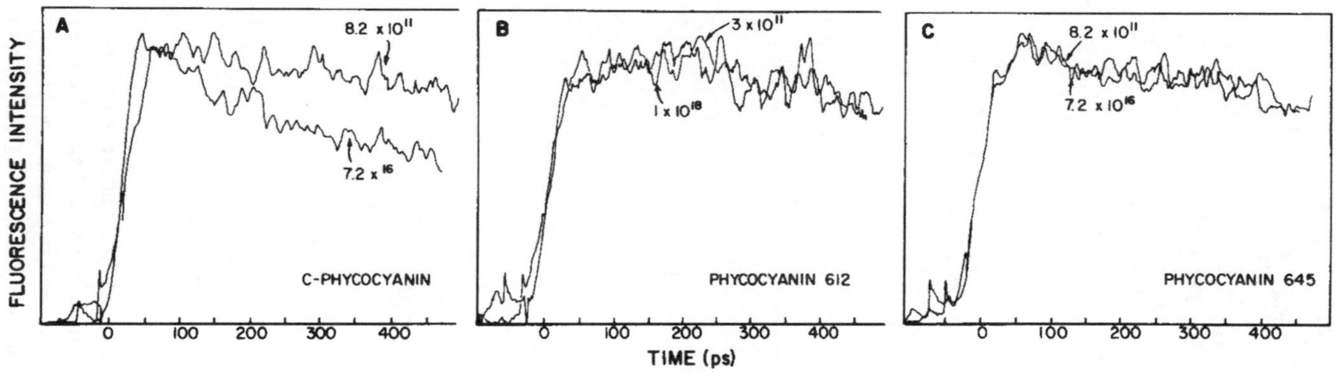

FIGURE 7. Effects of fluence on the decay of C-phycocyanin (panel A), phycocyanin 612 (panel B), and phycocyanin 645 (panel C). (Reproduced from *The Biophysical Journal*, 47, 787, 1985, by copyright permission of the Biophysical Society.)

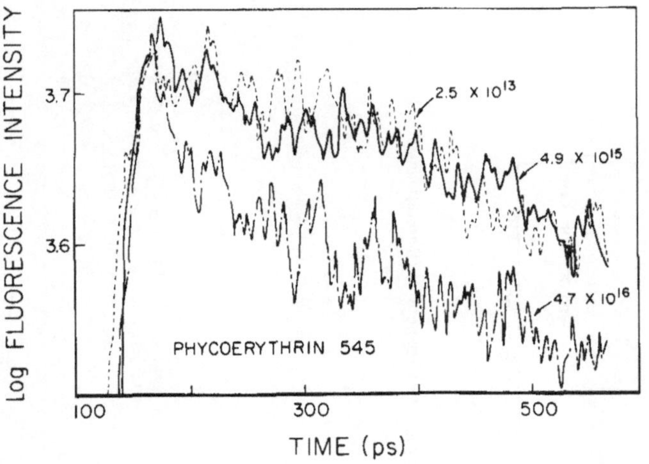

FIGURE 8. Effect of fluence on the decay of phycoerythrin 545 (*Rhodomonas lens*). (Reproduced from *The Biophysical Journal*, 47, 787, 1985, by copyright permission of the Biophysical Society.)

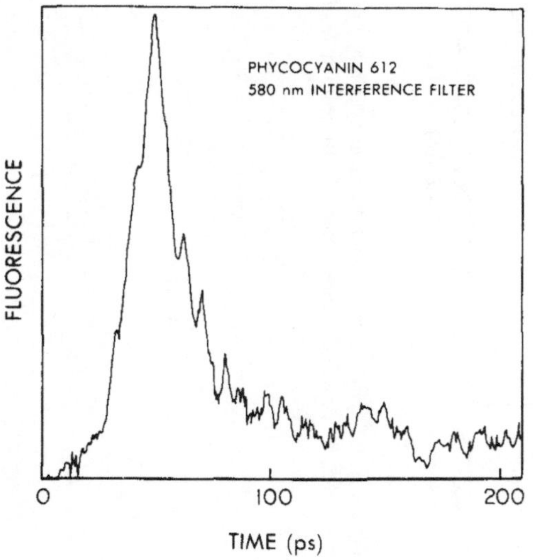

FIGURE 9. Spectrum of the 10-psec component in the decay of phycocyanin 612 (*Hemiselmis virescens*). This experiment filters out fluorescence emission above 590 nm. Unfiltered emission is shown in Figure 7. The protein was excited at 532 nm.

bands can readily give rise to the 593- and 637-nm fluorescence emission bands, respectively. The observation of emission from the s chromophore was due to some leakage of energy not transferred to the f chromophores. The kinetics results were very supportive of the s and f chromophore concept, and the decay of the s chromophores seemed to fit well with the rise times of the f chromophores. This was the first experimental measurement of a spectrum of a chromophore group which had kinetics properties that clearly established these chromophores as the s type. This result conclusively established that the fast decays are produced from s to f transfers perhaps together with f to f transfer, but not by f to f transfers alone.

VIII. PICOSECOND STUDIES ON ISOLATED BILIPROTEIN SUBUNITS

Biliproteins are generally composed of two subunits, α and β, and some are structurally more complex. The α and β subunits can be separated by fully denaturing the biliproteins and then separating the subunits by chromotography. Once separated the individual subunits can be returned to a nondenaturing solvent.

There have been two picosecond fluorescence studies performed on the individual subunits of C-phycocyanin.[42,75] Switalski and Sauer[42] found that the β subunit of C-phycocyanin (*Anabaena variabilis*) had a biexponential decay of 1.3 and 0.3 to 0.4 nsec. The α subunit of C-phycocyanin (*A. variabilis*[42] and *Spirulina platensis*[75]) was the more interesting case. This subunit had only one phycocyanobilin chromophore, whereas β had two. A single chromophore cannot possess both s and f states. Since biexponential behavior of biliproteins was usually viewed as proof that s and f chromophores existed, it was perhaps anticipated that this subunit would give only the decay associated with radiative emission from an f-type chromophore. Both Hefferle et al.[75] and Switalski and Sauer,[42] however, have clearly established that these α subunits were best fitted by a biexponential. Hefferle et al.[75] reported 250-psec and 1.3-nsec decays. Switalski and Sauer[42] found 440- to 500-psec and 1.5 nsec decays. In this case, the fast decay varied between 440 and 500 psec depending on excitation wavelength.

Both reports discussed two possible explanations for the biexponential decay of α subunits.[42,75] The separated α subunits might aggregate in nondenaturing solvents and thus have two or more chromophores in different environments. The addition of 1.0 *M* NH_4SCN which dissociated biliproteins had no effect on the decay. However, this concentration of thiocyanate when applied to biliproteins produced monomers (αβ)[39] and might not fully dissociate an α_n aggregate. A second proposal[42,75] was that chromophores on individual subunits have different environments and thus different conformations with individual decay kinetics. These experiments were critically of interest because they reflected possible interpretations of results from phycobilisomes and individual biliproteins. It must be remembered that harsh treatment was applied to obtain the isolated subunits and that returning them to nondenaturing buffers without the other subunit that they usually associate with can result in a very artificial situation.

IX. QUANTUM YIELDS AND LIFETIMES

The energy transfer ability of biliproteins depends on quantum yields and fluorescence lifetimes of donor pigments (Equation 1). There have been a large number of measurements of these quantities with a fairly wide distribution of results. For C-phycocyanin, the quantum yields vary over a very wide range: *Synechocystis* sp.[79] at 0.53, *Anacystis nidulans*[44] at 0.51, *Schizothrix calcicola*[44] at 0.81, *Tolypothrix tenuis*[81] at 0.41, *Nostoc* sp.[73] at 0.75, *Phormidium laminosum*[80] at 0.22, and *F. diplosiphon*[29] at 0.52. Quantum yields for the various biliproteins and their subunits are listed in Table 3. The subunits were separated under denaturing conditions and then returned to nondenaturing solvents.

The fluorescence quantum yields from intact phycobilisomes have been determined.[32] Phycobilisomes were found to have quantum yields similar to those for isolated biliproteins: *F. diplosiphon* at 0.68; *Nostoc* sp. at 0.65, and *Porphyridium cruentum* at 0.60. Grabowski and Gantt[32] have probed the effects of phycobilisome structure on fluorescence quantum by growing these three organisms in red, green, and white light. The two cyanobacteria had markedly less C-phycoerythrin in red light conditions, and *F. diplosiphon* adapted most extensively. The fluorescence quantum yields on all nine phycobilisome preparations showed no effect of these structural variations.

Fluorescence lifetimes have been obtained for various biliproteins. The variations from

Table 3
QUANTUM YIELDS FOR VARIOUS BILIPROTEINS

Biliprotein	Fluorescence quantum yield (range)	Ref.
C-Phycocyanin	0.22—0.81	29, 44, 72, 79—81
R-Phycocyanin	0.59	29
R-Phycoerythrin	0.82—0.84	44, 81
B-Phycoerythrin	0.85—0.98	29, 44, 79
C-Phycoerythrin	0.45—0.88	29, 44, 48, 72
Allophycocyanin	0.53—0.68	29, 41
b-Phycoerythrin	0.48	29
α subunit, allophycocyanin	1.09	41
β subunit, allophycocyanin	0.51	41
α subunit, C-phycoerythrin	0.51	48
β subunit, C-phycoerythrin	0.56	48

Table 4
FLUORESCENCE LIFETIMES FOR VARIOUS BILIPROTEINS

Biliproteins	Range of fluorescence lifetimes (nsec)	Ref.
C-Phycocyanin	1.0—2.3	29, 44, 72, 74, 77, 80—85
R-Phycocyanin	2.0	29
C-Phycoerythrin	1.0—3.5	29, 44, 72
R-Phycoerythrin	2.3—3.1	44, 81, 86
B-Phycoerythrin	2.0—3.8	29, 44, 85
b-Phycoerythrin	2.9	29
Allophycocyanin	1.6—2.7	29, 72, 73
Phycoerythrocyanin	2.2	74
Allophycocyanin I	1.9	72
Allophycocyanin B	2.7	72
Phycocyanin 645	1.4—1.6	54, 76, 77
Phycocyanin 612	2.3	77
Phycoerythrin 545	1.5	77

report to report were not as substantial as those for the fluorescence quantum yields of C-phycocyanin. For C-phycocyanins, the lifetimes were obtained from several organisms and measured by different techniques: for *A. nidulans*,[44,82] 1.8 nsec; for *S. calciola*,[44] 2.3; for *Phormidium luridum*,[83] 1.3; for *T. tenuis*,[81] 2.1; for *Nostoc* sp.,[72] 2.1; for *P. laminosum*,[80] 1.5; for an unidentified organism,[84] 1.1; for *Agmenellum quadruplicatum*,[85] 1.0 to 1.1; for *F. diplosiphon*,[29] 2.2; for *M. laminosus*,[74] 1.7; and for *P. luridum*,[77] 2.3. A compilation of fluorescence lifetimes for various biliproteins showed that they fall in a fairly narrow range (Table 4). The fluorescence decay of the final emitter from intact phycobilisomes had a similar lifetime as the isolated biliproteins (Table 5).

Vernotte and Moya[83] have studied the effect of temperature on the fluorescence quantum yield and fluorescence lifetime of C-phycocyanin from the cyanobacterium *P. luridum*. At ambient temperatures, the monomers and hexamers of C-phycocyanin had identical lifetimes, but the quantum yield of hexamers exceeded that for monomers. As the temperature was raised from about 5 to about 55°C, the lifetimes of both aggregates gradually declined. At temperatures above 60°C, however, a very rapid rise in lifetime was observed. The lower-temperature behavior was ascribed to internal conversion, and the large increases at higher temperatures were assigned to direct fluorescence from s chromophores. Above 60°C, the protein was being denatured, and the energy transfer efficiency between s and f chromophores was declining. By 60°C the quantum yield of the C-phycocyanin was quite low.

Table 5
FLUORESCENCE LIFETIMES OF FINAL EMITTERS, ALLOPHYCOCYANIN 680, OF PHYCOBILISOMES

Organism	Fluorescence lifetime (nsec)	Ref.
Mastigocladus laminosus	1.5	66
Rhodella violacea	1.9—2.0	65
Porphyridium cruentum	1.8 and 1.0	63
Anacystis nidulans	1.9	67

Hefferle et al.[75] also examined biliprotein fluorescence quantum yields and lifetimes as a function of temperature. They used allophycocyanin and C-phycocyanin isolated from the cyanobacterium *Spirulina platensis* and applied picosecond fluorescence techniques. The quantum yields dropped as temperature increased for both monomers and trimers of both biliproteins, but the decline in quantum yields of allophycocyanin trimers was much less than that of the monomeric form. The picosecond fluorescence decay of the biliproteins was essentially biexponential, having an s and f and f to f transfer component and the radiative decay of the f chromophores. They studied these components at various temperatures and reached conclusions similar to those of Vernotte and Moya.[83] A proposal was made that the high-temperature effects were caused by protein unfolding which changed the distance between chromophores and their orientations.[75] As noted in Equation 1, these two factors could cause a decrease in the efficiency of energy transfer. This work carefully showed that the fastest decays retained polarization and were assigned to the s to f and f to f transfers.

X. EXCITON ANNIHILATION

Laser beams of fairly high intensities were used to produce measurable signals in the picosecond time-resolved studies. As noted above, some biliproteins may show fluorescence decays that were nonlinear because of these high intensities.[57,62,64,72,73,77]

Singlet-singlet exciton annihilation is described below for an isolated aggregate:

$$A + h\nu \rightarrow A^1 \tag{15}$$

$$A^1 + A^1 \rightarrow A + A^2 \tag{16}$$

A is the ground state, A^1 is the first excited state, and A^2 is a higher state. As noted in Equation 16, multiple excitations on the same aggregate are needed for an annihilation event. Following the very detailed analysis of this problem by Wong et al.[72], Doukas et al.,[73] and Pellegrino et al.[64] for several different biliproteins, fluences are routinely kept at 10^{13} photons/cm^2 or below to avoid conflicts in interpretations between annihilation and s to f transfer.

Guard-Friar et al.[77] in addition noted that multiple excitations as suggested by Equation 16 required several chromophores. The lack of annihilation for cryptomonad phycocyanins 612 and 645 was readily attributed to a small number of chromophores (eight) on the isolated aggregates. By extension, other aggregates low in chromophore content like monomers of allophycocyanin (two), monomers of C-phycocyanin (three), and monomers of C-phycoerythrin (six) should be less susceptible to annihilation than larger aggregates. This proposal was emphasized by the failure of various theoretical treatments to explain the annihilation kinetics of certain biliproteins. Since these equations were derived assuming large numbers of chromophores or other factors not applicable to biliproteins,[87,88] they did not accurately reproduce the biliprotein situations of limited chromophores.

Exciton annihilation may yield data of importance in the study of biliproteins. Phycobilisomes from *Nostoc* sp. and *Porphyridium cruentum* showed exciton annihilation at higher intensities.[62,64] For the latter alga, allophycocyanin was the only biliprotein in those phycobilisomes to exhibit the effect.[62] Purified allophycocyanin conversely showed only a small annihilation effect upon 530-nm excitation, but purified C-phycoerythrin and C-phycocyanin were much more susceptible to annihilation.[72] This result for the purified biliproteins was readily explained from the studies on the correspondence between annihilation and chromophore content.[77] Allophycocyanin is isolated as a trimer having a total of only six chromophores. The difference between the annihilation behaviors of purified allophycocyanin and allophycocyanin in the phycobilisomes was readily related to the phycobilisome structure-function relationship. The Gantt phycobilisome model[30] presented allophycocyanins as links between the more abundant biliproteins in the rods of the phycobilisomes and the chlorophyll *a* in the membranes. All the excitation energy harvested by the various phycocyanins, phycoerythrins, or phycoerythrocyanins must be funneled through the various allophycocyanins in the core of the phycobilisomes. The allophycocyanin content of phycobilisomes was generally only around 10% and was further differentiated into two allophycocyanins (which were 660-nm emitters) and smaller amounts of two types of allophycocyanin 680. Pellegrino et al.[64] therefore noted that the energy transfer from the other biliproteins to allophycocyanins accounted for its augmented annihilation in the phycobilisomes. The efficiency of energy transfer in the phycobilisomes at high enough fluences finally overloaded the ability of the allophycocyanin core to function without exciton collisions occurring.

XI. SPECTRAL DECONVOLUTION

The functional segregation of chromophores into s and f types has been well documented. Some consideration has been given to the idea that there are multiple s and f chromophores within a given biliprotein aggregate. One method to determine the absorption spectra of these various chromophores is by deconvolution of the absorption spectra of the biliproteins. There are various methods for approaching this task. One method is using the mirror image of the fluorescence emission spectrum and successively subtracting this from the absorption spectrum. A second major approach is to divide the absorption spectrum into a number of Gaussian- or Lorentzian-shaped curves.

Zickendraht-Wendelstadt et al.[48] employed the mirror image technique on C-phycoerythrin from the cyanobacterium *Pseudanabaena* W1173. On the α subunit of C-phycoerythrin there were two chromophores. The deconvolution method yielded two absorption bands which represented s and f chromophores. The β subunit had its three phycoerythrobilin chromophores divided by deconvolution into two s and one f types. The native C-phycoerythrin spectrum was fitted by using the results for α and β, and there were three curves in the deconvolution: one curve had a single s chromophore from the β subunit; another curve had two identical s chromophores, one from the α and one from the β subunit; and the remaining curve represented two f chromophores, again with one from the α and the other from the β subunit.

Jung et al.[54] used the mirror image concept and applied it to cryptomonad phycocyanin 645. They obtained five components from their deconvolution. Two of these bands were assigned to band splitting from strong exciton coupling. The two phycocyanobilin chromophores that comprised the exciton pair were found on different β subunits. The pair was brought close enough together to produce the exciton interaction because of the proximity of the β subunits in the dimer and the relative locations of the chromophores on the subunits. The two exciton bands both showed similar kinetics when the protein was treated with $KMnO_4$ or $Hg(CH_3COO)_2$. The f chromophore of phycocyanin 645 was then generated from this exciton pair, since its low-energy band was at the lowest energy of all the chromophores.

This was the first evidence presented, other than the shape of a CD spectrum, that seriously raised the question of exciton-exciton interaction contributing to the spectrum of a biliprotein. In this case, splitting apparently generated the f chromophore that produced the 660-nm emission for this protein.

Another possible case of a spectral shift produced by exciton interaction was that of the phycobilisomes of the cyanobacterium *Nostoc* sp. Gray and Gantt[89] found an absorption band which was lacking in dissociated phycobilisomes. One possible origin for such a band was strong coupling of chromophores which specifically interact only in intact phycobilisomes.

The spectrum of allophycocyanin has been discussed in terms of strong coupling of dipoles. The 650-nm absorption band was proposed by some as being possibly produced by dipole-dipole interaction.[40,90] Others have argued that a variation in phycocyanobilin conformation can readily produce the 650-nm band.[91,92] Theoretical calculations on the chromophores of allophycocyanin from the cyanobacterium *Anabaena cylindrica* have demonstrated that the conformational proposal was possible.[93] Mimuro et al.[92] have used Gaussian deconvolution of the allophycocyanin spectrum to develop a detailed model for understanding its genesis. They assigned s and f states to allophycocyanin based on their deconvolution results.

Csatorday et al.[90] have also studied allophycocyanin by deconvolution techniques. Allophycocyanin monomers were readily fitted to one Gaussian and one Gaussian-Lorentzian band. Since monomers have two chromophores, this was a satisfactory deconvolution. Trimers were deconvoluted into four bands for both their absorption and CD spectra. Of course, four was then the minimum number necessary for a good fit. In order to achieve a good CD fit it was necessary to introduce negative bands. Once this was done, there was quite excellent agreement in wavelength maxima between the bands in the deconvolution of CD and absorption spectra. In addition, the negative band in the deconvoluted CD spectrum of the trimer beautifully explained a slight negative sliver in the high-energy side of the experimental CD spectrum.[94] This deconvolution eradicated a serious challenge to the proposal that the 650-nm band was generated through dipole-dipole interaction. To produce the needed spectral shift, strong coupling was probably needed, but strong coupling required both a negative and a positive CD component. It was generally considered that the CD spectrum of allophycocyanin had only two positive bands in the visible region. The realization that the negative bands were present but mainly obscured by overlapping positive bands of noninteracting chromophores added greatly to the possible acceptance of this model. The original basis for proposing the exciton interaction model was the observation that the 650-nm band was totally lacking in the monomer spectrum and the need to explain the flat and very low fluorescence polarization spectrum of trimers.[40,95] It should be noted that an interaction model does not preclude a chromophore conformation change also occurring when the trimers dissociated to monomers. For C-phycocyanin, a large conformational change accompanied the hexamer to monomer dissociation.[71]

Zilinskas et al.[96] performed the first Gaussian analysis of the absorption spectrum of allophycocyanin. They were mainly concerned with the spectral differences between allophycocyanin and the two forms (I and B) of allophycocyanin 680. Allophycocyanins I and B shared Gaussian bands at 679 and 667 to 671 nm. These bands were both missing from the Gaussian deconvolution analysis of allophycocyanin, and thus they represented the chromophores responsible for the 680-nm emission from these proteins and from intact phycobilisomes.

XII. PHOTOCHEMICAL HOLE BURNING

Photochemical hole burning can be observed when a photolabile substance is irradiated with a narrow band of light. Certain molecules that can undergo photochemistry can be bleached, and a hole appears in their absorption spectrum. The hole is actually a small notch in the absorption band. These experiments must be performed at very low temperatures, 2 to 20° K and result in high-resolution spectral data.

Friedrich et al.[97] have used photochemical hole burning to study a mixture of C-phycocyanin and allophycocyanin from the cyanobacterium *S. platensis.* Preliminary to the hole burning experiment the low-temperature absorption of the material was recorded. At 5.3° K, the C-phycocyanin absorption spectrum was divided into two relatively sharp bands at 595 and 627.5 nm, and the allophycocyanin absorption at 650 nm was clearly observed. At room temperature, only a single broad band was observed. Gray and Gantt[89] observed a very similar low-temperature absorption for C-phycocyanin from the cyanobacterium *Nostoc* sp. They reported two low-temperature bands at 598 and 631 nm. Friedrich et al.[97] also examined the low-temperature spectrum of free tetrapyrrole as a model for the behavior of phycocyanobilin that was not coupled to the apoprotein. While in the case of C-phycocyanin the spectrum split into two bands with no large-scale shifting at low temperatures, the model tetrapyrrole underwent a large bathochromic shift. They suggested that this shift resulted from a change in tetrapyrrole protonation and that the tetrapyrroles on the protein were protected against protonation changes. Since C-phycocyanin monomers each have three chromophores, the 595-nm band was assumed to represent one chromophore and the 627.5-nm band two overlapping chromophores.

When the laser irradiated the C-phycocyanin-allophycocyanin mixture, the allophycocyanin and the higher-wavelength C-phycocyanin band showed a hole, but the low-wavelength C-phycocyanin band did not. The narrowness of the holes and other characteristics of the spectra suggested that the chromophores were held rigidly by the apoprotein. This agreed with the usually high fluorescence quantum yields of the biliproteins (Table 3). The phonon wing which was observed as a satellite to each hole showed relatively weak coupling and suggested that little of the excitation energy was transformed to heat. This meant that most of the energy was available for energy transfer. The very fact that the holes were observed demonstrated that a small amount of excitation energy was used in some type of photochemical reaction and thus unavailable for energy transfer. The lack of a hole on the 595-nm band demonstrated that it was an independent electronic transition from the 627.5-nm band.

Friedrich et al.[98] have also studied C-phycoerythrin from the cyanobacterium *Phormidium persicinum* by photochemical hole burning at very low temperatures. As with C-phycocyanin,[97] the results indicated that chromophores were held rigidly by the apoprotein and that only a very small amount of excitation energy was lost to the environment as heat.

The use of free tetrapyrroles as models for biliprotein chromophores has been occasionally very useful. Braslavsky et al.[99] have reviewed the conformations and spectroscopic properties of certain tetrapyrroles.

XIII. LOW-TEMPERATURE FLUORESCENCE

Very often fluorescence and absorption spectroscopy were performed on purified biliproteins and biliprotein-containing organisms at very low temperatures, commonly at −196°C (the temperature of liquid nitrogen). One result of this procedure was that the spectral bands were significantly narrower, and better resolution of overlapping spectra was obtained. Many examples of this approach have appeared.[22,23,50,89,100-117]

Studies on low[108,110] and ambient[117] temperature spectroscopy of mixtures of phycoerythrin and chlorophyll derivatives have demonstrated that excitation energy transfer can occur from this biliprotein to chlorophyll. As noted previously, this does not occur in cyanobacteria or red algae because of the organization of the phycobilisomes (see Section III).

REFERENCES

1. **Förster, T.**, Zwischenmolekulare Energiewanderung und Fluoreszenz, *Ann. Phys.*, 2, 55, 1948.
2. **Förster, T.**, Transfer mechanisms of electronic excitation, *Discuss. Faraday Soc.*, 27, 7, 1959.
3. **Förster, T.**, Delocalized excitation and excitation transfer, in *Modern Quantum Chemistry, Part III*, Sinanoglu, O., Ed., Academic Press, New York, 1965, 93.
4. **Eisinger, J. and Dale, R. E.**, Interpretation of intermolecular energy transfer experiments, *J. Mol. Biol.*, 84, 643, 1974.
5. **Kasha, M.**, Energy transfer mechanisms and the molecular exciton model for molecular aggregates, *Radiat. Res.*, 20, 55, 1963.
6. **Kasha, M., Rawls, H. R., and El-Bayoumi, M. A.**, The exciton model in molecular spectroscopy, *Pure Appl. Chem.*, 11, 371, 1965.
7. **Tinoco, I., Jr.**, The exciton contribution to the optical rotation of polymers, *Radiat. Res.*, 20, 133, 1963.
8. **Philipson, K. D. and Sauer, K.**, Exciton interaction in a bacteriochlorophyll-protein from *Chloropseudomonas ethylica*. Absorption and circular dichroism at 77°K, *Biochemistry*, 11, 1880, 1972.
9. **Dratz, E. A., Schultz, A. J., and Sauer, K.**, Chlorophyll-chlorophyll interactions, *Brookhaven Symp. Biol.*, 19, 303, 1966.
10. **Simpson, W. T. and Peterson, D. L.**, Coupling strength for resonance force transfer of electronic energy in van der Waals solids, *J. Chem. Phys.*, 26, 588, 1957.
11. **Arnold, W. and Oppenheimer, J. R.**, Internal conversion in the photosynthetic mechanism of blue-green algae, *J. Gen. Physiol.*, 33, 423, 1950.
12. **Duysens, L. N. M.**, Transfer of light energy within the pigment systems present in photosynthesizing cells, *Nature*, 168, 548, 1951.
13. **French, C. S. and Young, V. K.**, The fluorescence spectra of red algae and the transfer of energy from phycoerythrin to phycocyanin and chlorophyll, *J. Gen. Physiol.*, 35, 873, 1952.
14. **Tomita, G. and Rabinowitch, E.**, Excitation energy transfer between pigments in photosynthetic cells, *Biophys. J.*, 2, 483, 1962.
15. **Brody, M. and Emerson, R.**, The quantum yield of photosynthesis in *Porphyridium cruentum*, and the role of chlorophyll *a* in the photosynthesis of red algae, *J. Gen. Physiol.*, 43, 251, 1959.
16. **Brody, S. S. and Brody, M.**, Induced changes in the efficiency of energy transfer in *Porphyridium cruentum*. I, *Arch. Biochem. Biophys.*, 82, 161, 1959.
17. **Ghosh, A. K. and Govindjee**, Transfer of the excitation energy in *Anacystis nidulans* grown to obtain different pigment ratios, *Biophys. J.*, 6, 611, 1966.
18. **Wang, R. T. and Myers, J.**, Reverse energy transfer from chlorophyll to phycobilin in *Anacystis nidulans*, *Plant Cell Physiol.*, Photosynthetic Organelles issue, 3, 1977.
19. **Ley, A. C.**, Effective absorption cross-sections in *Porphyridium cruentum*. Implications for energy transfer between phycobilisomes and photosystem II reaction centers, *Plant Physiol.*, 74, 451, 1984.
20. **Diner, B. A.**, Energy transfer from the phycobilisomes to photosystem II reaction centers in wild type *Cyanidium caldarium*, *Plant Physiol.*, 63, 30, 1979.
21. **Schreiber, U. and Vidaver, W.**, Photosynthetic energy transfer reversibly inhibited by hydrostatic pressure, *Photochem. Photobiol.*, 18, 205, 1973.
22. **Harnischfeger, G. and Codd, G. A.**, Factors affecting energy transfer from phycobilisomes to thylakoids in *Anacystis nidulans*, *Biochem. Biophys. Acta*, 502, 507, 1978.
23. **Schreiber, U., Rijgersberg, C. P., and Amesz, J.**, Temperature-dependent reversible changes in phycobilisome-thylakoid membrane attachment in *Anacystis nidulans*, *FEBS Lett.*, 104, 327, 1979.
24. **Schreiber, U.**, Cold-induced uncoupling of energy transfer between phycobilins and chlorophyll in *Anacystis nidulans*, *FEBS Lett.*, 107, 4, 1979.
25. **Gantt, E. and Lipschultz, C. A.**, Energy transfer in phycobilisomes from phycoerythrin to allophycocyanin, *Biochim. Biophys. Acta*, 292, 858, 1973.
26. **Glazer, A. N. and Bryant, D. A.**, Allophycocyanin B (λmax 671, 618 nm). A new cyanobacterial phycobiliprotein, *Arch. Microbiol.*, 104, 15, 1975.
27. **Zilinskas, B. A., Zimmerman, B. K., and Gantt, E.**, Allophycocyanin forms isolated from *Nostoc* sp. phycobilisomes, *Photochem. Photobiol.*, 27, 587, 1978.
28. **Lundell, D. J., Yamanaka, G., and Glazer, A. N.**, A terminal energy acceptor of the phycobilisome: the 75,000-dalton polypeptide of *Synechococcus* 6301 phycobilisomes — a new biliprotein, *J. Cell Biol.*, 91, 315, 1981.
29. **Grabowski, J. and Gantt, E.**, Photophysical properties of phycobiliproteins from phycobilisomes: fluorescence lifetimes, quantum yields, and polarization spectra, *Photochem. Photobiol.*, 28, 39, 1978.
30. **Gantt, E., Lipschultz, C. A., and Zilinskas, B.**, Further evidence for a phycobilisome model from selective dissociation, fluorescence emission, immunoprecipitation, and electron microscopy, *Biochem. Biophys. Acta*, 430, 375, 1976.

31. **Bekasova, O. D., Romanyuk, V. A., and Zvalinsky, V. I.,** Efficiency of excitation energy migration in phycobilisomes of red sea macroalgae, *Biofizika*, 26, 74, 1981.
32. **Grabowski, J. and Gantt, E.,** Excitation energy migration in phycobilisomes: comparison of experimental results and theoretical predictions, *Photochem. Photobiol.*, 28, 47, 1978.
33. **Gulyaev, B. A., Tetenkin, V. L., and Bekasova, O. D.,** The pigment orientation and the energy migration in blue-green algae and isolated phycobilisomes, *Izv. Akad. Nauk S.S.S.R. Ser. Biol.*, 188, 1981.
34. **Kume, N. and Katoh, T.,** Dissociation kinetics of *Anabaena* phycobilisomes, *Plant Cell Physiol.*, 23, 803, 1982.
35. **Goedheer, J. C. and Birnie, F.,** Fluorescence polarization and location of fluorescence maxima of C-phycocyanin, *Biochim. Biophys. Acta*, 94, 579, 1965.
36. **Jabłonski, A.,** Theory of the polarisation of photoluminescence of colored solutions, *Z. Phys.*, 96, 236, 1935.
37. **Vernotte, C.,** Separation et caracterisation spectroscopique du monomere et des polymeres de la *C. phycocyanine*, *Photochem. Photobiol.*, 14, 163, 1971.
38. **Teale, F. W. J. and Dale, R. E.,** Isolation and spectral characterization of phycobiliproteins, *Biochem. J.*, 116, 161, 1970.
39. **MacColl, R., Berns, D. S., and Koven, N. L.,** Effect of salts on C-phycocyanin, *Arch. Biochem. Biophys.*, 146, 477, 1971.
40. **MacColl, R., Csatorday, K., Berns, D. S., and Traeger, E.,** Chromophore interactions in allophycocyanins, *Biochemistry*, 19, 2817, 1980.
41. **Cohn-Bazire, G., Beguin, S., Rimon, S., Glazer, A. N., and Brown, D. M.,** Physico-chemical and immunological properties of allophycocyanins, *Arch. Microbiol.*, 111, 225, 1977.
42. **Switalski, S. C. and Sauer, K.,** Energy transfer among the chromophores of C-phycocyanin from *Anabaena variabilis* using steady state and time-resolved fluorescence spectroscopy, *Photochem. Photobiol.*, 40, 423, 1984.
43. **Feofilov, P. P.,** *The Physical Basis of Polarized Emission*, Consultants Bureau Enterprises, New York, 1961.
44. **Dale, R. E. and Teale, F. W. J.,** Number and distribution of chromophore types in native phycobilisomes, *Photochem. Photobiol.*, 12, 99, 1970.
45. **Fujimori, E. and Quinlan, K.,** Fluorescence, energy transfer, and SH-groups in photosynthetic pigments of red and blue-green algae, in *Photosynthetic Mechanisms of Green Plants*, Natl. Res. Counc. Publ. 1145, National Academy of Science, Washington, D.C., 1963, 519.
46. **Fujimori, E. and Pecci, J.,** Distinct subunits of phycoerythrin from *Porphyridium cruentum* and their spectral characteristics, *Arch. Biochem. Biophys.*, 118, 448, 1967.
47. **MacDowall, F. D. H., Bednar, T., and Rosenberg, A.,** Conformation dependence of intramolecular energy transfer in phycoerythrin, *Proc. Natl. Acad. Sci. U.S.A.*, 59, 1356, 1968.
48. **Zickendraht-Wendelstadt, B., Friedrich, J., and Rüdiger, W.,** Spectral characterization of monomeric C-phycoerythrin from *Pseudanabaena* W1173 and its α and β subunits: energy transfer in isolated subunits and C-phycoerythrin, *Photochem. Photobiol.*, 31, 367, 1980.
49. **MacColl, R., Edwards, M. R., Mulks, M. H., and Berns, D. S.,** Comparison of the biliproteins from two strains of the thermophilic cyanophyte *Synechococcus lividus*, *Biochem. J.*, 141, 419, 1974.
50. **Canaani, O. D. and Gantt, E.,** Circular dichroism and polarized fluorescence characteristics of blue-green algal allophycocyanins, *Biochemistry*, 19, 2950, 1980.
51. **MacColl, R. and Guard-Friar, D.,** Phycocyanin 612: a biochemical and photophysical study, *Biochemistry*, 22, 5568, 1983.
52. **MacColl, R., Habig, W., and Berns, D. S.,** Characterization of phycocyanin from *Chroomonas* species, *J. Biol. Chem.*, 248, 7080, 1973.
53. **MacColl, R., Berns, D. S., and Gibbons, O.,** Characterization of cryptomonad phycoerythrin and phycocyanin, *Arch. Biochem. Biophys.*, 177, 265, 1976.
54. **Jung, J., Song, P.-S., Paxton, R. J., Edelstein, M. S., Swanson, R., and Hazen, E. E., Jr.,** Molecular topography of the phycocyanin photoreceptor from *Chroomonas* species, *Biochemistry*, 19, 24, 1980.
55. **MacColl, R., Guard-Friar, D., and Csatorday, K.,** Chromatographic and spectroscopic analysis of phycoerythrin 545 and its subunits, *Arch. Microbiol.*, 135, 194, 1983.
56. **MacColl, R. and Guard-Friar, D.,** Phycocyanin 645. The chromophore assay of phycocyanin 645 from the cryptomonad protozoa *Chroomonas* species, *J. Biol. Chem.*, 258, 14327, 1983.
57. **Porter, G., Tredwell, C. J., Searle, G. F. W., and Barber, J.,** Picosecond time-resolved energy transfer in *Porphyridium cruentum*. I. In the intact alga, *Biochim. Biophys. Acta*, 501, 232, 1978.
58. **Brody, S. S., Porter, G., Tredwell, C. J., and Barber, J.,** Picosecond energy transfer in *Anacystis nidulans*, *Photobiochem. Photobiophys.*, 2, 11, 1981.
59. **Brody, S. S., Treadwell, C., and Barber, J.,** Picosecond energy transfer in *Porphyridium cruentum* and *Anacystis nidulans*, *Biophys. J.*, 34, 439, 1981.

60. **Yamazaki, I., Mimuro, M., Murao, T., Yamazaki, T., Yoshihara, K., and Fujita, Y.,** Excitation energy transfer in the light harvesting antenna system of the red alga *Porphyridium cruentum* and the blue-green alga *Anacystis nidulans:* analysis of time-resolved fluorescence spectra, *Photochem. Photobiol.*, 39, 233, 1984.
61. **Yamazaki, I., Tamai, N., Yamazaki, T., Mimuro, M., and Fujita, Y.,** Excitation energy transfer in phycobilin-chlorophyll. A system of algal intact cells, in *Ultrafast Phenomena*, Auston, D. H. and Eisenthal, K. B., Eds., Springer-Verlag, New York, 1984, 490.
62. **Searle, G. F. W., Barber, J., Porter, G., and Tredwell, C. J.,** Picosecond time-resolved energy transfer in *Porphyridium cruentum*. II. In the isolated light harvesting complex (phycobilisomes), *Biochim. Biophys. Acta,* 501, 246, 1978.
63. **Wendler, J., Holzwarth, A. R., and Wehrmeyer, W.,** Picosecond time-resolved energy transfer in phycobilisomes isolated from the red alga *Porphyridium cruentum, Biochim. Biophys. Acta,* 765, 58, 1984.
64. **Pellegrino, F., Wong, D., Alfano, R. R., and Zilinskas, B. A.,** Fluorescence relaxation kinetics and quantum yield from the phycobilisomes of the blue-green alga *Nostoc* sp. measured as a function of single picosecond pulse intensity, *Photochem. Photobiol.*, 34, 691, 1981.
65. **Holzwarth, A. R., Wendler, J., and Wehrmeyer, W.,** Picosecond time resolved energy transfer in isolated phycobilisomes from *Rhodella violacea* (Rhodophyceae), *Photochem. Photobiol.*, 36, 479, 1982.
66. **Hefferle, P., Nies, M., Wehrmeyer, W., and Schneider, S.,** Picosecond time-resolved fluorescence study of the antenna system isolated from *Mastigocladus laminosus* Cohn. I. Functionally intact phycobilisomes, *Photobiochem. Photobiophys.*, 5, 41, 1983.
67. **Suter, G. W., Mazzola, P., Wendler, J., and Holzwarth, A. R.,** Fluorescence decay kinetics in phycobilisomes isolated from the blue-green alga *Synechococcus* 6301, *Biochim. Biophys. Acta,* 766, 269, 1984.
68. **Gillbro, T., Sandström, A., Sundström, V., and Holzwarth, A. R.,** Polarized absorption picosecond kinetics as a probe of energy transfer in phycobilisomes of *Synechococcus* 6301, *FEBS Lett.*, 162, 64, 1983.
69. **Glazer, A.N., Yeh, S. W., Webb, S. P., and Clark, J. H.,** Disk-to-disk transfer as the rate-limiting step for energy flow in phycobilisomes, *Science,* 227, 419, 1985.
70. **Kobayashi, T., Degenkolb, E. O., Bersohn, R., Rentzepis, P. M., MacColl, R., and Berns, D. S.,** Energy transfer among the chromophores in phycocyanins measured by picosecond kinetics, *Biochemistry,* 18, 5073, 1979.
71. **MacColl, R. and Berns, D. S.,** Biliproteins: some relationships among aggregation states, spectra, and excitation-energy transfer, *Isr. J. Chem.*, 21, 296, 1981.
72. **Wong, D., Pellegrino, F., Alfano, R. R., and Zilinskas, B. A.,** Fluorescence relaxation kinetics and quantum yield from the isolated phycobiliproteins of the blue-green alga *Nostoc* sp. measured as a function of single picosecond pulse intensity. I, *Photochem. Photobiol.*, 33, 651, 1981.
73. **Doukas, A. G., Stefancic, V., Buchert, J., Alfano, R. R., and Zilinskas, B. A.,** Exciton annihilation in the isolated phycobiliproteins from the blue-green alga *Nostoc* sp. using picosecond absorption spectroscopy, *Photochem. Photobiol.*, 34, 505, 1981.
74. **Hefferle, P., Nies, M., Wehrmeyer, W., and Schneider, S.,** Picosecond time-resolved fluorescence study on the antenna system isolated from *Mastigocladus laminosus* Cohn. II. Constituent biliproteins in various forms of aggregation, *Photobiochem. Photobiophys.*, 5, 325, 1983.
75. **Hefferle, P., John, W., Scheer, H., and Schneider, S.,** Thermal denaturation of monomeric and trimeric phycocyanins studied by static and polarized time-resolved fluorescence spectroscopy, *Photochem. Photobiol.*, 39, 221, 1984.
76. **Holzwarth, A. R., Wendler, J., and Wehrmeyer, W.,** Studies on chromophore coupling in isolated phycobiliproteins. I. Picosecond fluorescence kinetics of energy transfer in phycocyanin 645 from *Chroomonas* sp., *Biochim. Biophys. Acta,* 724, 388, 1983.
77. **Guard-Friar, D., MacColl, R., Berns, D. S., Wittmershaus, B., and Knox, R. S.,** Picosecond fluorescence of cryptomonad biliproteins. Effects of excitation intensity on the fluorescence decay times of phycocyanin 612, phycocyanin 645 and phycoerythrin 545, *Biophys. J.*, 47, 787, 1985.
78. **Hanzlik, C. A., Hancock, L. E., Knox, R. S., Guard-Friar, D., and MacColl, R.,** Picosecond fluorescence spectroscopy of the biliprotein phycocyanin 612: direct evidence for fast energy transfer, *J. Lumin.*, 34, 99, 1985.
79. **Latimer, P., Barnister, T. T., and Rabinowitch, E.,** Quantum yields of fluorescence of plant pigments, *Science,* 124, 585, 1956.
80. **Seibert, M. and Connolly, J. S.,** Fluorescence properties of C-phycocyanin isolated from a thermophilic cyanobacterium, *Photochem. Photobiol.*, 40, 267, 1984.
81. **Barber, D. J. W. and Richards, J. T.,** Energy transfer in the accessory pigments R-phycoerythrin and C-phycocyanin, *Photochem. Photobiol.*, 25, 565, 1977.
82. **Brody, S. S. and Brody, M.,** A quantitative assay for the number of chromophores on a chromoprotein; its application to phycoerythrin and phycocyanin, *Biochim. Biophys. Acta,* 50, 348, 1961.

83. **Vernotte, C. and Moya, I.,** Action de la temperature sur la duree de vie fluorescence et le rendement de fluorescence de la C-phycocyanine en solution. *Photochem. Photobiol.*, 17, 245, 1973.
84. **Merkelo, H., Hartman, S. R., Mar, T., Singhal, G. S., and Govindjee,** Mode-locked lasers: measurements of very fast radioactive decay in fluorescent systems, *Science*, 164, 301, 1969.
85. **Priestle, J. P., Jr., Rhyne, R. H., Jr., Salmon, J. B., and Hackert, M. L.,** Phycobiliproteins: comparison of solution and single crystal fluorescence for C-phycocyanin and B-phycoerythrin, *Photochem. Photobiol.*, 35, 827, 1982.
86. **MacDowall, F. and Walker, M.,** Fluorescence lifetime of phycoerythrin, *Photochem. Photobiol.*, 7, 109, 1968.
87. **Swenberg, C. E., Geacintov, N. E., and Pope, M.,** Molecular quenching of excitons and fluorescence in the photosynthetic unit, *Biophys. J.*, 16, 1447, 1976.
88. **Paillotin, G., Swenberg, C. E., Breton, J., and Geacintov, N. E.,** Analysis of picosecond laser-induced fluorescence phenomena in photosynthetic membranes utilizing a master equation approach, *Biophys. J.*, 25, 513, 1979.
89. **Gray, B. H. and Gantt, E.,** Spectral properties of phycobilisomes and phycobiliproteins from the blue-green alga — *Nostoc* sp., *Photochem. Photobiol.*, 21, 121, 1975.
90. **Csatorday, K., MacColl, R., Csizmadia, V., Grabowski, J., and Bagyinka, C.,** Exciton interaction in allophycocyanin, *Biochemistry*, 23, 6466, 1984.
91. **Murakami, A., Mimuro, M., Ohki, K., and Fujita, Y.,** Absorption spectrum of allophycocyanin isolated from *Anabaena cylindrica:* variation of the absorption spectrum induced by changes of the physico-chemical environment, *J. Biochem.*, 89, 79, 1981.
92. **Mimuro, M., Murakami, A., and Fujita, Y.,** Studies on spectral characteristics of allophycocyanin isolated from *Anabaena cylindrica:* curve-fitting analysis, *Arch. Biochem. Biophys.*, 215, 266, 1982.
93. **Sugimoto, T., Kikushima, M., Saito, M., and Suzuki, H.,** Models for the change in chromophore structure of allophycocyanin corresponding to its observed reversible change in absorbance, *J. Phys. Soc. Jpn.*, 53, 873, 1984.
94. **Brown, A. S., Foster, J. A., Voynow, P. V., Franzblau, C., and Troxler, R. F.,** Allophycocyanin from the filamentous cyanophyte, *Phormidium luridum, Biochemistry*, 14, 3581, 1975.
95. **MacColl, R., Csatorday, K., Berns, D. S., and Traeger, E.,** The relationship of the quaternary structure of allophycocyanin to its spectrum, *Arch. Biochem. Biophys.*, 208, 42, 1981.
96. **Zilinskas, B. A., Greenwald, L. S., Bailey, C. L., and Kahn, P. C.,** Spectral analysis of allophycocyanin I, II, III and B from *Nostoc* sp. phycobilisomes, *Biochim. Biophys. Acta*, 592, 267, 1980.
97. **Friedrich, J., Scheer, H., Zickendraht-Wendelstadt, B., and Haarer, D.,** High-resolution optical studies on C-phycocyanin via photochemical hole burning, *J. Am. Chem. Soc.*, 103, 1030, 1981.
98. **Friedrich, J., Scheer, H., Zickendraht-Wendelstadt, B., and Haarer, D.,** Photochemical hole burning: a means to observe high resolution optical structures in phycoerythrin, *J. Chem. Phys.*, 74, 2260, 1981.
99. **Braslavsky, S. E., Holzwarth, A. R., and Schaffner, K.,** Solution conformations, photophysics, and photochemistry of bile pigments; bilirubin and biliverdin dimethyl esters and related tetrapyrroles, *Angew. Chem. Int. Ed. Engl.*, 22, 656, 1983.
100. **Scheer, H. and Kufer, W.,** Studies on plant bile pigments. IV. Conformational studies on C-phycocyanin from *Spirulina platensis, Z. Naturforsch.*, 32c, 513, 1977.
101. **Gantt, E., Lipschultz, C. A., Grabowski, J., and Zimmerman, B. K.,** Phycobilisomes from blue-green and red algae. Isolation criteria and dissociation characteristics, *Plant Physiol.*, 63, 615, 1979.
102. **MacColl, R., O'Connor, G., Crofton, G., and Csatorday, K.,** Phycoerythrocyanin: its spectroscopic behavior and properties, *Photochem. Photobiol.*, 34, 719, 1981.
103. **Isono, T. and Katoh, T.,** Cylindrical phycobilisomes from a blue-green alga, *Anabaena variabilis, Plant Cell Physiol.*, 23, 1347, 1982.
104. **Isono, T. and Katoh, T.,** Subparticles of *Anabaena* phycobilisomes. I. Reconstitution of allophycocyanin cores and entire phycobilisomes, *Plant Cell Physiol.*, 24, 357, 1983.
105. **Rijgersberg, C. P. and Amesz, J.,** Fluorescence and energy transfer in phycobiliprotein-containing algae at low temperature, *Biochim. Biophys. Acta*, 593, 261, 1980.
106. **Khanna, R., Graham, J.-R., Myers, J., and Gantt, E.,** Phycobilisome composition and possible relationship to reaction centers, *Arch. Biochem. Biophys.*, 224, 534, 1983.
107. **Grabowski, J.,** Chlorophyllin and phycocyanin spectra at room and low temperature, *Photosynthetica*, 6, 291, 1972.
108. **Grabowski, J. and Frackowiak, D.,** Chlorophyllin and phycoerythrin spectra at room and low temperature, *Photosynthetica*, 6, 142, 1972.
109. **Goedheer, J. C.,** Fluorescence action spectra of algae and bean leaves at room and at liquid nitrogen temperatures, *Biochim. Biophys. Acta*, 102, 73, 1965.
110. **Frackowiak, D. and Grabowski, J.,** Low temperature spectra of biliproteins, *Photosynthetica*, 5, 146, 1971.

111. **Peterson, R. B., Dolan, E., Calvert, H. E., and Ke, B.,** Energy transfer from phycobiliproteins to photosystem I in negative cells and heterocysts of *Anabaena variabilis*, *Biochim. Biophys. Acta*, 634, 237, 1981.

112. **Krasnovsky, A. A., Jr. and Kovalev, Y. V.,** Low temperature delayed fluorescence of phycobilins in red alga, *Biofizika*, 26, 724, 1981.

113. **Harnischfeger, G. and Herold, B.,** Aspects of energy-transfer between phycobilins and chlorophyll in *Chroomonas* spec. (Cryptophycea), *Ber. Dtsch. Bot. Ges.*, 94, 65, 1981.

114. **Lichtlé, C., Jupin, H., and Duval, J. C.,** Energy transfers from photosystem II to photosystem I in *Cryptomonas rufescens* (Cryptophyceae), *Biochim. Biophys. Acta*, 591, 104, 1980.

115. **Cho, F. and Govindjee,** Low-temperature (4-77°K) spectroscopy of *Anacystis;* temperature dependence of energy transfer efficiency, *Biochim. Biophys. Acta*, 216, 151, 1970.

116. **Sineshchekov, V. A., Muslimov, I. A., and Bekasova, O. D.,** Energy migration in phycobilisomes, *Mol. Biol.*, 18, 447, 1984.

117. **Frackowiak, D. and Grabowski, J.,** Excitation energy transfer between biliproteins and chlorophyllide *a*, *Photosynthetica*, 4, 236, 1970.

Chapter 7

PHYSIOLOGY

I. BILIPROTEIN CONTENT

In the cyanobacterium *Anacystis nidulans,* Myers and Kratz[1] have calculated that C-phycocyanin represented 24.1% of the dry weight of the cell. This suggested that biliproteins are important to the functioning of the organism and that the cell must expend considerable effort in the synthesis of these pigments. An important environmental factor in the control of biliprotein synthesis is the intensity and color of the growth light.

II. EFFECT OF GROWTH LIGHT ON BILIPROTEINS

A. Background

Early studies on the red algae and cyanobacteria showed that striking changes occurred in the colors of certain organisms depending on the type of growth irradiance. The involvement of changes in biliprotein content in these color variations was demonstrated. In addition, the significance of the biliprotein in establishing the organism in habitats of different spectral composition and intensity was an early topic of biliprotein research.[2-12]

Boresch[5] and Kylin[10] obtained absorption spectra which clearly demonstrated the spectral distinctions between red algal and cyanobacterial phycoerythrins.

B. Growth in White Light

When cyanobacteria,[13-16] red algae,[17-20] or cryptomonads[21,22] were grown in high-intensity white light, the photosynthetic pigments — biliproteins and chlorophylls — were reduced in amount compared to growth at lower intensities. This is a reasonable response for the cells, since at higher light intensities less pigment would be needed for maintenance of photosynthetic activity required to sustain growth and reproduction.

The response of these organisms to growth light intensity in regard to the various pigment ratios was less clear. Myers and Kratz[1] reported that the C-phycocyanin to chlorophyll *a* ratio first rose as intensity increased and then a subsequent intensity increase caused the ratio to drop below the original level. These changes were both quite small. Others have reported a lower biliprotein to chlorophyll ratio at higher growth intensities,[14,15,23,24] but Raps et al.[16] have surprisingly found the reverse, a significantly higher ratio at higher light intensities. Brown and Richardson[25] studied seven species of cyanobacteria and red algae where the biliprotein to chlorophyll ratio was examined as a function of light intensity. The ratio of C-phycoerythrin to chlorophyll *a* in the cyanobacterium *Phormidium persicinum* dropped sharply as intensity increased, but the C-phycocyanin to chlorophyll *a* ratio in the red alga *Porphyridium aerugineum* after an initial drop showed a dramatic rise as the intensity of growth light increased. Other organisms had more gradual or very level responses to intensity.[25]

These adaptations were more than just productions of different amounts of pigments. The number of phycobilisomes[26] in the cyanobacterium *Spirulina platensis* decreased as light intensity increased. Vierling and Alberte[24] found that the number of photosystem I reaction centers per cell also decreased in higher-intensity white light for the cyanobacterium *A. nidulans*.

In the red algae, responses to growth light intensity were also variable. Koch[17] found that the B-phycoerythrin to chlorophyll *a* ratio was higher in low-intensity growth light. This work was also interesting in that it presented an early visible absorption spectrum for B-phycoerythrin *(P. cruentum)*, showing an absorption shoulder near 500 nm which is now associated with phycourobilin. Brody and Emerson[18] found the same intensity dependence for the same organism. Nonetheless two subsequent reports on this same red alga indicated that the B-phycoerythrin to chlorophyll *a* ratio underwent no significant change during large variations in growth light intensities.[20,25] For a different red alga, *Porphyridium* sp., the intensity of the growth illumination did not affect the phycoerythrin to chlorophyll ratio,[19] but for the red alga *Griffithsia pacifica* the R-phycoerythrin to chlorophyll *a* ratio dropped as intensity increased.[27] The number of phycobilisomes per segment of thylakoid membrane dropped with a very similar dependence on growth intensity as the biliprotein to chlorophyll ratio.[27] Döhler[28] has studied the growth of *A. nidulans* in different light conditions and related them to differing enzyme activities.

Two species of cryptomonad showed declines in their biliprotein to chlorophyll ratios as light intensity increased. However, the change in *Cryptomonas* was much larger than the change in *Chroomonas* sp.[21,22]

The great variations observed on the ratios of biliprotein to chlorophyll were products of several factors. The way the experiments were performed can be important. Both the physical setup of the equipment and the cell cycle can contribute to varying degrees of internal shading which would alter intensity results. The amounts of available nutrients can also affect the influence of intensity on pigment ratio (see Section V.A).

Despite inconsistencies there is a clear generalization which these results have established. The effect of increasing the intensity of white growth light was to diminish both biliprotein and chlorophyll, and this change in the content for both types of photosynthetic pigments can readily occur in the absence of a variation in pigment ratio. For example, Ley and Butler[20] studied the effect of high and low white light intensity on *P. cruentum*, varying the irradiance intensity by over 100-fold. The absorption and therefore the pigment content per cell was three- to five-fold higher at the lower irradiance, but the relative pigment proportions were virtually identical. The pigment ratio changes which are frequently observed in this type of experiment are, therefore, best viewed as secondary or individualistic phenomena. A result for one organism or even one group of organisms is obviously an invalid criterion for development of generalizations in this complex area.

C. Growth in Colored Light

When cyanobacteria and red algae are grown in colored light, the changes in pigment ratios can be quite spectacular. A very interesting result occurred when the organisms were grown in light that was specifically absorbed by chlorophyll *a*, either blue or red light. Jones and Myers[29] irradiated the cyanobacterium *A. nidulans* with red or blue growth light and found a strong decline in the chlorophyll *a* content per cell when compared to cells grown in white light or light absorbed mainly by C-phycocyanin. The red and blue wavelengths were those strongly absorbed by chlorophyll *a*, and after growth at these wavelengths the C-phycocyanin to chlorophyll *a* ratio was greatly increased. Varying the intensities of blue (436 nm) or red (680 nm) light did not seriously change the pigment ratios. Using the red alga *Porphyridium* sp., Guérin-Dumartrait et al.[19] also found that red growth light lowered the chlorophyll *a* content and had little effect on the phycoerythrin content. Ley and Butler[20] found that red and blue growth lights produced a large increase in the phycoerythrin to chlorophyll *a* ratio of *P. cruentum*. Conversely, growth in green light mainly absorbed by b- and B-phycoerythrins had little impact on the pigment ratio. Likewise, Jones and Myers[29] found that growth illumination in the wavelength region of biliprotein absorption did not lower the C-phycocyanin content. Jones and Myers[29] noted that red and blue lights were mainly directed at photosystem I chlorophyll *a*. A reduction in its content was analogous to a light intensity effect. The reduction of pigment in red or blue light would tend toward maintaining comparable rates for the two photosystems. Bennett and Bogorad[30] also found

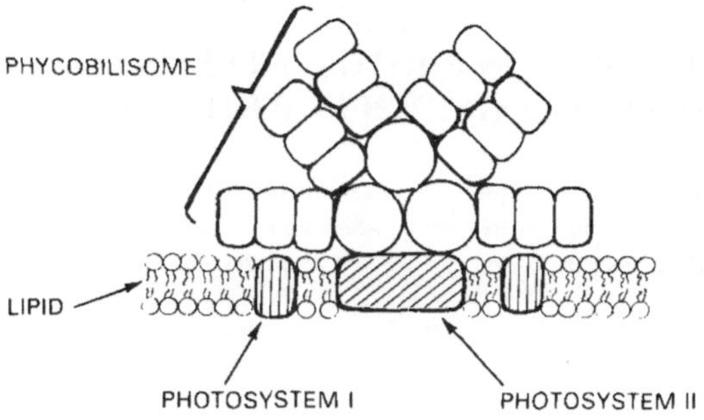

FIGURE 1. Schematic drawing showing the possible spacial relationships among a phycobilisome and photosystems I and II. Excitation energy in this model would pass from the phycobilisome via allophycocyanin 680 to photosystem II. In order to maintain functional balance between the two photosystems, excitation energy from photosystem II can spillover into photosystem I. Note that the relative chlorophyll contents of the two photosystems are not reflected in the cross-section areas in this drawing.

that red light increased the biliprotein to chlorophyll *a* ratio in the cyanobacterium *Fremyella diplosiphon*. Other reports have also appeared on the effects of the spectra of growth irradiation on the photosynthetic pigments.[14-18]

These studies could be and are called ''chromatic adaptations'', but for clarity this designation will be avoided here. In certain cyanobacteria, the amounts of C-phycoerythrin and usually C-phycocyanin were varied by light quality. In this process the C-phycoerythrin content was augmented by the green light. In some organisms the amount of C-phycocyanin was increased by growth in red-orange light. The process is called ''complementary chromatic adaptation'' and will be discussed in Section III.

D. Growth Light and Photosynthesis

When various parameters were measured for cells grown in lights of different light quality, a general observation was that the cells adapted so that these growth wavelengths were used with increased efficiency for photosynthesis.[15,31-34] Brody and Emerson[31] found that cells of *P. cruentum* grown in green light had a higher quantum yield for photosynthesis in green light than did cells grown in blue light. Likewise, cells grown in blue light had a higher quantum yield for photosynthesis in these wavelengths than cells grown in green light.

Ley and Butler[20] have studied the effects of growing *P. cruentum* in lights of different light quality on its photosynthetic operation. They grew cells in high-intensity white, low-intensity white, green, red, and blue lights. The blue and red lights were absorbed mostly by chlorophyll *a*, and the green and white were chiefly absorbed by the biliproteins. In white or green light which was absorbed by the biliproteins, the ratio of biliprotein to chlorophyll *a* did not change significantly, but in lights absorbed by chlorophyll *a* — the red and blue lights — the relative amounts of pigments were quite different from those in cells grown under green and white lights, although all the growth lights had similar intensities. The photosynthetic apparatus responded to optimize utilization of the irradiance it was receiving. Measurements on *P. cruentum* showed that in low-intensity white light, the light absorbed by phycoerythrin went into photosystem II and 95% of the light absorbed by chlorophyll *a* went into photosystem I. A schematic drawing shows the general spacial relationships between a phycobilisome and photosystems I and II (Figure 1). When grown in red light, chlorophyll *a* would transmit an excess of energy to photosystem I if the situation were unchanged from that of growth in white light. The cells responded in two ways to

Table 1
ENERGY DISTRIBUTION IN PHOTOSYNTHESIS FOR *PORPHYRIDIUM CRUENTUM* CELLS GROWN IN WHITE OR RED LIGHT[20]

Growth light[a]	Percent transfer of light absorbed at 550 nm to photosystem I	Percent transfer of light absorbed at 435 nm to photosystem I	Yield of energy transfer for photosystem II to I
White	10	95	0.55
Red	5	60	0.23

[a] The intensities in this experiment are similar for both types of irradiation.

keep the two photosystems in balance (Table 1). More chlorophyll *a* transferred to photosystem II, and perhaps the biliprotein content was increased; this increased the amount of red light going to photosystem II. The second response was to lower the fraction of energy being transferred from photosystem II to I. Energy transfer in the white and green light regimes were similar, since both kinds of light were strongly absorbed by the biliproteins. The high efficiency of energy transfer from photosystem II to I and the very large fraction of chlorophyll *a* in photosystem I tended to maintain the needed balance (Table 1). Ley and Butler[20] pointed out that these variations allow *P. cruentum* to adapt to environments in which the spectral composition and intensity of light vary greatly. They observed that the ability to respond to these factors may be part of the reason that red algae are an abundant group in the oceanic and intertidal regions. In Chapter 8 the relationships among phycobilisomes and photosystems I and II are explored in more detail.

III. COMPLEMENTARY CHROMATIC ADAPTATION

Tandeau de Marsac[35] has surveyed 44 strains of cyanobacteria to determine whether the spectra of the growth light would influence the synthesis of C-phycoerythrin and C-phycocyanin. Three types of response to green and red lights were recognized. First, 12 strains did not show a change in the ratios of biliproteins when grown in red or green light. Second, C-phycoerythrin alone was affected in seven strains. In these organisms, C-phycoerythin was synthesized at a higher rate in green light. It was observed that even in red growth light some C-phycoerythin was still being made. Third, in the remaining 25 strains, synthesis of both the C-phycoerythrin and C-phycocyanin was tied to the type of growth illumination. The rate of C-phycoerythrin synthesis was greater in green than in red light, while for C-phycocyanin the situation was reversed. In red light, the synthesis of C-phycoerythin ceased, but in green light C-phycocyanin was still made to some extent. This latter type of growth response depended on the nature of the growth light and other experimental factors. A typical result for growth of the cyanobacterium *Calothrix parietina* showed a vastly reduced C-phycoerythrin content in cells grown in red light. Cells grown in white light were observed to have an intermediate C-phycoerythrin to C-phycocyanin ratio when compared to red and green light-grown cells (Figure 2). The C-phycoerythrin to C-phycocyanin ratios varied tenfold from 0.090 for red to 1.02 for green light-grown cells.

Chromatic adaptation in the cyanobacterium *Tolypothrix tenuis* has been studied extensively.[36-43] This organism is the type in which light regulated both C-phycoerythrin and C-phycocyanin syntheses. A major focus of these studies was to determine the spectra of the pigments controlling the chromatic effects. Not only do these studies contribute to understanding the phenomenon of chromatic adaptation, but they also provide a vehicle to study the general problem of biliprotein synthesis. Hattori and Fujita[37] developed an excellent system for chromatic studies of these cells. Specific biliproteins can be synthesized in the dark if the dark period is preceded by a suitable preillumination in the absence of a suitable

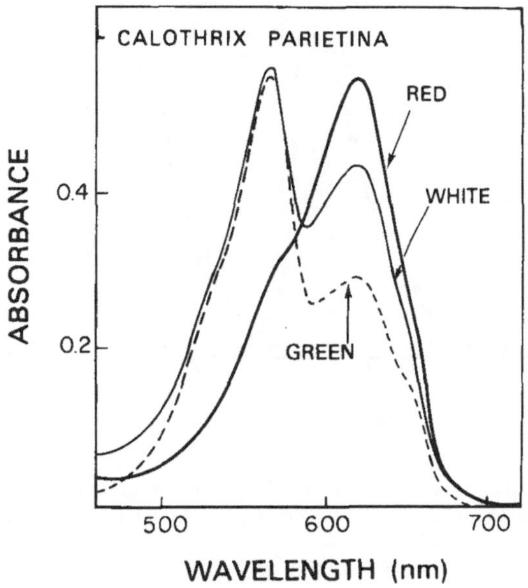

FIGURE 2. Absorption spectra of aqueous extracts of cells chromatically adapted by red, green, and white growth lights for the cyanobacterium *Calothrix parietina*. The three spectra are drawn so that the visible maxima are all roughly equal for comparison purposes.

source of nitrogen. When the cells were given the necessary form of nitrogen, the synthesis of biliproteins commenced in the dark. Preillumination decreased the level of biliproteins. The nature of the biliproteins produced depended on the spectral distribution of the final preillumination that the cells were given. Fujita and Hattori[39] found that preillumination with red light for 5 min produced cells that could synthesize C-phycocyanin in the subsequent dark period. Using cells which contained only C-phycocyanin, i.e., no C-phycoerythrin, the synthesis of C-phycocyanin and C-phycoerythrin could be evaluated after preillumination with green light. The action spectra corresponding to the red and green light effects were determined (Figure 3). The maximum conversion rate for C-phycoerythrin synthesis was at 541 nm, and the formation of C-phycocyanin was achieved best at 641 nm. Subsequently, Diakoff and Scheibe[42] investigated chromatic adaptation in this cyanobacterium. As was previously found by Fujita and Hattori,[37] after preillumination, addition of the proper source of nitrogen in the dark induces the synthesis of biliproteins. There was a 4- to 5-hr delay after nitrogen addition, and synthesis typically proceeded for several hours. When the preillumination was with red light, C-phycocyanin was produced, whereas preillumination with green light caused both C-phycocyanin and C-phycoerythrin to be synthesized. The action spectra of Diakoff and Scheibe[42] were similar to those of Fujita and Hattori.[39] The maximum wavelength for stimulating C-phycoerythrin synthesis was 550 nm, whereas 660-nm light caused cessation of C-phycoerythrin synthesis. A new finding in these action spectra was the detection of action maxima at 350 to 360 nm. These UV action maxima were much smaller than those in the visible region.

A proposal was made that this chromatic adaptation effect might be controlled by a photopigment similar to phytochrome which occurs in higher plants.[42] Phytochrome, unlike this red/green receptor, was active in the red and far-red. Ohki and Fujita[43] performed experiments on *T. tenuis* designed to test whether a pool of photoreactive precursors could be the substances responding to the preillumination by red or green light. They found evidence against accumulated precursors and acknowledged the existence of a photoreversible control pigment proposed by Diakoff and Scheibe.[42]

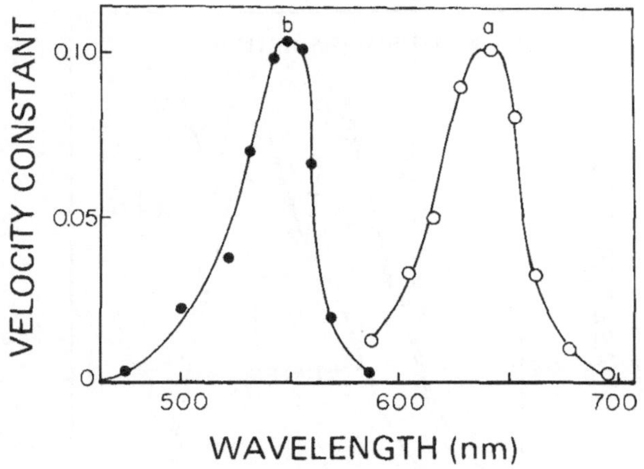

FIGURE 3. Action spectrum of chromatic adaptation in *Tolypothrix tenuis*. Curve a shows the response for the induction of C-phycocyanin, and curve b is for the induction of C-phycoerythin. The intensity of incident light is 20% higher for curve a than for curve b. (From Fujita, Y. and Hattori, A., *Plant Cell Physiol.*, 3, 209, 1962. With permission.)

Since biliproteins synthesized in the dark in *Anabaena variabilis* were not able to transfer energy to photosystem II chlorophyll *a*,[44] Ohki and Gantt[45] proposed the questions of whether the dark-synthesized biliproteins of *T. tenuis* were organized into phycobilisomes and if so whether they were attached to the thylakoid membranes. They found that structurally and functionally complete phycobilisomes were produced in the dark, and they were observed by electron microscopy to be attached to the thylakoid membranes. They noted that this would allow a minimal delay in the commencement of light harvesting for photosynthesis when light did become available.

Haury and Bogorad[46] investigated biliprotein synthesis in the cyanobacterium *F. diplosiphon,* also addressing chromatic adaptation. They employed a red or fluorescent white light system to enrich the cells in C-phycocyanin or C-phycoerythrin content, respectively. When white light-grown cells were transferred to red light growth conditions, the amount of C-phycocyanin increased and the C-phycoerythrin content remained fairly constant. When red light-grown cells which contained very little C-phycoerythrin were transferred to white light growth, C-phycocyanin synthesis continued, but the rate of C-phycoerythrin synthesis was much greater. Action spectra for complementary chromatic adaptation for these experiments on *F. diplosiphon* were quite different from those on *T. tenuis* in the blue region of the spectrum. In these experiments the formation of C-phycoerythrin was strongly stimulated by wavelengths centered at 387 nm as well as a band at 550 nm, and C-phycocyanin synthesis was promoted by blue light action maximum (463 nm) as well as a red light action maximum at 641 nm. The action spectra were quite complex, and other bands might very likely be contained in these spectra. The term ''adaptochromes'' was used for these control pigments. Haury and Bogorad[46] speculated that either two photoreceptors or a single metalloporphyrin was responsible for both bands.

Vogelmann and Scheibe[47] repeated the measurement of the action spectra for complementary chromatic adaptation in *F. diplosiphon* using the same techniques employed for *T. tenuis.* Now the synthesis of C-phycoerythrin had an action spectra much like that for *T. tenuis* and unlike that previously determined for *F. diplosiphon* by Haury and Bogorad.[46] Two possibilities have been brought forward to explain these variant results. The preillumination used by Vogelmann and Scheibe may have bleached a photoreceptor, or the Haury and Bogorad method of continuous illumination may have contributed a component related to the photosynthetic activity of the cell. In any event, Vogelmann and Scheibe[47] emphasized that their experiments suggested that both *F. diplosiphon* and *T. tenuis* were under the control of similar systems for C-phycoerythrin synthesis.

Vogelmann and Scheibe[47] and Diakoff and Scheibe[42] have noted the functional similarities between the pigment controlling C-phycoerythrin in these cyanobacteria and phytochrome of higher plants. They proposed that the photoreversible pigment can exist in two spectroscopic forms corresponding to the 650- and the 540-nm band maxima. Also, it has been reported[30,42,47] that other developmental processes of the cyanobacteria may be controlled by the same or similar pigment. This ''master'' pigment,[30] for example, would be involved in the red/green development effect for *Nostoc muscorum* A.[48] Lazaroff and Schiff[48] noted that when grown in the dark, this cyanobacterium was a mass of undifferentiated cells, but a short pulse of weak light caused the development of specialized cells and filaments. The action spectrum for the differentiation had an action maximum at 650 nm, and the action spectrum for reversing the effect had an action maximum at 530 nm. Robinson and Miller[49] have studied a red/green effect on *N. commune* having maxima at 640 and 520 nm. This effect was completely reversible, as the culture responded only to the color of the last irradiance. Red light produced 95% motile trichomes, and green light had 97% aseriate colonies and sheathed filaments. *F. diplosiphon*[30] had shorter filaments when grown in red light and less cylindrical cells when grown in white light.

Tandeau de Marsac et al.[50] have studied chromatic adaptation in the cyanobacterium *Synechocystis* sp. 6701 which unlike *F. diplosiphon* and *T. tenuis* has C-phycoerythrin but not C-phycocyanin under chromatic control. After cells were partially depleted of biliproteins by nitrate starvation in the light, nitrate was added and biliproteins were synthesized in the dark. The action spectrum for synthesis had a maximum at 540 nm; for stopping C-phycoerythrin synthesis the maximum was at 640 nm.

Bennett and Bogorad[30] have demonstrated that the C-phycoerythrin content in *F. diplosiphon* was decreased in red light. The C-phycoerythrin was not degraded, however, but it was lost due to cell lysis and growth. A decrease in cell filament length also occurred on transfer to red light, and this change was believed to involve cellular breakage. Gendel et al.[51] have used fluorescence spectroscopy and radiolabeling of polypeptides to study C-phycoerythrin synthesis during the early stages of complementary chromatic adaptation. Results indicated that chromatic adaptation involved other proteins in addition to C-phycoerythrin and that it was controlled at the level of transcription.

Bryant[52] and Bryant and Cohen-Bazire[53] have found in addition to the two usual subunits of C-phycocyanin that two additional subunits of this biliprotein may be synthesized in red light when C-phycocyanin and C-phycoerythrin were controlled by chromatic adaptation. The second pair of subunits was found in 24 out of the 31 strains of cyanobacteria having this property. This effect was not observed for cyanobacteria which did not exhibit chromatic adaptation or had only their C-phycoerythrin being controlled.[52] Purification and characterization of all four C-phycocyanin subunits from red light-grown *Pseudanabaena* 7409 showed that each is a unique gene product.[53]

The method used by the cells to accommodate the changes in C-phycoerythrin and C-phycocyanin content occurring during chromatic adaptation was to vary both the length of the phycobilisome rods and the ratio of C-phycoerythrin to C-phycocyanin in these rods (see also Chapter 5). To stabilize the varying amounts of biliproteins in these rods, different linker polypeptides were also modulated in red or green light-grown cells. In *Nostoc* sp., both C-phycoerythrin and C-phycocyanin were under chromatic control. In white light-grown cells the phycobilisomes had for each rod one C-phycocyanin hexamer and two C-phycoerythrin hexamers. However, in red light-grown cells they had two C-phycocyanin hexamers and no C-phycoerythrin was retained. The linkers at 34,000 and 32,000 mol wt in the white light phycobilisomes were replaced with a single 34,500-mol wt linker in red light-grown cells. Apparently, this latter linker was required to stabilize the connection of the two C-phycocyanin disks, while the two white light linkers were needed to stabilize the connection of two C-phycoerythrin disks and to connect the C-phycocyanin disk to the C-phycoerythrin.[54] For the cyanobacterium *Synechocystis* sp. 6701 the white light-grown cells had phycobilisomes with two C-phycocyanin and two C-phycoerythrin disks in each rod. There were linkers at 30,500, 31,500, and 33,500 mol wt to stabilize these rods. When grown in red light the C-phycocyanin content did not change, but the C-phycoerythrin content did drop sharply. The two C-phycocyanin disks and the 33,500-mol wt linker that stabilize their connection were retained in the red light phycobilisome rods, but the other two linkers declined simultaneously with the reduced C-phycoerythrin content.[55] Since additional C-phycocyanin was not added to the rods for this cyanobacterium, no new linkers were observed. The first demonstration that linkers were affected during chromatic adaptation was the gel electrophoresis results of Tandeau de Marsac and Cohen-Bazire.[56]

Neither allophycocyanins nor red algae have been mentioned in this section. Electron micrographs and energy transfer studies on chromatically adapted phycobilisomes suggested little or no changes in the allophycocyanin content of red or green light-grown cells. The available data on red algae suggested that perhaps this phylum did not behave like cyanobacteria in chromatic adaptation experiments. Further research on the red algae in this regard would be most interesting especially since the adaptation of red algal seaweeds to green light habitats in the ocean is an important research area.

Cobley and Miranda[57] have isolated UV-induced mutants of *F. diplosiphon*. These mutants varied in pigment content and exhibited various deficiencies in the ability to chromatically adapt.

IV. FAR-RED LIGHT MUTANTS

When the cyanobacterium *Anacystis nidulans* was transferred to growth in far-red light at wavelengths above 650 nm, its growth was quite poor and its chlorphyll *a* content diminished.[58] Myers et al.[59] found that a continuation of these growth conditions for extended periods eventually resulted in higher growth rates and higher levels of chlorophyll *a*. Colonies from this far-red-adapted material were obtained having three distinct colors: blue, green, and yellow. These pigment mutants had different C-phycocyanin to chlorophyll *a* ratios. The mutants all displayed better growth in far-red illumination than did the wild type.

Myers et al.[60] measured the amounts of chlorophyll *a* associated with photosystems I and II for these various far-red-induced pigment mutants. A very constant result of 118 ± 11 chlorophylls per reaction center I and 52 ± 9 chlorophylls per reaction center II was obtained for all mutants in two different growth lights. These results were obtained under experimental conditions that eliminated the influences of spillover on chlorophyll distribution.

Khanna et al.[61] have studied the photosynthetic properties of these pigment mutants in more detail. They reported that the number of phycobilisomes probably remained in proportion to the number of photosystem II reaction centers for a pigment mutant and wild-type cells. The ratios of phycobilisomes to reaction center II were calculated as follows: wild type, white light growth — 0.7; wild type, far-red light growth — 1.2; and mutant, far-red light growth — 0.8.

The total number of phycobilisomes in all cases for the wild type and far-red mutants of *A. nidulans* was 2.1 to 2.3×10^4 per cell.[61] The distribution of hemi-ellipsoidal phycobilisomes of the red alga *P. cruentum* has also been studied.[62,63] In a single cell it was estimated that there were approximately 6.6×10^5 phycobilisomes per cell.[62] There were more of the hemi-discoidal phycobilisomes per unit area, however, than hemi-ellipsoidal phycobilisomes (Table 2). The number of phycobilisomes per unit area was greater in far-red than in white light, apparently an adjustment to a harsh growth environment.

Table 2
NUMBER OF PHYCOBILISOMES PER CELL FOR *ANACYSTIS NIDULANS* AND *PORPHYRIDIUM CRUENTUM*

Organism	Growth light	Phycobilisomes per cell	Phycobilisomes per μm^2 of thylakoid[a]
Porphyridium cruentum	White	6.6×10^5	450[b]
Anacystis nidulans	White	2.2×10^4	980
A. nidulans (far-red mutant)	Far-red	2.1×10^4	1540
A. nidulans (wild type)	Far-red	2.3×10^4	1700

[a] The finding of significantly fewer phycobilisomes per unit area of thylakoid for hemi-ellipsoidal phycobilisomes is in agreement with earlier measurements on the two phycobilisome types.

[b] Dilworth and Gantt[62] reported a range of 370 to 560 phycobilisomes per square micrometer.

V. SOME EFFECTS OF HABITAT

A. Marine Ecology and Phycoerythrin

Biliprotein-containing organisms live in a plethora of habitats. They are abundant in oceans, intertidal zones, fresh water, and terrestrial regions and also flourish in waters which have extremes of pH, salinity, or temperature. Undoubtedly, the most studied habitat for these organisms is their vertical distribution in the ocean. The vast majority of red algae as well as many cryptomonads and a few cyanobacteria are marine. Certain red algae (seaweeds) exist at fairly fixed distances below the sea surface, at the bottom or in a very stratified water column, and can adapt to the light intensity and spectral compositions which are characteristic of their location. As light penetrates the surface of the water it is attenuated. In addition, the red portion of the spectra and to a lesser extent the blue are depleted much more extensively than the green wavelengths in much of the oceanic waters. This selective loss results in blue-green light penetrating deepest below the surface. In a particular water type, the following distribution was obtained for the percent (%) of incident surface light that penetrated to 10 m: 685 nm — 0; 630 nm — 3; 565 nm — 35; 535 nm — 38; 510 nm — 38; 460 nm — 28; and 435 nm — 18.[64] Of course there are great variations in marine waters, and Jerlov[65] has categorized several types which vary greatly in their tendencies to transmit light. Coastal waters do not always have the blue-green color of the other types. It is notable that phycoerythrins are spectroscopically extremely well suited to absorb light in the blue-green region. Recently, an organism has been discovered by Littler et al.[66] at a record depth of 268 m below the surface. This organism which was tentatively identified as a coralline red alga lived in light attenuated to 0.0005% of surface irradiance. In many cases in the vertical distribution of the three major types of benthic algae the red algae were found at a greater depth than the green or brown algae.

It was sometimes observed that the red algae growing beneath the ocean surface were more red in color due to increased phycoerythrin content. For example, *Chondrus crispus* was grown at 1 and at 10 m below the surface. This was accomplished by tying the red algae to a line and using a surface float and a weight to maintain vertical position. Ramus et al.[67] found that at 10 m the absorbance at 678 nm due to chlorophyll *a* doubled, while the absorbance at 565 nm which was mainly due to phycoerythrin increased more than four times the 1-m value. The red alga *Porphyra umbilicalis* showed a very similar increase in its phycoerythrin to chlorophyll *a* ratio at the greater depth. These organisms, therefore, have adapted their phycoerythrin content to utilize better the blue-green light which reached the greater depth, and increased concentrations of both pigment types were clearly demonstrated. A second type of experiment on these same red algae was to grow them near the surface in an essentially intertidal environment. Here they experienced a wide variation in exposure to the sun depending on the tide. Under these conditions, algae directly grown in the sun had a much lower content of pigment than those shaded by water part of the time. No changes in pigment ratios were reported. Rhee and Briggs[68] also examined the phycoerythrin to chlorophyll ratio of *C. crispus* and found the ratio to be quite stable down to 10 m and then to increase sharply as the depth approached 13 m. It is interesting that the ratio change was not gradual.

Results discussed above (in Sections I to III) are in accord with the concept that low light growth favors higher biliprotein and chlorophyll content and that green light may produce a higher content of phycoerythrin in certain organisms. In addition to light being able to regulate quantitatively levels of photosynthetic pigments, there is some limited evidence that light also may in select circumstances actually change the spectroscopic characteristics of a phycoerythrin. Haxo et al.[69] noted that it was a possibility that R-phycoerythrin might have a variable visible absorption spectrum depending on the growth conditions. Yu et al.[70] extended this concept by apparently showing that R-phycoerythrin from *Callithamnion roseum* can have a different phycoerythrobilin to phycourobilin ratio depending on growth light intensity. The algae grown at the lowest light intensity had the greatest proportion of phycourobilin. Phycourobilin absorbed maximally at 490 to 500 nm and was thus ideally positioned to harvest solar radiation below the sea surface. A possible advantage for marine light harvesting by phycourobilin was also suggested by the relatively recent discoveries of CU-phycoerythrins containing both phycourobilin and phycoerythrobilin in some marine cyanobacteria.[71-73] Alberte et al.[73] studied the photosynthetic characteristics of various marine cyanobacteria. In the group that was studied, the phycoerythrin-containing algae had higher photosynthetic efficiencies at low light intensity than organisms lacking phycoerythrin, and *Synechococcus* WH7803, which had a urobilin component on its phycoerythrin, was the most efficient.

How important is the presence of phycoerythrin in determining the vertical distribution of various marine organisms? First, other factors entirely unrelated to solar energy harvesting are quite vital. Since many red algae require a surface to which they attach and the composition of the attachment surface varies with depth, the viability of the algae can be affected. The distribution of nutrients can also vary with depth, and obviously such variations will affect growth and the competitive edge of various algae. In addition, many other external factors as well as the gross morphology of the organisms will cause different responses to varying habitats. For example, Ramus et al.[74] studied two green and two red algae and concluded that light alone did not determine the vertical distribution of these algae.

A quandary has existed in the literature for a long time concerning the response of marine algae to light intensity vs. the spectral composition of light. Red algae grow well in deep marine waters. Is light harvesting a critical factor? Does the response of red algae to low light intensity, to the deeper penetration of blue-green light, or to both determine the pigment response or viability of the organisms? Answers to these questions cannot yet be given in absolutes because as already described (in Sections II.B and II.C), these organisms even under laboratory-controlled conditions responded in quite individualistic ways. It is clear from laboratory and field studies that intensity effects are quite likely to be important. Many of the cyanobacteria and red algae studied increase their overall photosynthetic pigment content at low light intensities to harvest more of the less available energy. This effort and the fact that the red algae have some species living in deep waters below 200 m where light is severely attenuated expounds the idea that their pigment composition may be well suited to establish life under these extreme conditions of solar energy starvation.

Intensity effects cause increased pigment content, but pigment ratios may also be affected in marine environments with increasing depth. Are these ratios varied by light intensity, spectrum, or both? In dealing with this question a complicating factor is that all red algae and cyanobacteria will not react in the same way. Complementary chromatic adaptation, for example, has been studied extensively for a few dozen cyanobacteria. Three distinct classes of responses are cataloged, and variability within a class is possible. For red algae, variability in different species is also to be expected. It has already been mentioned (Section II.C) that the biliprotein to chlorophyll *a* ratios in different light intensities may be either higher, lower, or constant depending on the species. Goedheer[75] has provided an extremely perceptive analysis for part of this problem. He pointed out that even with the same organism, light-induced pigment variations occurred from laboratory to laboratory. Using *A. nidulans* the following results were obtained[75] at high-intensity illumination: a deficiency of CO_2 or low growth temperatures yielded low C-phycocyanin to chlorophyll *a* ratios; ample CO_2 and moderate growth temperatures resulted in little change in biliprotein to chlorophyll ratios; and high growth temperatures gave high C-phycocyanin to chlorophyll *a* ratios. Light was far from being an independent variable in these experiments. Therefore, variable results in light intensity research were probably tied to experimental protocols. In certain situations, a pigment ratio change at elevated light intensity may only reflect factors like nutrient availability. In field studies, environments of differing light intensities may well differ in nutrient content or temperature. Nutrition needs will fluctuate depending on optimal growth rates at different light intensities.

Dring[76] performed a theoretical analysis on the photosynthetic abilities of various benthic marine algae as a function of depth and water type. It was concluded from an extensive selection of data that light intensity rather than spectra determines the adaptation of pigment. It was noted that the concept of light quality determining phycoerythrin levels or the success of red algal adaptation in deep water was "the mean tricks Nature sometimes plays on us, such as dressing up one form of adaptation to look like another." Ramus and van der Meer[77] have studied a mutant of the red alga *Gracilaria tikvahiae* which was deficient in phycoerythrin content. The mutant grew as well as the wild type, and it was pointed out that this did not support the concept of phycoerythrin being important in promoting algal growth. Briefly, a few points on this topic can be offered. First, there is some evidence that the phycoerythrin to chlorophyll *a* ratio may change apparently in response to light quality for certain red algae under natural conditions. Ramus et al.[67] found that pure intensity changes produced alterations in biliprotein and chlorophyll contents but none in their ratio. However, when the same red algae were subjected to lower intensity and blue-green light, the phycoerythrin content increased sharply relative to chlorophyll *a*.[67] Ramus et al.[74] considered adaptation to both intensity and spectral composition to occur simultaneously. Second, many phycoerythrin-rich red algae are already at the absolute limitation of their potential to adapt to growth under blue-green light. Phycobilisomes have allophycocyanin cores and probably must have at least one hexamer of C- or R-phycocyanin per rod adjacent to this core. The rest of the phycobilisome can be all phycoerythrin, but a limit of about three disks is all that can be added to a single stack. Addition of more phycoerythrin disks in order to increase the ratio of phycoerythrin to other biliproteins is probably impossible for many algae. These algae then are always at their highest level of adaptation in terms of pigment content when grown in low-intensity blue-green light. Near the surface they do not change in gross pigment content, but this only indicates that they can exist in ideal habitats as well as in locations where the available light is extremely limited. A majority of white light is also harvested by the phycoerythrin of these algae. Third, as was demonstrated for the red alga *Porphyridium cruentum* by Ley and Butler,[20] the adjustments made to light quality may be more subtle than simple overall pigment ratio alterations. The pigment mutant of *G. tikvahiae* would be interesting to investigate to determine if the amount of energy sent to photosystem I from II is reduced or if more chlorophyll *a* is directed to photosystem II relative to the wild type.

In *P. cruentum* grown in white or green light the phycoerythrin harvests most of the light energy for photosystem II and even has a large amount of excitation energy for spillover to photosystem I (Table 1).

In considering the possibilities for generalizations to explain how organisms responded to the environment and the relative advantages of certain pigment characteristics, it must be kept in mind that many of these organisms have complex and highly individualistic behaviors. In addition, experiments in natural habitats are difficult to control, and experiments in the laboratory may oversimplify important natural processes. The application of our increased understanding of the complete photosynthetic apparatus and how it responds to environmental stresses seems the appropriate way to proceed.

Cox et al.[78] have studied the cyanobacterium *Gloeocapsa* strain NS4 which was isolated from inside a cave entrance. The low light intensities of growth seemed to have produced very large phycobilisomes which could be an adaptation to low light. Larkum et al.[79] found that red algae were dominant in a submarine cave due to their ability to grow at low light intensity.

Vesk and Jeffrey[80] have studied some marine organisms grown in blue-green light which simulated light available at the greater oceanic depths. The cryptomonad *Chroomonas* sp. showed increases in both its phycoerythrin 545 and chlorophyll *a* and c_2 content over white light growth. A small increase in the phycoerythrin 545 to chlorophyll *a* ratio was also observed. As an aside, most cryptomonads named *Chroomonas* sp. contain phycocyanin 645 and not phycoerythrin.

B. Thermophiles

Although phycoerythrins are commonly distributed in marine algae and provide these organisms with advantageous light-harvesting abilities needed for coping with various low light intensity and blue-green light quality habitats, there appears to be no phycoerythrin-containing organism that lives at very high temperatures. Brock[81] has studied a cyanobacterium *Synechococcus* that grows in hot springs at 73 to 75°C. Edwards and Gantt[82] have shown that this thermophilic organism had the typical hemi-discoidal phycobilisomes. Both the C-phycocyanin[83,84] and allophycocyanin[85] isolated from thermophilic cyanobacteria had higher temperatures for thermal denaturation than biliproteins from mesophilic cyanobacteria. The thermophilic cyanobacterium *Mastigocladus laminosus* had phycoerythrocyanin together with the C-phycocyanin and allophycocyanin.[86]

C. Habitats of High Salt and Dehydration

Satoh et al.[87] have studied an intertidal red alga *Porphyra perforata*. They noted that at low tide, intertidal organisms were completely exposed to air and, in addition to drying, were subjected to increases in extracellular salt as the water around them evaporated. This salinity change produced osmotic pressure differences between the organism and its environment. The alga at low tide was likewise subjected to higher light intensities than it was when submerged in water. Satoh et al.,[87] Satoh and Fork,[88,89] and Öquist and Fork[90] discovered various ways that the red alga may use to accomodate these large and regular changes in environment. High salinity was shown to decrease the amount of light energy reaching photosystem II.[87] Satoh and Fork[88,89] proposed that at higher light intensities where photosynthesis was saturated, the excess excitation energy might be converted into heat. Öquist and Fork[90] have shown that desiccation causes almost all the energy absorbed to be funneled into photosystem I. This was accomplished by increasing the percentage of transfer from photosystem II to I and by increasing the amount of light harvested by photosystem I.

Kao et al.[91] have studied a cyanobacterium *Coccochloris elabens* which grows in salt ponds at about 30% (w/v) NaCl. The C-phycocyanin of this organism was isolated and

Table 3
SPECIFIC ABSORPTIVITIES USED TO DERIVE BEER'S LAW EQUATIONS 1 to 17[a]

	Absorptivities[b]						
Wavelength (nm)	**CPE**	**CPC**	**APC**	**RPC**	**BPE**	**PEC**	**RPE**
650	0	1.8	7.3	0.93	0	0.26	0
620	0	7.0	4.0		0	1.5	
615			3.8	7.0	0		0
573		3.4	1.7			8.5	
565	12.7	2.8	1.4	4.0			8.2
545		1.5	0.72	4.2	10.0		

[a] Absorptivities are calculated for 1-mg/mℓ solutions of biliprotein, and absorption is measured in a 1-cm light path at ambient temperatures.

[b] CPE, C-phycoerythrin; CPC, C-phycocyanin; APC, allophycocyanin; RPC, R-phycocyanin; BPE, B-phycoerythrin; PEC, phycoerythrocyanin; and RPE, R-phycoerythrin.

several properties characterized. Blumwald and Tel-Or[92] have found that the cyanobacterium *Synechococcus* 6311 had an increased biliprotein content when adapted to growth at higher salt.

VI. BILIPROTEIN ASSAYS

In many of the experiments on the effects of environmental factors on these organisms, it is necessary to calculate the resulting biliprotein and chlorophyll contents. One method is to extract the cells with an organic solvent to obtain chlorophyll and break the cells open to obtain the highly water-soluble biliproteins. When an aqueous solution of the biliproteins is obtained, it is possible to calculate the concentration of each biliprotein using an appropriate expression for Beer's law. The specific absorptivities for the biliproteins are known, but there are differences in the values among the reporting groups. C-Phycocyanin is especially difficult because of its tendency to form various aggregates with different absorption maxima and absorptivities. Using discretion, certain selected values (Table 3) were used to derive equations that can be used to calculate the amounts of biliproteins from mixtures characteristic of specific organisms of the cyanobacteria and red algae:

C-Phycoerythrin, C-phycocyanin, and allophycocyanin

$$C(CPE) = 0.00251\ A_{650} - 0.0321\ A_{620} + 0.0787\ A_{565} \tag{1}$$

$$C(CPC) = -0.0911\ A_{650} + 0.166\ A_{620} \tag{2}$$

$$C(APC) = 0.159\ A_{650} - 0.0410\ A_{620} \tag{3}$$

B-Phycoerythrin, C-phycocyanin, and allophycocyanin

$$C(BPE) = 0.00219\ A_{650} - 0.0220\ A_{620} + 0.100\ A_{545} \tag{4}$$

$$C(CPC) = -0.0911\ A_{650} + 0.166\ A_{620} \tag{5}$$

$$C(APC) = 0.159\ A_{650} - 0.0410\ A_{620} \tag{6}$$

R-Phycoerythrin, R-phycocyanin, and allophycocyanin

$$C(RPE) = 0.0138\ A_{650} - 0.0715\ A_{615} + 0.122\ A_{565} \quad (7)$$

$$C(RPC) = -0.0799\ A_{650} + 0.153\ A_{615} \quad (8)$$

$$C(APC) = 0.147\ A_{650} - 0.0196\ A_{615} \quad (9)$$

Phycoerythrocyanin, C-phycocyanin, and allophycocyanin

$$C(PEC) = 0.00499\ A_{650} - 0.0638\ A_{620} + 0.129\ A_{573} \quad (10)$$

$$C(CPC) = -0.0922\ A_{650} + 0.181\ A_{620} - 0.0291\ A_{573} \quad (11)$$

$$C(APC) = 0.160\ A_{650} - 0.0423\ A_{620} + 0.00258\ A_{573} \quad (12)$$

B-Phycoerythrin, R-phycocyanin, and allophycocyanin

$$C(BPE) = 0.0230\ A_{650} - 0.0631\ A_{615} + 0.100\ A_{545} \quad (13)$$

$$C(RPC) = -0.0799\ A_{650} + 0.154\ A_{615} \quad (14)$$

$$C(APC) = 0.147\ A_{650} - 0.0196\ A_{615} \quad (15)$$

C-Phycocyanin and allophycocyanin

$$C(CPC) = -0.0911\ A_{650} + 0.166\ A_{620} \quad (16)$$

$$C(APC) = 0.159\ A_{650} - 0.0410\ A_{620} \quad (17)$$

where C is the concentration in milligrams per milliliter and A is the absorbance at the wavelength indicated in nanometers.

Jeffrey and Humphrey[93] have derived Beer's law equations for chlorophylls *a* and c_2 dissolved in 90% acetone:

$$C(Chl\,a) = 11.43\ A_{663} - 0.64\ A_{630} \quad (18)$$

$$C(Chl\,c_2) = -3.63\ A_{663} + 27.09\ A_{630} \quad (19)$$

where the specific absorptivity at the maxima for chlorophyll *a* was 87.67 and for chlorophyll c_2 was 40.4. All values are for a 1-cm light path and ambient temperature.

VII. ISOLATION OF PHOTOREVERSIBLE PIGMENTS

Since complementary chromatic adaptation and the conversion of aseriate to filamentous growth in certain cyanobacteria are likely to be under the control of a red/green photoreversible pigment, its isolation would be beneficial in the study of these processes. Phytochrome, the photoreversible pigment from higher plants which may be analogous but not identical to the reputed cyanobacterial pigment, occurred in very minute amounts. Therefore it is possible that the cyanobacterial pigment or pigments may be present in exceedingly

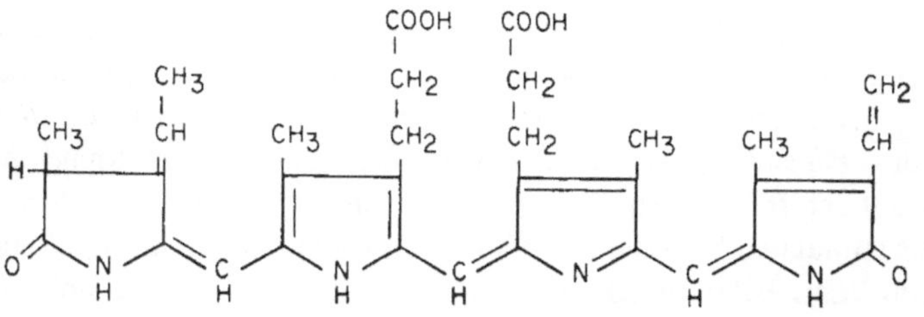

FIGURE 4. Chromophore structure of bilin of phytochrome. The structure is not meant to suggest any particular configuration.

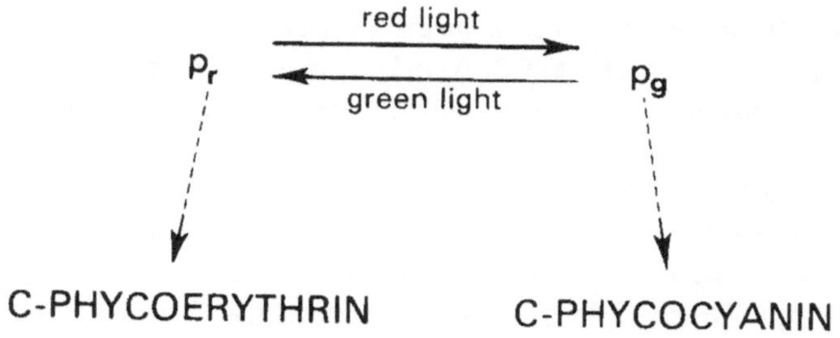

FIGURE 5. Suggested response of the proposed cyanobacterial photoreversible pigment to red and green lights. (From Scheibe, J., *Science*, 176, 1037, 1972. Copyright 1972 by the American Association for the Advancement of Science. With permission.)

small quantities. It is noteworthy that the chromophore of phytochrome was very similar in structure to phycocyanobilin (Figure 4).[94-96] The phycocyanobilin chromophore was in the Z,Z,Z configuration and can be converted to a Z,Z,E isomer.[97] Rüdiger et al.[97] discussed the possible relevance of these geometric isomers to phytochrome activity. Thümmler and Rüdiger[98] have also studied some of the properties of the Z,Z,Z and Z,Z,E configurations of the chromophores of phytochrome and C-phycocyanin. The designations E and Z refer to the configuration around the double bond connecting the individual pyrrole groups in the chromophore. Since the sought-after photoreversible pigment had activity with maxima at 650 and 540 nm, the biliproteins with C-phycoerythrin at 565 nm, C-phycocyanin at 620 nm, and allophycocyanin at 650 nm would greatly overlap and with their higher concentrations would probably tend to overwhelm the spectra of the photoreversible pigment. Also it is possible that the biliproteins are not photoreversible.

Scheibe[99] has prepared extracts from the cyanobacterium *T. tenuis* that have photoreversible behavior. This cyanobacterium had the ability to undergo complementary chromatic adaptation. In one experiment using duplicate samples, one cyanobacterium was placed in the sample compartment and the other in the reference compartment of a spectrophotometer. One was irradiated with red light and the other with green light, and the visible absorption difference spectrum was recorded. The lights were reversed, the spectra were recorded, and the spectrum obtained by subtracting the second from the first difference spectrum showed an absorption maximum at 650 nm. A modified type of experiment showed that the spectral effect can be saturated at certain light intensities. An argument against this absorption change being due to biliprotein bleaching was the photoreversibility. The repeated alteration of red and green light each reversed the effect of the other. The photoconvertible pigment had two forms which were interconverted depending on whether red or green light was used for irradiation (Figure 5). When irradiated with green light, the absorption at 650 nm increased and the absorption at 520 nm decreased. When irradiated with red light, the reverse was observed.

Björn and Björn[100,101] and Björn[102,103] have reported on a group of photoreversible pigments from cyanobacteria that they termed phycochromes. Aqueous extracts of various algae were separated on an isoelectric focusing column. Difference spectra for green minus red light-irradiated fractions were recorded. Red/green photoreversibility was found in some of the bands. Lazaroff and Schiff[48] have shown that in the cyanobacterium *N. muscorum* A, filament development was stimulated by red light (650-nm absorption maximum), and this effect was reversed by green light. Björn and Björn[101] have isolated a photoreversible pigment, phycochrome *c*, from this cyanobacterium and obtained its action spectrum for absorbance changes at 650 nm. The maximum efficiency for restoration of absorbance at 650 nm was at 580 nm, and the maximum for decreasing absorbance at 650 nm was at 630 nm. A proposal was made that these were the three spectral forms of phycochrome *c*, having maxima at 580, 630, and 650 nm. Björn[102] has isolated a different photoreversible pigment, phycochrome *d*, from the cyanobacterium *Tolypothrix distorta*. Further studies[103] have been carried out on phycochrome *b*, and it was shown to exist in two forms with maxima at 500 and 570 nm.

Ohad et al.[104,105] have also isolated a photoreversible pigment from the cyanobacteria *N. muscorum* and *F. diplosiphon*. Isoelectric focusing fractions containing a mixture of pigments were irradiated with red light and showed a small loss in 620-nm absorbance and a hyperchromicity at 560 nm. Green light irradiation produced the reversal of the spectrosocpic changes. These active fractions contained C-phycoerythrin, C-phycocyanin, and allophycocyanin. Antibodies to C-phycocyanin, C-phycoerythrin, and allophycocyanin were each adsorbed to protein A-Sepharose,® and portions of the photoreversible fractions were passed through each column. The antibodies to each biliprotein are totally specific, e.g., antibodies to purified C-phycocyanin will not react with C-phycoerythrin or allophycocyanin. Photoreversible samples passed through columns containing antibodies to C-phycocyanin or C-phycoerythrin retained their activity, but samples passed through the antiallophycocyanin column became completely inactive. It was concluded that allophycocyanin is a photoresponsive pigment.[104] Later, Ohad et al.[105] pointed out that the small absorbance changes (approximately 2% of the total absorbance) might suggest that a minor component or a special state of allophycocyanin was the active element. They studied the temperature dependence of photoreversibility and found it to be complex. It is not clear, however, whether some of these effects could be produced by perturbing the equilibrium between monomers and trimers of allophycocyanin.

Ohki and Fujita[106,107] have made a contribution to this area of research which both extended and complicated the issue of extraction of photoreversible pigments from the cyanobacteria. Their extracts from the cyanobacterium *T. tenuis* as well as the whole cells themselves did not exhibit any red/green photoreversible absorbance changes. This type of photoreversibility was induced when the extracts were treated with 0.4 *M* guanidine hydrochloride. Guanidine hydrochloride is an effective protein denaturant when used at higher concentrations. At a 0.1 *M* concentration it had the surprising effect of increasing the aggregation of C-phycocyanin, but by 0.5 *M* concentration the guanidine produced considerably less aggregation.[108] Ohki and Fujita[106] then examined purified C-phycocyanin and allophycocyanin in the absence of guanidine hydrochloride. These biliproteins lacked photoreversible behavior, but as with the crude extracts, the biliproteins were made responsive by 0.4 *M* guanidine hydrochloride. The visible absorption spectrum of allophycocyanin was altered by the guanidine hydrochloride treatment in that the 650-nm maximum was reduced and the shoulder at 620 nm remained unchanged or perhaps rose slightly. It has been shown that this type of spectral change can result from a partial dissociation of allophycocyanin trimers to monomers.[109,110] The 650-nm band was characteristic of the trimeric aggregation state.[111] The green- minus the red-induced difference spectrum of guanidine hydrochloride-treated allophycocyanin was similar to that obtained for extracts by Scheibe[99] and the phycochrome *c* of Björn and Björn.

The red/green difference spectrum for guanidine hydrochloride-treated C-phycocyanin was very different from that of allophycocyanin.

Ohki and Fujita[107] further discovered that photobleaching of *T. tenuis* cells enabled extracts of the treated cells to be photoreversible. Since Scheibe photobleached the cells used for his extracts, this photoreversibility was then understood.

de Kok et al.[112] and de Kok[113] have developed another method to produce photoreversible biliproteins. In the presence of 75% ethylene glycol, C-phycocyanin and allophycocyanin *(A. nidulans)*, when irradiated with red light, showed a decline in red absorption and a rise in green. When followed by green light irradiation this change was partially restored. As with guanidine hydrochloride, allophycocyanin in the presence of 75% ethylene glycol showed an absorption spectrum that demonstrated a conversion of trimers to monomers.

These results have left the question of the nature of the photoreversible pigment still a mystery. C-Phycocyanin and particularly allophycocyanin have, when properly induced, very similar photoreversible behavior to the pigment controlling the filament formation and chromatic adaptation in certain cyanobacteria. In normal cells there could be forces active in converting a small fraction of the biliproteins to the red/green active form. It seems more likely that biliprotein photoreversibility is coincidental and the photoreversible pigments controlling the development are unique gene products. The induction of photoreversibility in C-phycocyanin and allophycocyanin demonstrated another facet of the role of apoprotein-chromophore interaction in determining the properties of biliproteins.

VIII. NITROGEN CHLOROSIS AND HETEROCYSTS

When cyanobacteria were deprived of a usable form of nitrogen, e.g., nitrate or ammonium ion, they lost color and turned yellow. This process was named nitrogen chlorosis. Allen and Smith[114] have quantitated the loss of C-phycocyanin during this process. The earlier literature dating back to 1910 was surveyed in their report. Allen and Smith[114] showed that the C-phycocyanin content of *A. nidulans* began to decline as soon as nitrate was removed from the growth medium and was completely depleted in 15 hr. There was only a slight change in the chlorophyll *a* and carotenoid content during this period. Biliproteins were therefore the major nitrogen source of cyanobacteria during nitrogen starvation. Upon reintroduction of nitrate, C-phycocyanin reappeared and returned to a normal level in about 8 hr. Once C-phycocyanin was fully restored, cell division also could commence. Lau et al.[115] further studied the degradation of C-phycocyanin during nitrate starvation. They observed by sodium dodecyl sulfate (SDS) gel electrophoresis that both subunits of C-phycocyanin were lost during starvation. Moreover, it was found that *de novo* synthesis of C-phycocyanin was halted during starvation. Allen[116] has reviewed more recent developments in this field.

A certain group of filamentous cyanobacteria possess a very effective method for survival during nitrogen starvation. These filaments developed specialized cells, heterocysts, during periods of combined nitrogen deprivation. These heterocysts occurred at very regular intervals along the filament. Haselkorn[117] and Neuer et al.[118] have reviewed various properties of these morphologically distinct heterocysts. Haselkorn[117] defined heterocysts as anaerobic factories for nitrogen fixation when the atmosphere was aerobic. The heterocysts must be highly oxygen free because the enzymes required to convert N_2 to a usable oxidation state were strongly poisoned by O_2. Photosystem II in the vegetative cells resulted in O_2 evolution, but photosystem II activity was absent in heterocysts although photosystem I was maintained. Haselkorn noted that reports on the content of biliproteins in these heterocysts were inconsistent, ranging from no residual C-phycocyanin to readily measurable amounts. Once a heterocyst was formed, it converted atmospheric nitrogen to a usable form and then was able to transfer this nitrogen along the filament from cell to cell, resulting in the entire filament overcoming the chlorotic state. Normal color and function was regained in the vegetative cells, and nitrogen fixation continued in the heterocysts.

Peterson et al.,[119] noting several earlier reports that found biliproteins in heterocysts, also found C-phycocyanin and what appeared to be allophycocyanin in heterocysts of *Anabaena variabilis*. They reported significantly less biliprotein content than in vegetative cells and confirmed previous findings of no photosystem II activity. Since biliproteins in vegetative cells were closely coupled to photosystem II, these findings raised questions concerning their function in the heterocysts. Peterson et al.[119] found two answers: they stimulated photosystem I activity and promoted the conversion of acetylene to ethylene.

Yamanaka and Glazer[120] have studied the biliproteins in heterocysts of *Anabaena* 7119. One question concerned the origin of heterocyst biliprotein. Was it a remnant of the vegetative state or does biliprotein synthesis occur in conjunction with heterocyst formation? It was determined that the biliproteins were those remaining from the vegetative state. The biliproteins, notably C-phycocyanins, were found as 9, 13, and 16S aggregates in the heterocysts. Allophycocyanin was absent. The lack of allophycocyanin was surprising, since the heterocyst-biliprotein aggregates showed fluorescence emission at 670 nm. If this low-energy emission was from C-phycocyanin, its properties were changed in some way in the heterocysts to compensate for the loss of allophycocyanin. It appeared that the phycobilisome cores were degraded during the establishment of the heterocysts, but portions of the rods were retained and altered.

Yamanaka and Glazer[120] speculated on the genesis of the conflicting reports on biliprotein content in cyanobacterial heterocysts. When heterocysts were first formed, the precursor cells were in a very chlorotic state. The first heterocysts, since they do not synthesize biliproteins, would have to be correspondingly very low in biliprotein content. Subsequently, the heterocysts rectified the nitrogen deficiency and normal pigmentation was restored. When new heterocysts formed from the vegetative cells, they were produced from cells having biliproteins and thus so did the resulting heterocysts. Therefore, depending on when the observations on the biliprotein content in heterocysts were made, wide variability would be found in the measurements.

Specific enzymes are probably needed for both the degradation of biliproteins during nitrogen chlorosis and for the adaptation of the phycobilisomes to the allophycocyanin-free state needed in second-generation heterocysts. Foulds and Carr[121] have isolated one such enzyme from the cyanobacterium *A. cylindrica*. This proteolytic enzyme degraded both C-phycocyanin purified from the same source and C-phycocyanin from *Anacystis nidulans*. During the period of heterocyst development in *Anabaena cylindrica* the activity of this enzyme increased substantially. Wood and Haselkorn[122] have found a similar activity in heterocyst-forming cells and studied it in some detail.

IX. BILIN SYNTHESIS

A. *Cyanidium caldarium*

Frequently, a single organism is responsible for the bulk of the progress in a general field of inquiry. The initial research, for example, on the structure and function of phycobilisomes relied on the particularly advantageous properties of these organelles from the red alga *Porphyridium cruentum*. Most of the information on the biosynthesis of phycocyanobilin came from research on the red alga *C. caldarium*.

Allen[123] has presented some of the general characteristics of the organism. It was a dominant species in many acidic hot springs, and one of the strains used in the study was isolated from water corresponding to about 0.1 *N* sulfuric acid at temperatures in a range of 70 to 75°C. It grew well under laboratory conditions at pH values near 2 and ambient temperatures. Its photosynthetic pigments included C-phycocyanin, allophycocyanin, and chlorophyll *a*.[123,124] Allen[123] noted that cultures grown in the dark in a sugar-containing medium had a pale yellow color and were devoid of chlorophyll *a* and biliproteins. In the

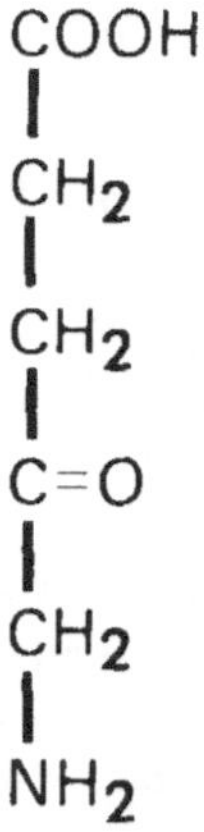

FIGURE 6. Structure of 5-aminolevulinic acid.

light the pigments were formed. They were light-intensity sensitive, and very high light intensities greatly depleted the cells of their photosynthetic pigments. Allen[123] concluded correctly that the growth characteristics of the organism would allow its use in studying bilin formation.

Nichols and Bogorad[124] have prepared some UV light-induced pigment mutants of *C. caldarium:* III-C contained no C-phycocyanin or allophycocyanin in the light but did have chlorophyll *a*, GGB had biliproteins when grown in the light but no chlorophyll *a*, and GGB-Y lacked the ability to form biliproteins and chlorophyll *a* when grown in the light. Using the GGB mutant, a preliminary action spectrum for biliprotein synthesis showed that a heme protein could be a precursor in the formation of biliproteins. Nichols and Bogorad[125] studied the action spectrum of biliprotein formation in greater detail. Blue light (420 nm) was most effective in biliprotein formation, but the results were greatly dependent on the age of the cultures. Light in the range of 550 to 650 nm for older cells was also found to be effective. The heme precursor appeared to be most active in younger cells, and a possible precursor absorbing in the range of 550 to 650 nm was likely to be involved more in older cells.

Schneider and Bogorad[126] have studied the action spectra for biliprotein and chlorophyll *a* synthesis in more detail using the III-C and GGB mutants. Data from these mutants suggested that different photocontrol mechanisms exist for chlorophyll *a* and biliprotein. In another mutant (III-D-2) which synthesized both types of pigment, the action spectra for both were not readily separable. Schneider and Bogorad[126] suggested that the synthesis of the two pigments might be coordinated through certain common intermediates.

Another important tool that will be very useful in the study of bilin synthesis is the disovery by Troxler and Bogorad[127] that when 5-aminolevulinic acid (Figure 6), a porphyrin precursor, was included in the growth medium, porphobilinogen, porphyrins, and a blue bilin were excreted by those cells into the growth medium. Studies have been made on the structure of *C. caldarium* and its mutants[128,129] and the various immunochemical,[130] physical, and biochemical properties of its biliproteins.[131-133]

B. Excreted Bilin

The 5-aminolevulinic acid-induced excretion of a blue-colored tetrapyrrole[127] was a clearly important lead in the study of the biliprotein chromophores. Troxler and Lester[134] used radiolabeled 5-aminolevulinic acid to follow its incorporation into phycocyanobilin. They

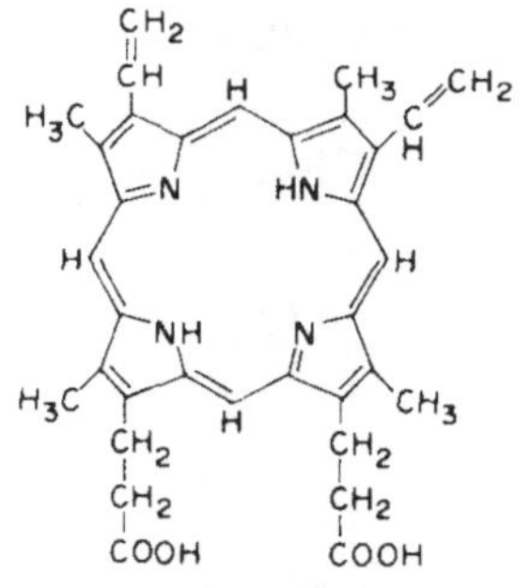

FIGURE 7. Structure of protoporphyrin IX.

also showed that the excreted blue pigment had an absorption spectrum identical to that obtained from the bilin cleaved from C-phycocyanin in boiling methanol. Troxler and Brown[135] demonstrated that for various other biliprotein-containing organisms 5-aminolevulinic acid was primarily incorporated into amino acids and not the tetrapyrroles. This underlines the unique utility of *C. caldarium* in bilin biosynthesis. Troxler et al.[136] studied the structure of the excreted bilin in detail using several experimental protocols. It was clearly demonstrated to be identical to phycocyanobilin. This was a unique occurrence of phycocyanobilin biosynthesis in the absence of concurrent apoprotein synthesis.

C. A Heme Precursor

An extensive body of information has accumulated that strongly suggests that heme was a direct precursor in the synthesis of phycocyanobilin. The cornerstone of this proof was the work by Brown et al.[137] in which radiolabeled heme was incorporated directly into the phycocyanobilin of *C. caldarium*. The first problem in conducting these experiments was that heme was insoluble in pH 2, the normal pH for healthy algal growth. Cells that were grown in the dark were pelleted, washed, and resuspended at pH 8. Under these conditions heme was rapidly taken up (30 min), and at this time cells were again pelleted, washed, and resuspended in the proper acid growth medium. At this point, two experiments were performed: one in the dark and the other in light. The dark experiments consisted of placing the heme-containing cells into medium including 5-aminolevulinic acid and glucose. After 90 hr in the dark at 37°C the medium became very dark in color from the excreted phycocyanobilin and porphyrins. The bilin was extracted and purified. When radiolabeled heme was used, a significant incorporation of radioactivity was found in phycocyanobilin. The second experiment consisted of the heme-containing cell being transferred to the light at 37°C in medium which lacked glucose. After 64 hr, cells were pelleted, washed, and broken in a French® pressure cell. The biliproteins were then refluxed in methanol for 16 hr to cleave phycocyanobilin from the apoproteins. Chlorophyll *a* was prepared from these cells by an acetone-water extraction. When radiolabeled heme was used, the radiolabel was incorporated into phycocyanobilin. No radiolabel was found in chlorophyll *a*. It was conceivable that heme might lose its iron atom and that the resulting protoporphyrin IX (Figure 7) would then act as a precursor for phycocyanobilin. This pathway was rejected, since chlorophyll *a* would also be synthesized from the radiolabeled protoporphyrin IX, but as was previously noted, no radiolabeled chlorophyll *a* was observed. Heme was therefore a direct precursor of phycocyanobilin.

Prior to the heme-uptake experiments, other experimental results strongly supported a precursor role for heme. Troxler[138] used the chlorophyll-free GGB mutant of *C. caldarium* to measure the production of CO and phycocyanobilin after administration of radiolabeled

5-aminolevulinic acid. It should be noted that CO is also a by-product of bile pigment synthesis from heme in mammals. When a single carbon was radiolabeled in 5-aminolevulinic acid-5-^{14}C, the heme synthesized from it had eight radiolabeled carbons. If phycocyanobilin was subsequently derived form the radiolabeled heme, one of the radioactive carbons would be given off as CO and the remaining seven would be retained on the bilin; there would be equimolar concentrations of both products. In the GGB example, equimolar amounts of CO and bilin were formed, and in addition, as was predicted for a porphyrin precursor, the specific radioactivity of the bilin was almost precisely seven times that found in CO. Very comparable results were also found for wild-type cells, but GGB-Y and III-C mutants which cannot make phycocyanobilin did not generate CO. Troxler and Dokos[139] demonstrated that in five cyanobacteria and one other red alga that CO and bilin were also synthesized in equimolar amounts. These results were consistent with a porphyrin precursor for the bilins. Troxler et al.[140] treated cells of *C. caldarium* with $^{18}O_2$. The method of incorporation of ^{18}O into the bilin indicated that the bilin was synthesized by a mechanism identical to bile pigment synthesis from mammalian heme. Brown et al.[141] essentially repeated the ^{18}O incorporation experiments mentioned above under more controlled conditions. The mechanism of phycocyanobilin synthesis was shown to be similar to that found for bilirubin synthesis from heme.

Subsequent to the heme-uptake experiments, Beale and Chen[142] and Brown et al.[143] studied the effects of compounds that inhibited the insertion of iron into protoporphyrin IX. The effect of this inhibition was to stop phycocyanin but not chlorophyll *a* synthesis in the dark. This again endorsed heme as a percursor for phycocyanobilin synthesis in *C. caldarium*. A rather surprising result was obtained in the light; Brown et al.[143] reported that under these more natural growth conditions, both chlorophyll *a* and phycocyanobilin were inhibited by *N*-methylprotoporphyrin IX, a specific inhibitor of ferrochelatase activity. The synthesis of chlorophyll *a* utilized a magnesium insertion which should not be affected. The inhibition of both pigments was entirely parallel, as if the pigments were being affected in a concerted manner. Brown et al.[143] postulated that perhaps C-phycocyanin and chlorophyll *a* syntheses were coordinated so that a specific inhibition of only one effectively halted synthesis of the other.

D. A Biliverdin Precursor

In mammals, heme is synthesized from protoporphyrin IX and degraded to biliverdin and then bilirubin for excretion. The structures of phycocyanobilin and biliverdin are similar, indicating that heme undergoes ring opening in the identical position during synthesis of both tetrapyrroles. Phycocyanobilin synthesis would require reduction of two double bonds and an isomerization in addition to heme ring opening. A few different pathways were available starting from heme and ending with phycocyanobilin. A series of results have very strongly demonstrated that ring opening of protoheme occurred before the isomerization or reduction steps and that therefore biliverdin IX was a direct precursor of phycocyanobilin. As might be anticipated, all these studied utilized the unicellular red alga *C. caldarium* for their proofs.[144-149]

Koest and Benedikt[144] found that biliverdin IX and phycocyanobilins were both excreted into the medium when the cells were incubated with 5-aminolevulinic acid. Beale and Cornejo[145] cultured cells with radiolabeled biliverdin and an inhibitor of protoheme formation. After a period in the dark, cells were harvested and protoheme and phycocyanobilin (after cleavage from protein) were obtained and purified. Very little radiolabel was found in the protoheme, and large amounts were found incorporated into phycocyanobilin. They concluded that biliverdin was a direct precursor of phycocyanobilin. The lack of label in protoheme excluded the breakdown of biliverdin-forming products which might then form radiolabeled protoheme. Brown et al.[146] also administered radiolabeled biliverdin and found

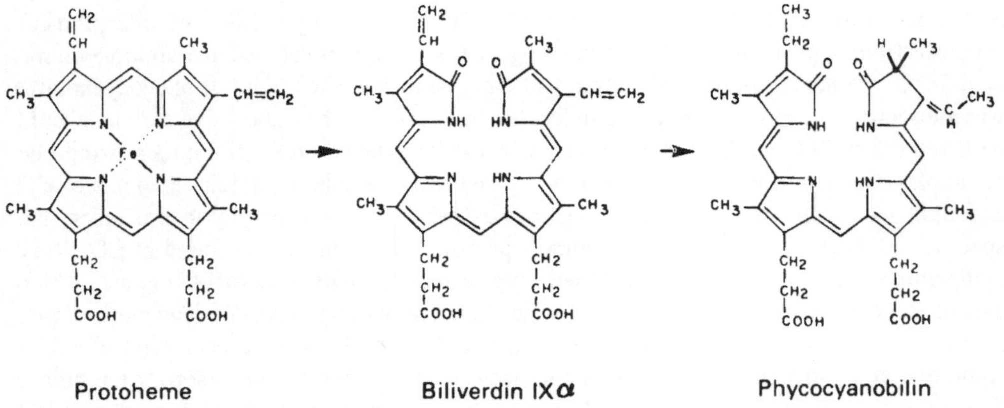

FIGURE 8. Structures of heme, biliverdin, and phycocyanobilin. The phycocyanobilin is in the Z,Z,Z configuration.

incorporation into phycocyanobilin when cells were grown in the light. Brown and Holroyd[147] determined that mesoheme and mesobiliverdin were not precursors, which again pointed toward biliverdin as a direct precursor of phycocyanobilin.

The conversion of biliverdin to phycocyanobilin required the operation of specific enzymes. Beale and Cornejo[148] incubated biliverdin with cell-free extracts and found that biliverdin was transformed into phycocyanobilin. Evidence was presented that the transformation was enzymatic. Beale and Cornejo[149] found that the reaction required O_2 and a reduced pyridine nucleotide and concluded that the enzyme was a mixed-function oxidase. A summary of the basic events (Figures 8 and 9) following additions of 5-aminolevulinic acid illustrates that the details of the last two reductions are still not characterized although substantial progress has clearly been achieved.

E. Other Approaches

Using intact cells, Csatorday et al.[150] studied the fluorescence properties of the chlorophyll-free mutant GGB of *C. caldarium*. In the dark, cells accumulated protoporphyrin IX and zinc protoporphyrin IX. In the light, both porphyrins disappeared and biliproteins were produced. Lewis et al.[151] showed that chlorophyll *a* and phycocyanobilin in the cyanobacterium *Anacystis nidulans* were formed by a common pathway that diverged after protoporphyrin IX. Cosner and Troxler[152] prepared protoplasts of *A. nidulans* by lysozyme digestion of the cell wall. The resulting protoplasts had greater permeability and were treated with radiolabeled 5-aminolevulinic acid. Imides obtained by chromic acid degradation of phycocyanobilin were shown to contain the radiolabel.

X. *APLYSIA*

Studies of protein-free bilins were sharply limited to phycocyanobilin and to a lesser extent phycoerythrobilin. A natural system that collected these bilins in a protein-free form is the sea hare *Aplysia*. This marine organism eats red algae, cleaves their chromophores, and stores them in an ink gland. Rüdiger[153] has isolated one component from this ink and has proven that it has the structure of phycoerythrobilin monomethylester. Results from our laboratory have shown that phycourobilin may also be an ingredient of the ink (Figures 10 and 11); these results confirmed some earlier reports. A red-colored component was partially purified and shown to have an absorption maximum at 490 nm and a fluorescence emission maximum at 510 nm. The phycourobilin-designate would probably be derived from the R-phycoerythrin of the red algae ingested by the *Aplysia*.

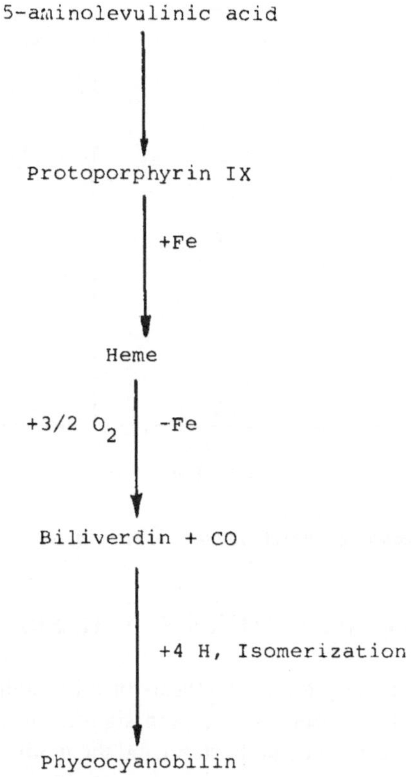

FIGURE 9. Proposed route of biosynthesis of phycocyanobilin from 5-aminolevulinic acid.

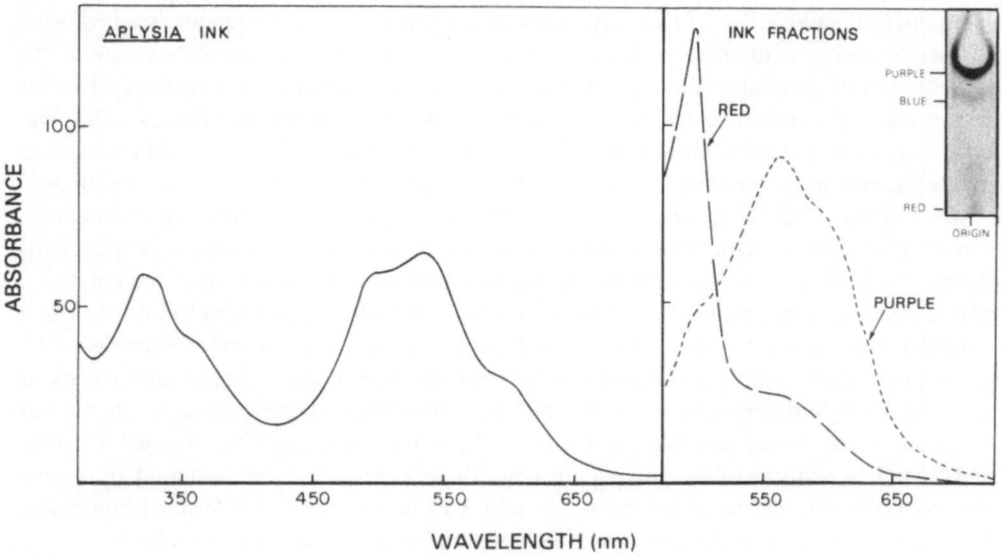

FIGURE 10. Absorption spectra of *Aplysia* ink and two components. The ink was fractionated by thin-layer chromatography. The red fraction has absorption characteristics resembling a urobilin. A typical chromatography experiment is shown in the inset. The solvent for chromatography was 50% benzene, 35% methanol, and 15% ethyl acetate. The plates were precoated with silica gels, and the components were suspended in pH 6.0, 0.1-ionic strength sodium phosphate buffer after chromatography.

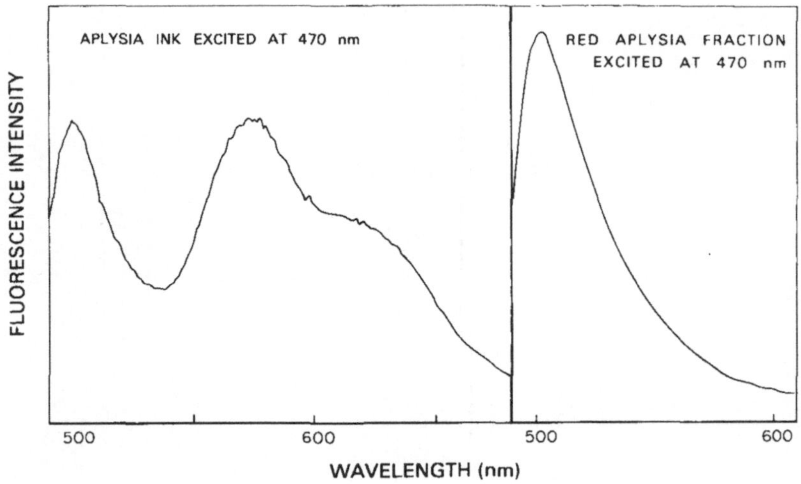

FIGURE 11. Fluorescence spectra of *Aplysia* ink and the red fraction from thin-layer chromatography.

XI. BILIPROTEIN SYNTHESIS

A principal vehicle used to study the biosynthesis of biliproteins has been the effects of light. Two types of systems have been used to gain significant data. Red/green light photoreversibility (Section III) of certain cyanobacteria and the unique properties of the red alga *C. caldarium* (Section IX.A) have generated insights into the action spectra for synthesis and have suggested the possibility of a photoreversible control pigment for the cyanobacteria.

Belford et al.[154] have performed the first cell-free translation of messenger RNAs yielding biliprotein subunit polypeptides. RNA was isolated from disrupted cells of *C. caldarium* and the poly(A)-rich and poly(A)-deficient RNA fractions were prepared by chromatography on oligo(dT)-cellulose. Reticulocyte lysates were prepared, and translation was studied using radioactive amino acid incorporation. Both poly(A^+) and poly(A^-)-mRNAs caused the incorporation of radiolabeled methionine into protein. SDS gel electrophoresis experiments did not show prominent bands in the region expected for biliprotein subunits. This was unusual, since C-phycocyanin was the major protein of the alga. To purify and concentrate any biliproteins in the mixture, the in vitro translation products were precipitated with anti-C-phycocyanin or antiallophycocyanin antisera. In poly(A^+)-mRNA-directed translations, immune precipitation with these antiserums did not reveal the characteristic bands of the biliprotein subunit polypeptides. Anti-C-phycocyanin immune precipitates from the poly(A^-)-mRNA-directed translations showed bands on gels at 15,900 and 19,700 mol wt. These molecular weights were within experimental error of those of the α and β subunits of C-phycocyanin. Antiallophycocyanin immunoprecipitates from poly(A^-)-directed translations contained a single major band of 17,500 mol wt. The second allophycocyanin subunit was not detected, and it was possible that both α and β subunit polypeptides migrated as single bands. Limited analysis of the polypeptides by Edman degradation demonstrated agreement between the in vitro products and the amino acid sequences of the *C. caldarium* biliproteins. These in vitro biliprotein subunits were the apoproteins and lacked tetrapyrroles.

Shortly thereafter, Steinmüller et al.[155] also translated the subunits of C-phycocyanin and allophycocyanin from poly(A^-)-mRNA. They used two red algae in these experiments, *C. caldarium* and *P. aerugineum*. For both algae, the translation products from poly(A^+)-mRNA did not form immunoprecipitates of C-phycocyanin or allophycocyanin. Steinmüller et al.[156] observed that light promoted biliprotein synthesis, and glucose had the reverse property. Light and glucose were shown to function primarily by altering the levels of translatable mRNA.

Egelhoff and Grossman[157] examined the effects of chloramphenicol and cycloheximide on the incorporation of radiolabeled amino acid into phycobilisome polypeptides. Cycloheximide blocked incorporation of amino acid into some of the nonpigmented phycobilisome proteins, whereas chloramphenicol blocked incorporation into biliprotein subunit polypeptides. This provided indirect evidence that phycocyanin subunit polypeptides were synthesized on 70S ribosomes in chloroplasts, and nonpigmented polypeptides were synthesized on 80S ribosomes in the cytoplasm. These results suggested but did not prove that biliproteins were synthesized in the chloroplast. Similar results were found for *C. caldarium, P. aerugineum,* and *P. cruentum,* indicating that this was a general phenomenon.

Under normal circumstances the synthesis of bilin and apoprotein seemed to be highly coordinated events. Schuster et al.[158] inhibited tetrapyrrole synthesis by administering levulinic acid to *C. caldarium.* This chemical treatment also depressed the rate of apoprotein synthesis, but some bilin-free apoprotein was obtained and purified from this process.

Lemaux and Grossman[159] isolated and studied a gene for the β subunit of C-phycocyanin from the endosymbiont of *Cyanophora paradoxa.* A DNA probe for a chromophore-containing region was obtained. They suggested a possible utility for the probe in identifying other genes for the biliprotein subunits. Subsequently, the genes for the C-phycocyanin α subunit and the α and β subunits of allophycocyanin were identified for this organism.[160] Conley et al.[161] have studied red light vs. green light transcripts of the C-phycocyanin genes from the cyanobacterium *F. diplosiphon.*

The genes for the two subunits of C-phycocyanin from the cyanobacterium *Agmenellum quadruplicatum* PR-6 have been isolated.[162,163] A set of oligonucleotide probes were prepared by de Lorimier et al.[162] and Pilot and Fox[163] based on the knowledge of the partial sequence of the amino acids of the C-phycocyanin subunits. The various probes were hybridized against total RNA, and the one that hybridized best was selected for use as a hybridization probe.[162] The sequence homologous to the probe was cloned, and ultimately, genes corresponding to both C-phycocyanin subunits were obtained by both groups. Although the amino acid sequencing had not been completed for C-phycocyanin from this cyanobacterium, its partial amino acid sequence was known and results from the two completely sequenced C-phycocyanins were available for comparison.[164-166] Cloned DNA sequences showed good homology to the sequences of the proteins, leaving little doubt that the cloned DNA contained the genes of the C-phycocyanin subunits. The genes for the two subunits were in the same orientation and were separated by a noncoding sequence of 105 or 108 base pairs.[162,163] Both the studies by de Lorimier et al.[162] and those by Belford et al.[154] indicated that posttranslational cleavage at the N terminus of C-phycocyanin subunit polypeptides did not occur because the DNA sequences did not contain a signal peptide and the N terminal amino acid of both subunits was methionine.

The genes for the allophycocyanin subunit polypeptides have also been isolated.[167] Bryant et al.[167] have performed nucleotide sequencing on genes from the chloroplast-like cyanelles of *Cyanophora paradoxa.* The deduced amino acid sequences of these subunits had 81 to 85% homology to sequences of heterologous allophycocyanins obtained by amino acid sequencing. The amino acid sequence of allophycocyanin from *Cyanophora paradoxa* has not yet been obtained. A 40% homology was found between the deduced amino acid sequences of the α and β subunits.

REFERENCES

1. **Myers, J. and Kratz, W. A.,** Relations between pigment content and photosynthetic characteristics in a blue-green alga, *J. Gen. Physiol.*, 39, 11, 1956.
2. **Gaidukov, N.,** Über den Einfluss farbigen Lichtes auf die Färbung lebender Oscillarien, *Abh. Preuss. Akad. Wiss.*, 5, 1, 1902.
3. **Gaidukov, N.,** Weitere Untersuchungen über den Einfluss farbigen Lichtes auf die Färbung der Oscillarien, *Ber. Dtsch. Bot. Ges.*, 21, 484, 1903.
4. **Boresch, K.,** Über die Einwirkung farbigen Lichtes auf die Färbung von Cyanophyceen, *Ber. Dtsch. Bot. Ges.*, 37, 25, 1919.
5. **Boresch, K.,** Die komplementäre chromatische adaptation, *Arch. Protistenkd.*, 44, 1, 1921.
6. **Kylin, H.,** Über die Farbe der Florideen und Cyanophyceen, *Sven. Bot. Tidskr.*, 6, 531, 1912.
7. **Engelmann, T. W.,** Farbe und Assimilation, *Bot. Ztg.*, 41, 1, 1883.
8. **Engelmann, T. W.,** Untersuchungen über die quantitativen Beziehungen zwischen Absorption des Lichtes und Assimilation in Pflanzenzellen, *Bot. Ztg.*, 42, 82, 1884.
9. **Oltmanns, F.,** Über die Cultur- und Lebensbedingungen der Meeresalgen, *Jahr. Wiss. Bot.*, 23, 349, 1891.
10. **Kylin, H.,** Einige Bemerkungen über Phykoerythrin und Phykocyan, *Hoppe-Seyler's Z. Physiol. Chem.*, 197, 1, 1931.
11. **Harder, R.,** Über die Bedeutung von Lichtintensität und Wellenlänge für die Assimilation farbiger Algen, *Z. Bot.*, 15, 305, 1923.
12. **Lubimenko, V.,** Sur la quantité de la chlorophylle chez les algues marines, *C. R.*, 179, 1073, 1924.
13. **Halldal, P.,** Pigment formation and growth in blue-green algae in crossed gradients of light intensity and temperature, *Physiol. Plant.*, 11, 401, 1958.
14. **Zhevner, V. D., Gusev, M. V., and Shestakov, S. V.,** Change in composition and content of pigments of the blue-green algae as a function of the spectral composition of light and degree of illumination, *Mikrobiologiya*, 34, 209, 1965.
15. **Ghosh, A. K. and Govindjee,** Transfer of the excitation energy in *Anacystis nidulans* grown to obtain different pigment ratios, *Biophys. J.*, 6, 611, 1966.
16. **Raps, S., Wyman, K., Siegelman, H. W., and Falkowski, P. G.,** Adaptation of the cyanobacterium *Microcystis aeruginosa* to light intensity, *Plant Physiol.*, 72, 829, 1983.
17. **Koch, W.,** Untersuchungen an bakterienfreien Massenkulturen der einzelligen Rotalge *Porphyridium cruentum* Naegeli, *Arch. Mikrobiol.*, 18, 232, 1953.
18. **Brody, M. and Emerson, R.,** The effect of wavelength and intensity of light on the proportion of pigments in *Porphyridium cruentum*, *Am. J. Bot.*, 46, 433, 1959.
19. **Guérin-Dumartrait, E., Hoarau, J., Leclerc, J.-C., and Sarda, C.,** Effets de quelques conditions d'eclairement, notamment de la lumiere rouge, sur la composition pigmentaire et la structure de *Porphyridium* sp. (Lewin), *Phycologia*, 12, 119, 1973.
20. **Ley, A. C. and Butler, W. L.,** Effects of chromatic adaptation on the photochemical apparatus of photosynthesis in *Porphyridium cruentum*, *Plant Physiol.*, 65, 714, 1980.
21. **Faust, M. A. and Gantt, E.,** Effect of light intensity and glycerol on the growth, pigment composition, and ultrastructure of *Chroomonas* sp., *J. Phycol.*, 9, 489, 1973.
22. **Thinh, L.-V.,** Effect of irradiance on the physiology and ultrastructure of the marine cryptomonad, *Cryptomonas* strain Lis (Cryptophyceae), *Phycologia*, 22, 7, 1983.
23. **Öquist, G.,** Light-induced changes in pigment composition of photosynthetic lamellae and cell-free extracts obtained from the blue-green alga *Anacystis nidulans*, *Physiol. Plant.*, 30, 45, 1974.
24. **Vierling, E. and Alberte, R. S.,** Functional organization and plasticity of the photosynthetic unit of the cyanobacterium *Anacystis nidulans*, *Physiol. Plant.*, 50, 93, 1980.
25. **Brown, T. E. and Richardson, F. L.,** The effect of growth environment on the physiology of algae: light intensity, *J. Phycol.*, 4, 38, 1968.
26. **Van Eykelenburg, C.,** The ultrastructure of *Spirulina platensis* in relation to temperature and light intensity, *Antonie van Leeuwenhoek; J. Microbiol. Serol.*, 45, 369, 1979.
27. **Waaland, J. R., Waaland, S. D., and Bates, G.,** Chloroplast structure and pigment composition in the red alga *Griffithsia pacifica:* regulation by light intensity, *J. Phycol.*, 10, 193, 1974.
28. **Döhler, G.,** Photosynthetische carboxylierungsreaktionen verschieden pigmentierter *Anacystis* — Zellen, *Planta*, 131, 129, 1976.
29. **Jones, L. W. and Myers, J.,** Pigment variations in *Anacystis nidulans* induced by light of selected wavelengths, *J. Phycol.*, 1, 7, 1965.
30. **Bennett, A. and Bogorad, L.,** Complementary chromatic adaptation in a filamentous blue-green alga, *J. Cell Biol.*, 58, 419, 1973.
31. **Brody, M. and Emerson, R.,** The quantum yield of photosynthesis in *Porphyridium cruentum*, and the role of chlorophyll *a* in the photosynthesis of red algae, *J. Gen. Physiol.*, 43, 251, 1959.

32. **Brody, M. and Brody, S. S.,** Induced changes in the photosynthetic efficiency of *Porphyridium cruentum*. II, *Arch. Biochem. Biophys.*, 96, 354, 1962.
33. **Brody, S. S. and Brody, M.,** Induced changes in the efficiency of energy transfer in *Porphyridium cruentum*. I, *Arch. Biochem. Biophys.*, 82, 161, 1959.
34. **Yocum, C. S. and Blinks, L. R.,** Light-induced efficiency and pigment alterations in red algae, *J. Gen. Physiol.*, 41, 1113, 1958.
35. **Tandeau de Marsac, N.,** Occurrence and nature of chromatic adaptation in cyanobacteria, *J. Bacteriol.*, 130, 82, 1977.
36. **Hattori, A. and Fujita, Y.,** Formation of phycobilin pigments in a blue-green alga, *Tolypothrix tenuis*, as induced by illumination with colored lights, *J. Biochem.*, 46, 521, 1959.
37. **Hattori, A. and Fujita, Y.,** Effect of pre-illumination on the formation of phycobilin pigments in a blue-green alga, *Tolypothrix tenuis*, *J. Biochem.*, 46, 1259, 1959.
38. **Fujita, Y. and Hattori, A.,** Effect of chromatic lights on phycobilin formation in a blue-green alga, *Tolypothrix tenuis*, *Plant Cell Physiol.*, 1, 293, 1960.
39. **Fujita, Y. and Hattori, A.,** Photochemical interconversion between precursors of phycobilin chromoproteids in *Tolypothrix tenuis*, *Plant Cell Physiol.*, 3, 209, 1962.
40. **Fujita, Y. and Hattori, A.,** Formation of phycoerythrin in pre-illuminated cells of *Tolypothrix tenuis* with special reference to nitrogen metabolism, *Plant Cell Physiol.*, 1, 281, 1960.
41. **Fujita, Y. and Hattori, A.,** Changes in composition of cellular material during formation of phycobilin chromoproteids in a blue-green alga, *Tolypothrix tenuis*, *J. Biochem.*, 52, 38, 1962.
42. **Diakoff, S. and Scheibe, J.,** Action spectra for chromatic adaptation in *Tolypothrix tenuis*, *Plant Physiol.*, 51, 382, 1973.
43. **Ohki, K. and Fujita, Y.,** Photocontrol of phycoerythrin formation in the blue-green alga *Tolypothrix tenuis* growing in the dark, *Plant Cell Physiol.*, 19, 7, 1978.
44. **Ohki, K. and Katoh, T.,** Incompetence of dark-synthesized phycocyanin in excitation transfer to photosystem II chlorophyll, *Planta*, 129, 249, 1976.
45. **Ohki, K., and Gantt, E.,** Functional phycobilisomes from *Tolypothrix tenuis* (Cyanophyta) grown heterotrophically in the dark, *J. Phycol.*, 19, 359, 1983.
46. **Haury, J. F. and Bogorad, L.,** Action spectra for phycobiliprotein synthesis in a chromatically adapting cyanophyte, *Fremyella diplosiphon*, *Plant Physiol.*, 60, 835, 1977.
47. **Vogelmann, T. C. and Scheibe, J.,** Action spectra for chromatic adaptation in the blue-green alga *Fremyella diplosiphon*, *Planta*, 143, 233, 1978.
48. **Lazaroff, N. and Schiff, J.,** Action spectrum for developmental photo-induction of the blue-green alga *Nostoc muscorum*, *Science*, 137, 603, 1962.
49. **Robinson, B. L. and Miller, J. H.,** Photomorphogenesis in the blue-green alga *Nostoc commune* 584, *Physiol. Plant.*, 23, 461, 1970.
50. **Tandeau de Marsac, N., Castets, A.-M., and Cohen-Bazire, G.,** Wavelength modulation of phycoerythrin synthesis in *Synechocystis* sp. 6710, *J. Bacteriol.*, 142, 310, 1980.
51. **Gendel, S., Ohad, I., and Bogorad, L.,** Control of phycoerythrin synthesis during chromatic adaptation, *Plant Physiol.*, 64, 786, 1979.
52. **Bryant, D. A.,** The photoregulated expression of multiple phycocyanin species. A general mechanism for the control of phycocyanin synthesis in chromatically adapting cyanobacteria, *Eur. J. Biochem.*, 119, 425, 1981.
53. **Bryant, D. A. and Cohen-Bazire, G.,** Effects of chromatic illumination on cyanobacterial phycobilisomes. Evidence for the specific induction of a second pair of phycocyanin subunits in *Pseudanabaena* 7409 grown in red light, *Eur. J. Biochem.*, 119, 415, 1981.
54. **Zilinskas, B. A. and Howell, D. A.,** Role of the colorless polypeptides in phycobilisome assembly in *Nostoc* sp., *Plant Physiol.*, 71, 379, 1983.
55. **Gingrich, J. C., Blaha, L. K., and Glazer, A. N.,** Rod substructure in cyanobacterial phycobilisomes: analysis of *Synechocystis* 6701 mutants low in phycoerythrin, *J. Cell Biol.*, 92, 261, 1982.
56. **Tandeau de Marsac, N. and Cohen-Bazire, G.,** Molecular composition of cyanobacterial phycobilisomes, *Proc. Natl. Acad. Sci. U.S.A.*, 74, 1635, 1977.
57. **Cobley, J. G. and Miranda, R. D.,** Mutations affecting chromatic adaptation in the cyanobacterium *Fremyella diplosiphon*, *J. Bacteriol.*, 153, 1486, 1983.
58. **Myers, J., Graham, J.-R., and Wang, R. T.,** On spectral control of pigmentation in *Anacystis nidulans* (Cyanophyceae), *J. Phycol.*, 14, 513, 1978.
59. **Myers, J., Graham, J.-R., and Wang, R. T.,** Spontaneous pigment mutants of *Anacystis nidulans* selected by growth under far-red light, *Arch. Microbiol.*, 124, 143, 1980.
60. **Myers, J., Graham, J.-R., and Wang, R. T.,** Light harvesting in *Anacystis nidulans* studied in pigment mutants, *Plant Physiol.*, 66, 1144, 1980.
61. **Khanna, R., Graham, J.-R., Myers, J., and Gantt, E.,** Phycobilisome composition and possible relationship to reaction centers, *Arch. Biochem. Biophys.*, 224, 534, 1983.

62. **Dilworth, M. F. and Gantt, E.,** Phycobilisome-thylakoid topography on photosynthetically active vesicles of *Porphyridium cruentum*, *Plant Physiol.*, 67, 608, 1981.
63. **Wanner, G. and Köst, H.-P.,** Investigations on the arrangement and fine structure of *Porphyridium cruentum* phycobilisomes, *Protoplasma*, 102, 97, 1980.
64. **Seybold, A.,** Über die Lichtenergiebilanz submerser Wasserpflanzen, vornehmlich der Meeresalgen, *Jahrb. Wiss. Bot.*, 79, 593, 1934.
65. **Jerlov, N. G.,** *Marine Optics*, Elsevier, Amsterdam, 1976.
66. **Littler, M. M., Littler, D. S., Blair, S. M., and Norris, J. N.,** Deepest known life discovered on an uncharted seamount, *Science*, 227, 57, 1985.
67. **Ramus, J., Beale, S. I., Mauzerall, D., and Howard, K. L.,** Changes in photosynthetic pigment concentration in seaweeds as a function of water depth, *Mar. Biol.*, 37, 223, 1976.
68. **Rhee, C. and Briggs, W. R.,** Some responses of *Chondrus crispus* to light. I. Pigmentation changes in the natural habitat, *Bot. Gaz.*, 138, 123, 1977.
69. **Haxo, F., O'hEocha, C., and Norris, P.,** Comparative studies of chromatographically separated phycoerythrins and phycocyanins, *Arch. Biochem. Biophys.*, 54, 162, 1955.
70. **Yu, M.-H., Glazer, A. N., Spencer, K. G., and West, J. A.,** Phycoerythrins of the red alga *Callithamnion*. Variation in phycoerythrobilin and phycourobilin content, *Plant Physiol.*, 68, 482, 1981.
71. **Shimura, S. and Fujita, Y.,** Phycoerythrin and photosynthesis of the pelagic blue-green alga *Trichodesmium thiebautii* in the waters of Kuroshio, Japan, *Mar. Biol.*, 31, 121, 1975.
72. **Parry, D. L.,** Cyanophytes with R-phycoerythrins in association with seven species of ascidians from the Great Barrier Reef, *Phycologia*, 23, 503, 1984.
73. **Alberte, R. S., Wood, A. M., Kursar, T. A., and Guillard, R. R. L.,** Novel phycoerythrins in marine *Synechococcus* spp. Characterization and evolutionary and ecological implications, *Plant Physiol.*, 75, 732, 1984.
74. **Ramus, J., Beale, S. I., and Mauzerall, D.,** Correlation of changes in pigment content with photosynthetic capacity of seaweeds as a function of water depth, *Mar. Biol.*, 37, 231, 1976.
75. **Goedheer, J. C.,** Spectral properties of the blue-green alga *Anacystis nidulans* grown under different environmental conditions, *Photosynthetica*, 10, 411, 1976.
76. **Dring, M. J.,** Chromatic adaptation of photosynthesis in benthic marine algae: an examination of its ecological significance using a theoretical model, *Limnol. Oceanogr.*, 26, 271, 1981.
77. **Ramus, J. and van der Meer, J. P.,** A physiological test of the theory of complementary chromatic adaptation. I. Color mutants of a red seaweed, *J. Phycol.*, 19, 86, 1983.
78. **Cox, G., Benson, D., and Dwarte, D. M.,** Ultrastructure of a cave-wall cyanophyte-*Gloeocapsa* NS4, *Arch. Microbiol.*, 130, 165, 1981.
79. **Larkum, A. W. D., Drew, E. A., and Crossett, R. N.,** The vertical distribution of attached marine algae in Malta, *J. Ecol.*, 55, 361, 1967.
80. **Vesk, M. and Jeffrey, S. W.,** Effect of blue-green light on photosynthetic pigments and chloroplast structure in unicellular marine algae from six classes, *J. Phycol.*, 13, 280, 1977.
81. **Brock, T. D.,** Life at high temperatures. Evolutionary, ecological, and biochemical significance of organisms living in hot springs is discussed, *Science*, 158, 1012, 1967.
82. **Edwards, M. R. and Gantt, E.,** Phycobilisomes of the thermophilic blue-green alga *Synechococcus lividus*, *J. Cell Biol.*, 50, 896, 1971.
83. **Hattori, A., Crespi, H. L., and Katz, J. J.,** Effect of side-chain deuteration on protein stability, *Biochemistry*, 4, 1213, 1965.
84. **Kao, O. H. W. and Berns, D. S.,** Similar C-phycocyanins from two strains of thermotolerant cyanophyte *Mastigocladus laminosus*, *Can. J. Microbiol.*, 23, 510, 1977.
85. **MacColl, R., Edwards, M. R., and Haaksma, C.,** Some properties of allophycocyanin from a thermophilic blue-green alga, *Biophys. Chem.*, 8, 369, 1978.
86. **Nies, M. and Wehrmeyer, W.,** Biliprotein assembly in the hemidiscoidal phycobilisomes of the thermophilic cyanobacterium *Mastigocladus laminosus* Cohn. Characterization of dissociation products with special reference to the peripheral phycoerythrocyanin-phycocyanin complexes, *Arch. Microbiol.*, 129, 374, 1981.
87. **Satoh, K., Smith, C. M., and Fork, D. C.,** Effects of salinity on primary processes of photosynthesis in the red alga *Porphyra perforata*, *Plant Physiol.*, 73, 643, 1983.
88. **Satoh, K. and Fork, D. C.,** A new mechanism for adaptation to changes in light intensity and quality in the red alga *Porphyra perforata*. I. Relation to state 1-state 2 transitions, *Biochim. Biophys. Acta*, 722, 190, 1983.
89. **Satoh, K. and Fork, D. C.,** A new mechanism for adaptation to changes in light intensity and quality in the red alga *Porphyra perforata*. III. Fluorescence transients in the presence of 3-(3,4 dichlorophenyl)-1,1-dimethylurea, *Plant Physiol.*, 71, 673, 1983.
90. **Öquist, G. and Fork, D. C.,** Effects of desiccation on the excitation energy distribution from phycoerythrin to the two photosystems in the red alga *Porphyra perforata*, *Physiol. Plant.*, 56, 56, 1982.

91. **Kao, O. H. W., Berns, D. S., and Town, W. R.,** The characterization of C-phycocyanin from an extremely halo-tolerant blue-green alga, *Coccochloris elabens*, *Biochem. J.*, 131, 39, 1973.
92. **Blumwald, E. and Tel-Or, E.,** Salt adapation of the cyanobacterium *Synechococcus* 6311 growing in a continuous culture (turbidostat), *Plant Physiol.*, 74, 183, 1984.
93. **Jeffrey, S. W. and Humphrey, G. F.,** New spectrophotometric equations for determining chlorophylls *a*, *b*, c_1, and c_2 in higher plants, algae and natural phytoplankton, *Biochem. Physiol. Pflanz.*, 167, 191, 1975.
94. **Grombein, S., Rüdiger, W., and Zimmermann, H.,** The structure of the phytochrome chromophore in both photoreversible forms, *Hoppe-Seyler's Z. Physiol. Chem.*, 356, 1709, 1975.
95. **Klein, G., Grombein, S., and Rüdiger, W.,** On the linkages between chromophore and protein in biliproteins. VI. Structure and protein linkage of the phytochrome chromophore, *Hoppe-Seyler's Z. Physiol. Chem.*, 358, 1077, 1977.
96. **Lagarias, J. C. and Rapoport, H.,** Chromopeptides from phytochrome. The structure and linkage of the P_R form of the phytochrome chromophore, *J. Am. Chem. Soc.*, 102, 4821, 1980.
97. **Rüdiger, W., Thümmler, F., Cmiel, E., and Schneider, S.,** Chromophore structure of the physiologically active form (P_{fr}) of phytochrome, *Proc. Natl. Acad. Sci. U.S.A.*, 80, 6244, 1983.
98. **Thümmler, F. and Rüdiger, W.,** Models for the photoreversibility of phytochrome. Z, E isomerization of chromopeptides from phycocyanin and phytochrome, *Tetrahedron*, 39, 1943, 1983.
99. **Scheibe, J.,** Photoreversible pigment: occurrence in a blue-green alga, *Science*, 176, 1037, 1972.
100. **Björn, G. S. and Björn, L. O.,** Photochromic pigments from blue-green algae: phycochromes a, b, c, *Physiol. Plant.*, 36, 297, 1976.
101. **Björn, G. S. and Björn, L. O.,** Action spectra for conversions of phycochrome c from *Nostoc muscorum*, *Physiol. Plant.*, 43, 195, 1978.
102. **Björn, G. S.,** Phycochrome d, a new photochromic pigment from the blue-green alga, *Tolypothrix distorta*, *Physiol. Plant.*, 42, 321, 1978.
103. **Björn, G. S.,** Action spectra for *in vivo* and *in vitro* conversions of phycochrome *b*, a reversibly photochromic pigment in blue-green alga, and its separation from other pigments, *Physiol. Plant.*, 46, 281, 1979.
104. **Ohad, I., Schneider, H.-J.A.W., Gendel, S., and Bogorad, L.,** Light-induced changes in allophycocyanin, *Plant Physiol.*, 65, 6, 1980.
105. **Ohad, I., Clayton, R. K., and Bogorad, L.,** Photoreversible absorbance changes in solutions of allophycocyanin purified from *Fremyella diplosiphon:* temperature dependence and quantum efficiency, *Proc. Natl. Acad. Sci. U.S.A.*, 76, 5655, 1979.
106. **Ohki, K. and Fujita, Y.,** Photoreversible absorption changes of guanidine-HCl-treated phycocyanin and allophycocyanin isolated from the blue-green alga *Tolypothrix tenuis*, *Plant Cell Physiol.*, 20, 483, 1979.
107. **Ohki, K. and Fujita, Y.,** *In vivo* transformation of phycobiliproteins during photobleaching of *Tolypothrix tenuis* to forms active in photoreversible absorption changes, *Plant Cell Physiol.*, 20, 1341, 1979.
108. **Berns, D. S. and Morgenstern, A.,** Two denaturant effects of guanidine salts on the protein C-phycocyanin, *Arch. Biochem. Biophys.*, 123, 640, 1968.
109. **MacColl, R., Csatorday, K., Berns, D. S., and Traeger, E.,** The relationship of the quaternary structure of allophycocyanin to its spectrum, *Arch. Biochem. Biophys.*, 208, 42, 1981.
110. **MacColl, R.,** Stability of allophycocyanin's quaternary structure, *Arch. Biochem. Biophys.*, 223, 24, 1983.
111. **MacColl, R., Csatorday, K., Berns, D. S., and Traeger, E.,** Chromophore interactions in allophycocyanin, *Biochemistry*, 19, 2817, 1980.
112. **de Kok, J., Braslavsky, S. E., and Spruit, C. J. P.,** Solvent-induced photoreversible reactions of C-phycocyanin from *Synechococcus* sp., *Photochem. Photobiol.*, 34, 705, 1981.
113. **de Kok, J.,** Photoreactions of the chromophore cations of denatured species of C-phycocyanin and allophycocyanin, *Photochem. Photobiol.*, 35, 849, 1982.
114. **Allen, M. M. and Smith, A. J.,** Nitrogen chlorosis in blue-green algae, *Arch. Mikrobiol.*, 69, 114, 1969.
115. **Lau, R. H., MacKenzie, M. M., and Doolittle, W. F.,** Phycocyanin synthesis and degradation in the blue-green bacterium *Anacystis nidulans*, *J. Bacteriol.*, 132, 771, 1977.
116. **Allen, M. M.,** Cyanobacterial cell inclusions, *Annu. Rev. Microbiol.*, 38, 1, 1984.
117. **Haselkorn, R.,** Heterocysts, *Annu. Rev. Plant Physiol.*, 29, 319, 1978.
118. **Neuer, G., Papen, H., and Bothe, H.,** Heterocyst biochemistry and differentiation, in *Photosynthetic Prokaryotes: Cell Differentiation and Function*, Papageorgiou, G. C. and Packer, L., Eds., Elsevier, New York, 1983, 219.
119. **Peterson, R. B., Dolan, E., Calvert, H. E., and Ke, B.,** Energy transfer from phycobiliproteins to photosystem I in vegetative cells and heterocysts of *Anabaena variabilis*, *Biochim. Biophys. Acta*, 634, 237, 1981.
120. **Yamanaka, G. and Glazer, A. N.,** Phycobiliproteins in *Anabaena* 7119 heterocysts, in *Photosynthetic Prokaryotes: Cell Differentiation and Function*, Papageorgiou, G. C. and Packer, L., Eds., Elsevier, New York, 1983, 69.

121. **Foulds, I. J. and Carr, N. G.,** A proteolytic enzyme degrading phycocyanin in the cyanobacterium *Anabaena cylindrica, FEMS Microbiol. Lett.*, 2, 117, 1977.
122. **Wood, N. B. and Haselkorn, R.,** Control of phycobiliprotein proteolysis and heterocyst differentiation in *Anabaena, J. Bacteriol.*, 141, 1375, 1980.
123. **Allen, M. B.,** Studies with *Cyanidium caldarium*, an anomalously pigmented chlorophyte, *Arch. Mikrobiol.*, 32, 270, 1959.
124. **Nichols, K. E. and Bogorad, L.,** Studies on phycobilin formation with mutants of *Cyanidium caldarium, Nature*, 188, 870, 1960.
125. **Nichols, K. E. and Bogorad, L.,** Action spectra studies of phycocyanin formation in a mutant of *Cyanidium caldarium, Bot. Gaz.*, 124, 85, 1962.
126. **Schneider, H. A. W. and Bogorad, L.,** Spectral response curves for the formation of phycobiliproteins, chlorophyll and δ-aminolevulinic acid in *Cyanidium caldarium, Z. Pflanzenphysiol.*, 94, 449, 1979.
127. **Troxler, R. F. and Bogorad, L.,** Studies on the formation of phycocyanin, porphyrins, and a blue phycobilin by wild-type and mutant strains of *Cyanidium caldarium, Plant Physiol.*, 41, 491, 1966.
128. **Mercer, F. V., Bogorad, L., and Mullens, R.,** Studies with *Cyanidium caldarium*. I. The fine structure and systematic position of the organism, *J. Cell Biol.*, 13, 393, 1962.
129. **Bogorad, L., Mercer, F. V., and Mullens, R.,** Studies with *Cyanidium caldarium*. II. The fine structure of pigment-deficient mutants, in *Photosynthetic Mechanisms of Green Plants*, Publ. National Research Council, National Academy of Science, Washington, D.C., 1963, 560.
130. **Bogorad, L.,** Studies of phycobiliproteins, *Rec. Chem. Prog.*, 26, 1, 1965.
131. **Kao, O. H. W., Edwards, M. R., and Berns, D. S.,** Physical-chemical properties of C-phycocyanin isolated from an acido-thermophilic eukaryote, *Cyanidium caldarium, Biochem. J.*, 147, 63, 1975.
132. **Brown, A. S. and Troxler, R. F.,** Properties and N-terminal sequence of allophycocyanin from the unicellular rhodophyte *Cyanidium caldarium, Biochem. J.*, 163, 571, 1977.
133. **Troxler, R. F., Foster, J. A., Brown, A. S., and Franzblau, C.,** The α and β subunits of *Cyanidium caldarium* phycocyanin. Properties and amino acid sequences at the amino terminus, *Biochemistry*, 14, 268, 1975.
134. **Troxler, R. F. and Lester, R.,** Biosynthesis of phycocyanobilin, *Biochemistry*, 6, 3840, 1967.
135. **Troxler, R. F. and Brown, A. S.,** Metabolism of δ-aminolevulinic acid in red and blue-green algae, *Plant Physiol.*, 55, 463, 1975.
136. **Troxler, R. F., Kelly, P., and Brown, S. B.,** Phycocyanobilin synthesis in the unicellular rhodophyte *Cyanidium caldarium, Biochem. J.*, 172, 569, 1978.
137. **Brown, S. B., Holroyd, J. A., Troxler, R. F., and Offner, G. D.,** Bile pigment synthesis in plants. Incorporation of haem into phycocyanobilin and phycobiliproteins in *Cyanidium caldarium, Biochem. J.*, 194, 137, 1981.
138. **Troxler, R. F.,** Synthesis of bile pigments in plants. Formation of carbon monoxide and phycocyanobilin in wild-type and mutant strains of the alga, *Cyanidium caldarium, Biochemistry*, 11, 4235, 1972.
139. **Troxler, R. F. and Dokos, J. M.,** Formation of carbon monoxide and bile pigment in red and blue-green algae, *Plant Physiol.*, 51, 72, 1973.
140. **Troxler, R. F., Brown, A. S., and Brown, S. B.,** Bile pigment synthesis in plants. Mechanism of ^{18}O incorporation into phycocyanobilin in the unicellular rhodophyte, *Cyanidium caldarium, J. Biol. Chem.*, 254, 3411, 1979.
141. **Brown, S. B., Holroyd, J. A., and Troxler, R. F.,** Mechanism of bile-pigment synthesis in algae. ^{18}O incorporation into phycocyanobilin in the unicellular rhodophyte, *Cyanidium caldarium, Biochem. J.*, 190, 445, 1980.
142. **Beale, S. I. and Chen, N. C.,** N-Methyl mesoporphyrin IX inhibits phycocyanin, but not chlorophyll synthesis in *Cyanidium caldarium, Plant Physiol.*, 71, 263, 1983.
143. **Brown, S. B., Holroyd, J. A., Vernon, D. I., Troxler, R. F., and Smith, K. M.,** The effect of N-methylprotoporphrin IX on the synthesis of photosynthetic pigments in *Cyanidium caldarium*. Further evidence for the role of haem in the biosynthesis of plant bilins, *Biochem. J.*, 208, 487, 1982.
144. **Koest, H.-P. and Benedikt, E.,** Biliverdin IXα, intermediate and end product of tetrapyrrole biosynthesis, *Z. Naturforsch.*, 37C, 1057, 1982.
145. **Beale, S. I. and Cornejo, J.,** Biosynthesis of phycocyanobilin from exogenous labeled biliverdin in *Cyanidium caldarium, Arch. Biochem. Biophys.*, 227, 279, 1983.
146. **Brown, S. B., Holroyd, J. A., and Vernon, D. I.,** Biosynthesis of phycobiliproteins. Incorporation of biliverdin into phycocyanin of the red alga, *Cyanidium caldarium, Biochem. J.*, 219, 905, 1984.
147. **Brown, S. B. and Holroyd, J. A.,** Biosynthesis of the chromophore of phycobiliproteins. A study of mesohaem and mesobiliverdin as possible intermediates and further evidence for an algae haem oxygenase, *Biochem. J.*, 217, 265, 1984.
148. **Beale, S. I. and Cornejo, J.,** Enzymatic transformation of biliverdin to phycocyanobilin by extracts of the unicellular red alga *Cyanidium caldarium, Plant Physiol.*, 76, 7, 1984.

149. **Beale, S. I. and Cornejo, J.,** Enzymatic heme oxygenase activity in soluble extracts of the unicellular red alga, *Cyanidium caldarium, Arch. Biochem. Biophys.*, 235, 371, 1984.
150. **Csatorday, K., MacColl, R., and Berns, D. S.,** Accumulation of protoporphyrin IX and Zn protoporphyrin IX in *Cyanidium caldarium, Proc. Natl. Acad. Sci. U.S.A.*, 78, 1700, 1981.
151. **Lewis, N. G., Walter, J. A., and Wright, J. L. C.,** Tetrapyrrole biosynthesis in *Anacystis nidulans;* incorporation of [1-^{13}C]-, [2-^{13}C]-, [1,2-^{13}C]- and [2-^{13}C, 2-$^{2}H_3$] acetate, *Phytochemistry*, 23, 1611, 1984.
152. **Cosner, J. C. and Troxler, R. F.,** Phycobiliprotein synthesis in protoplasts of the unicellular cyanophyte, *Anacystis nidulans, Biochim. Biophys. Acta*, 519, 474, 1978.
153. **Rüdiger, W.,** Über die Abwehrfarbstoffe von Aplysia-Arten. II. Die Struktur von Aplysioviolin, *Hoppe-Seyler's Z. Physiol. Chem.*, 348, 1554, 1967.
154. **Belford, H. S., Offner, G. D., and Troxler, R. F.,** Phycobiliprotein synthesis in the unicellular rhodophyte, *Cyanidium caldarium.* Cell-free tranlsation of the mRNAs for the α and β subunit polypeptides of phycocyanin, *J. Biol. Chem.*, 258, 4503, 1983.
155. **Steinmüller, K., Kaling, M., and Zetsche, K.,** *In-vitro* synthesis of phycobiliproteids and ribulose-1,5 bisphosphate carboxylase by non-poly-adenylated-RNA of *Cyanidium caldarium* and *Porphyridium aerugineum, Planta*, 159, 308, 1983.
156. **Steinmüller, K. and Zetsche, K.,** Photo- and metabolite regulation of the synthesis of ribulose bisphosphate carboxylase/oxygenase and the phycobiliproteins in the alga *Cyanidium caldarium, Plant Physiol.*, 76, 935, 1984.
157. **Egelhoff, T. and Grossman, A.,** Cytoplasmic and chloroplast synthesis of phycobilisome polypeptides, *Proc. Natl. Acad. Sci. U.S.A.*, 80, 3339, 1983.
158. **Schuster, A., Köst, H.-P., Rüdiger, W., and Eder, J.,** Investigations on the apoprotein of phycocyanin from *Cyanidium caldarium, Arch. Microbiol.*, 135, 30, 1983.
159. **Lemaux, P. G. and Grossman, A.,** Isolation and characterization of a gene for a major light-harvesting polypeptide from *Cyanophora paradoxa, Proc. Natl. Acad. Sci. U.S.A.*, 81, 4100, 1984.
160. **Lemaux, P. G. and Grossman, A. R.,** Major light-harvesting polypeptides encoded in polycistronic transcripts in a eukaryotic alga, *EMBO J.*, 4, 1911, 1985.
161. **Conley, P. B., Lemaux, P. G., and Grossman, A. R.,** Cyanobacterial light-harvesting complex subunits encoded in two red light-induced transcripts, *Science*, 230, 550, 1985.
162. **de Lorimier, R., Bryant, D. A., Porter, R. D., Liu, W.-Y., Jay, E., and Stevens, S. E., Jr.,** Genes for the α and β subunits of phycocyanin, *Proc. Natl. Acad. Sci. U.S.A.*, 81, 7946, 1984.
163. **Pilot, T. J. and Fox, J. L.,** Cloning and sequencing of the genes encoding the α and β subunits of C-phycocyanin from the cyanobacterium *Agmenellum quadruplicatum, Proc. Natl. Acad. Sci. U.S.A.*, 81, 6983, 1984.
164. **Offner, G. D., Brown-Mason, A. S., Ehrhardt, M. M., and Troxler, R. F.,** Primary structure of phycocyanin from the unicellular rhodophyte, *Cyanidium caldarium.* I. Complete amino acid sequence of the α subunit, *J. Biol. Chem.*, 256, 12,167, 1981.
165. **Frank, G., Sidler, W., Widmer, H., and Zuber, H.,** The complete amino acid sequence of both subunits of C-phycocyanin from the cyanobacterium *Mastigocladus laminosus, Hoppe-Seyler's Z. Physiol. Chem.*, 359, 1491, 1978.
166. **Troxler, R. F., Ehrhardt, M. M., Brown-Mason, A. S., and Offner, G. D.,** Primary structure of phycocyanin from the unicellular rhodophyte *Cyanidium caldarium.* II. Complete amino acid sequence of the β subunit, *J. Biol. Chem.*, 256, 12,176, 1981.
167. **Bryant, D. A., de Lorimier, R., Lambert, D. H., Dubbs, J. M., Stirewalt, V. L., Stevens, S. E., Jr., Porter, R. D., Tam, J., and Jay, E.,** Molecular cloning and nucleotide sequence of the α and β subunits of allophycocyanin from the cyanelle genome of *Cyanophora paradoxa, Proc. Natl. Acad. Sci. U.S.A.*, 82, 3242, 1985.

Chapter 8

BILIPROTEIN-THYLAKOID INTERACTION

I. LIGHT HARVESTING BY TWO PHOTOSYSTEMS

Engelmann[1,2] obtained early evidence that biliproteins were active in photosynthesis. Motile, aerotactic bacteria were used to detect O_2 evolution from various microorganisms. The wavelengths at which O_2 were detected suggested that the biliproteins were harvesting light energy for use in photosynthesis. The cyanobacteria, red algae, and cryptomonads all have chlorophyll *a* and perform the usual form of O_2-evolving photosynthesis:

$$CO_2 + H_2O \xrightarrow{\text{light}} O_2 + \text{carbohydrates}$$

The principal questions to be covered in this chapter concern the relationships among the biliproteins and the two pigment systems involved in photosynthesis. The vast literature that exists on the light reactions of photosynthesis will not be covered extensively, and only the limited examples needed to establish the functions of biliproteins and particularly phycobilisomes will be discussed. Emerson and Lewis[3] studied the quantum yields of photosynthesis in *Chroococcus* as a function of wavelength. The quantum yield was fairly constant from 570 to 690 nm, while the proportion of biliprotein to chlorophyll *a* was changing dramatically. For example, at 670 nm, chlorophyll *a* absorbed 94% of the total light, but between 560 and 640 nm, C-phycocyanin absorbed over 80%, while chlorophyll *a* absorbed less than 20% of the total. Therefore the efficient utilization of light absorbed by C-phycocyanin for photosynthesis was established. Emerson and Lewis[4] subsequently found that the quantum yield of photosynthesis was not constant over the entire spectrum. Although fairly constant from 580 to 685 nm, above 685 nm the quantum yield decreased rapidly. Also, at wavelengths shorter than 560 nm, another region of declining quantum yield was observed. It was noted that chlorophyll *a* continued to absorb light in the spectral region above 685 nm in which the quantum yield declined (''red-drop'' effect). This experiment proved to be quite important as a first step and along with later studies provided a basis for the concept of two pigment systems that were each responsible for their own light-driven reactions. The low quantum yields above 685 nm occurred because only a single photosystem (photosystem I) was heavily represented at these wavelengths, and excitation of photosystem II was also needed for efficient photosynthesis.

Haxo and Blinks[5] measured the absorption and photosynthetic action spectra for a number of marine red algae. A polarographic method of O_2 measurement was employed to obtain data for the action spectra. Since phycoerythrin absorbed maximally (545 to 565 nm) at wavelengths that were far removed from chlorophyll *a* absorbance, these red algae were selected as excellent test vehicles for the roles of antenna pigments in photosynthesis. The efficiency of the action spectra in wavelengths of phycoerythrin absorbance was always quite high, and the red-drop effect was very apparent in regions of chlorophyll *a* absorbance (Figure 1). These authors emphasized that the very low utilization of chlorophyll *a* absorbance in the red algae was an unexpected phenomenon.

Emerson et al.[6] produced strong support for the two pigment classes when they discovered that the red-drop effect could be overcome. The quantum yield of photosynthesis in the red algae could be improved when absorption by red light was supplemented with absorption at a second wavelength where a biliprotein or some other antenna pigment had strong absorbance. The current interpretation of these data would invoke the idea of two photo-

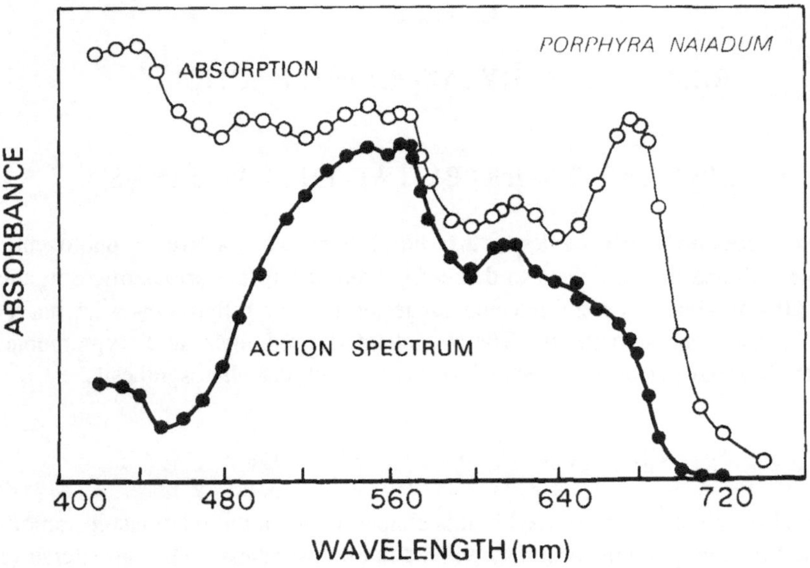

FIGURE 1. Action spectrum for O_2 evolution and absorption spectrum for a red alga. (From Haxo, F. T. and Blinks, L. R., *J. Gen. Physiol.*, 33, 389, 1950. With permission.)

systems being excited. This improvement of quantum yield was called the enhancement effect. Biliproteins and other antenna pigments were therefore shown to be essential parts of the efficient operation of the photosynthetic apparatus.

The action spectrum for photosynthesis of a cryptomonad alga has been measured by Haxo and Fork.[7] The cryptomonads do not have phycobilisomes but have a biliprotein and chlorophyll c_2 in addition to chlorophyll *a*. As was found for the cyanobacteria and red algae, cryptomonad biliproteins were quite active in photosynthesis. The red-drop effect was less apparent.

Duysens and Amesz[8] and Amesz and Duysens[9] obtained additional proof for two pigment systems in cyanobacteria and red algae. The two systems were found to operate serially. They obtained action spectra that were characteristic of each light reaction of the red alga *Porphyridium cruentum*. The action spectrum for photosystem I was obtained by employing a strong and constant background light at 562 nm and measuring the action spectrum with weak light. The 562-nm light saturated photosystem II, and photosystem I became limiting. Therefore the action spectrum obtained will be that of pigments contributing to photosystem I. The converse experiment using a strong background light at 680 nm yielded the action spectrum of photosystem II. With 680-nm background light, the ratio of activity at 560 to 680 nm was 7.0, and with a background light at 562 nm, the ratio was 1.4. Thus B-phycoerythrin contributed to the action spectra of both pigment systems, but the contribution of chlorophyll *a* to photosystem II was quite small.

Jones and Myers[10] measured action spectra for the cyanobacterium *Anacystis nidulans* as described above.[8] They discussed the possibility that light energy in photosystem II may spillover to photosystem I (see Section II.B). The action spectrum of photosystem II obtained against a 690-nm background was very much related to biliprotein and not to chlorophyll *a* absorption. Photosystem I activity was measured with a 620-nm background and exhibited components from both C-phycocyanin and chlorophyll *a*. These action spectra were not the actual spectra of only the specific pigments associated with individual photosystems, since certain light energy may be shared between photosystems. These experiments, therefore, also measure the biliprotein contributions to photosystem I that occur via transfer of energy from photosystem II to I (see Section II.B).

II. PHYCOBILISOMES AND CHLOROPHYLL *a*

A. Photosystems I and II Light Harvesting

Phycobilisomes are composed of a variety of biliproteins in which excitation energy migrates from the short-wavelength absorbers, phycoerythrin or phycoerythrocyanin when present, to phycocyanin to allophycocyanin to the two types of allophycocyanin 680. From the two types of allophycocyanin 680 the excitation energy is both energetically favorable and at the closest proximity for transfer, probably via Förster very weak coupling of dipoles, to chlorophyll *a* in the thylakoid membrane. Phycobilisomes transfer excitation energy directly to photosystem II for O_2 evolution, while chlorophyll *a* makes only a small contribution to this photosystem. A major question is how light harvested by the biliproteins reaches the reaction centers of photosystems I and II. Furthermore, why is the energy that is absorbed only by chlorophyll *a* apparently minimally utilized in photosynthesis, while light absorbed almost exclusively by biliproteins is quite effective? A guiding principle seems to be to balance the contributions from the two pigment systems into the reaction centers for optimal photosynthetic operation.

For cyanobacteria and red algae, various techniques will be discussed which suggested that the bulk of the light-harvesting chlorophyll *a* was associated with photosystem I. A much smaller quantity of chlorophyll *a* appeared to be directly tied to photosystem II as was the great majority of the biliproteins. Excitation energy, however, from phycobilisome absorption apparently, after first going to photosystem II, can readily be transferred into photosystem I (spillover). Gingras[11] has estimated for *P. cruentum* that approximately 90 to 95% of the energy absorbed at 695 nm was utilized by photosystem I.

The year 1977 was quite crucial in this field, with several overlapping discoveries being reported from a variety of laboratories.[12-18] Ley and Butler[12] studied the fluorescence spectra of the red alga *P. cruentum* at − 196°C. Photosystem I fluorescence was measurable at this low temperature. Fluorescence measurements, if properly analyzed, provided a unique method of determining the early events in the transfer of light energy. Excitation in the blue or red where chlorophyll *a* was the primary absorber was shown to be distributed almost exclusively to photosystem I. B- and b-phycoerythrin were the major biliproteins of this alga, and when they were excited by green light the excitation energy was given almost completely to photosystem II. Although the small amount of green light excitation going directly to photosystem I was difficult to measure due to overlapping emissions, it was found to be 5%, and a suggestion was made that 10% was a more reasonable figure. This amount of photosystem I excitation was probably a result of the chlorophyll *a* absorbing at these wavelengths. Ley and Butler[12] confirmed their fluorescence findings by measuring the action spectrum for O_2 evolution. A weak modulated monochromatic measuring beam was used in conjunction with a strong constant background illumination covering a broad spectrum. The evolution of O_2 (photosystem II activity) was stimulated by the biliproteins, but almost no O_2 evolved from regions where chlorophyll *a* was the exclusive absorber. They postulated that a model consistent with these data would have phycobilisomes in direct contact with the photosystem II units. Furthermore, it was found that some excitation energy from photosystem II was subsequently distributed to photosystem I. The partitioning of excitation energy insured optimal operation of the two photosystems. Another aspect of spillover from photosystem II to photosystem I was that certain measurements showed where the excitation finally arrived but not where it was originally transferred by the light-harvesting pigments.

Tel-Or and Malkin[13] examined the photochemical and fluorescence properties of the photosynthetic activity of the cyanobacterium *Phormidium luridum*. Based on quantum yields and pigment composition, the following distribution was obtained: photosystem I — 90% of chlorophyll *a* absorbance and 15% of the C-phycocyanin and photosystem II — 10% of the chlorophyll *a* absorbance and 85% of the C-phycocyanin. They noted that their results showed the importance of biliproteins in photosystem II. As demonstrated above,[12] this contribution of biliproteins to photosystem I may be a consequence of the redistribution of excitation energy originally found in photosystem II.

Wang et al.[14] studied the action spectra for both photosystems using the cyanobacterium *A. nidulans*. They found that 84% of the chlorophyll *a* molecules delivered their excitation energy to photosystem I. C-Phycocyanin utilized a small number of chlorophylls to transmit its light energy to photoreaction II.

Mimuro and Fujita[15] used two methods to examine the distribution of chlorophylls in the two photosystems of the cyanobacterium *Anabaena variabilis*. Both methods demonstrated that a very small portion of the chlorophyll *a* molecules were associated with photosystem II. The ratio of chlorophyll *a* in photosystem II to total chlorophyll *a* was quite small.

Larkum and Weyrauch[16] measured the photosynthetic action spectra of the red alga *Griffithsia monilis*. The amount of chlorophyll *a* associated with photosystem II was very much smaller than that associated with photosystem I. They suggested a close energy transfer linkage between phycobilisomes and photosystem II and that spillover of excitation energy could then occur to photosystem I.

Harnishfeger and Codd[17] studied the fluorescence at $-196°C$ for several cyanobacteria. They found that light harvested by the biliproteins was mainly transferred to photosystem II chlorophylls.

Ried et al.[18] studied the action spectra of photosystems I and II from several red algae. The action spectra for light reaction II showed that only 10 to 20% of the total chlorophyll *a* was associated with photosystem II together with the biliproteins. They measured action spectra with 550- or 699-nm background light, which will excite one or the other photosystem. The low-intensity measuring beam will cause a response characteristic of the other photosystem. Even so, the action spectrum using a 550-nm background must be considered light reaction I, and the 699-nm case, light reaction II. These action spectra reflected the pigment content of their respective photosystems, but spillover caused the pigments from photosystem II to be represented in the other action spectrum of the photosystem. They proposed two models to account for the action spectra; the phycobilisomes were associated entirely with photosystem II and spillover occurred from the chlorophylls of photosystem II or excitation energy from the phycobilisomes could be distributed to both systems in variable proportions. A photosystem I spectrum was calculated by correcting light reaction I. The calculated photosystem I action spectrum showed only contributions from chlorophyll *a* and carotenoids.

Larkum and Weyrauch,[16] in addition, observed that allophycocyanin, usually a minor biliprotein in terms of content, was active in light harvesting in *G. monilis*. Lemasson et al.[19] had previously shown in several cyanobacteria that the photosynthetic action spectra for O_2 evolution showed a band at 650 nm which was associated with allophycocyanin content.

Clearly, this evidence proved a close link between the biliproteins and photosystem II.[12-18] In two atypical situations, however, the biliproteins were switched over to direct coupling with photosystem I. Tel-Or and Malkin[13] studied spheroplasts (cells in which the outer envelope had been removed) of a cyanobacterium and found that the C-phycocyanin became exclusively associated with photosystem I. There was a concurrent loss of photosystem II activity. Peterson et al.[20] studied heterocysts of the cyanobacterium *A. variabilis*. Heterocysts, which act as centers for nitrogen fixation, are specialized cells formed along the filament of certain cyanobacteria. Photosystem II activity was absent from these cells, since the presence of O_2 would destroy the enzymes needed for fixation of N_2. The heterocyst biliproteins were found to be associated directly with photosystem I (see Chapter 7 for more details).

B. Spillover

The concept of two photosystems began with the proposals of Hill and Bendall.[21] Their suggestions together with other results were summarized in the "Z" scheme for electron transfer. Each of the two photosystems had a reaction center at which photochemistry could occur, and associated with each reaction center was an antenna pigment system. These antenna pigments harvested the light energy needed for the chemical reactions at each reaction center. It appeared that an essential factor in the proper functioning of the two photosystems was that they both be activated in an appropriate time frame. An important mechanism to insure efficient operation of the two photosystems was spillover. In cyanobacteria and red algae, the biliproteins — because of their high concentrations, absorbance spectra, and high absorptivities — harvested a high percentage of the available solar radiation. The phycobilisomes were closely coupled to photosystem II, while leaving most of the light-harvesting chlorophyll *a* associated with photosystem I. In many cases, the result of this arrangement was that a high percentage of light energy was intially directed into photosystem II. Spillover was the rerouting of excitation energy to photosystem I. Transfer of excitation energy from photosystem I to II was apparently not a consequential factor.

Murata et al.[22] performed fluorescence experiments at $-196°C$ that gave early credence to the spillover concept. At wavelengths where the photosystem II antenna pigments had high absorbances relative to chlorophyll *a,* a large quantum yield of fluorescence for photosystem I was observed for several organisms including red algae and cyanobacteria. It was suggested that the best explanation for these data was energy transfer from photosystem II to I. Working with the red alga *Porphyridium cruentum,* Murata[23] demonstrated that after preillumination with light absorbed by phycoerythrin the fluorescence at $-196°C$ showed decreases in photosystem II emissions and increases in the emission intensity of the band assigned to photosystem I. Preillumination of chlorophyll *a* did not alter the fluorescence from the cells. These observations were interpreted as indicating that phycoerythrin absorption either increased direct transfer of excitation energy to the photosystem I or increased spillover from photosystem II to I. At the same time as Murata et al.[22] were studying spillover for biliprotein-containing organisms, Bonaventura and Myers[24] were performing similar experiments with a green alga.

Ley and Butler[25] demonstrated that there was a 60 to 100% increase in the rate constant for energy transfer from photosystem II to I (spillover) in cells of *P. cruentum* that were previously exposed to light absorbed by photosytem II pigment. The nomenclature used for these experiments was *state 1* for cells equilibrated in light absorbed by the photosystem I pigment and *state 2* for cells equilibrated in light absorbed by photosystem II. Therefore a phycoerythrin-containing organism illuminated with green light was placed in state 2, which caused greater amounts of photosystem II excitation energy to be subsequently transferred to the chlorophyll *a* molecules associated with photosystem I. Spillover was observed in a certain time frame in which irradiation changes the distribution of excitation energy. At much longer times, in addition to spillover the process of changing the amounts of pigment molecules attached to a particular photosystem may become very prominent. *P. cruentum* was placed in state 1 by illumination of the cells for 5 min with blue light at room temperature and then during blue light treatment lowering the temperature to $-196°C$. State 2 was achieved in the same way using green irradiation.[25] The increased spillover in state 2 insured efficient photosynthetic operation even though there was minimal direct excitation of photosystem I chlorophylls.

Ley and Butler[26] measured the fraction of photosystem II excitation energy transferred to photosystem I as a function of the status of the photosystem II reaction centers. When these reaction centers were completely open the yield of excitation energy transfer was about 0.5. Therefore under growth conditions where the phycobilisome was the principle light-harvesting pigment, both photosystems I and II were maintained for optimal photosynthetic activity. When the reaction centers of photosystem II were entirely closed the yield for transfer went upt o 0.95.

Ried and Reinhardt[27] have examined several details in the redistribution of excitation energy between the two photosystems for nine red algae. A light pulse of 0.2 sec of medium intensity at 443 nm produced complete transfer from state 2 to 1. The doses of light needed to achieve half-maximum transitions from state 2 to 1 were standardized against the number of electron transport chains. They estimated that the transition for red algae required 2 to 4 photons per electron transport chain. The number of photons was corrected for spectral overlap of photosystems I and II and for absorption not used in photosynthesis.

Biggins et al.[28] examined the red alga *P. cruentum* in a state 1 or 2 to determine if reversible protein phosphorylation was involved in these transitions. The answer was negative. They noted that phosphorylation of light-harvesting proteins in higher plants was apparently an important part of the state 2-state 1 transitions, and they proposed that red algae must have different methods of accomplishing the distribution of excitation energy. Biggins[29] demonstrated that when the proteins of the cells were cross-linked with glutaraldehyde, the cells became fixed in whichever state (1 or 2) they were in during the cross-linking. This result suggested the chlorophyll *a*-protein complexes have specific positions or orientations that are characteristic of state 1 or 2. A model for state 1-state 2 transitions was developed that included the structure of the protein as a factor. Biggins et al.[28] proposed that the two photosystems were farther apart in state 1 than they were in state 2. Closer proximity would allow the increased energy transfer that was observed from photosystem II to I when the cells were in state 2.

Evidence on the process of excitation energy transfer from photosystem II to I was obtained through picosecond fluorescence spectroscopy of the cyanobacterium *Anacystis nidulans* and the red alga *P. cruentum*. Bruce et al.[30] placed both these organisms in state 2 or 1 by proper irradiation and then froze the cells in liquid nitrogen (−196°C). The fluorescence emission lifetime from photosystem II chlorohpyll *a* was 2.8 times longer for cells in state 1 for *A. nidulans* and 1.6 times longer for cells in state 1 for *P. cruentum*. They concluded that photosystem II chlorophyll *a* was involved in spillover. These increased lifetimes for state 1 cells compared to state 2 cells suggested that there was a significant uncoupling of photosystem II chlorophyll *a* from energy transfer to photosystem I chlorophyll *a* in state 1 vs. state 2.

Karukstis and Sauer[31] have used picosecond fluorescence spectroscopy to study the assignment of room temperature fluorescence emission bands to photosystems I and II.

In Chapter 7 other results relating to the response of photosynthesis to variations in growth light are considered.

C. Chlorophyll-Proteins

The above evidence (Section II.A) supported a model for the thylakoid membranes of red algae and cyanobacteria in which most of the chlorophyll *a* was associated with photosystem I. The small fraction of chlorophyll *a* in photosystem II acted as a conduit for funneling excitation energy from the phycobilisomes to the photosystem II reaction centers or for spillover to photosystem I. Confirmation of a structural nature for these assignments came from studies on isolated chlorophyll-protein complexes.

Ogawa et al.[32] have isolated chlorophyll-protein complexes from the cyanobacterium *Anabaena variabilis* by sonication and treatment with the detergent Triton® X-100. P700, which is the reaction center pigment of photosystem I, was located on one of the complexes together with 80% of the total chlorophyll *a*. Another complex which was believed to be related to photosystem II by its fluorescence properties had a much smaller chlorophyll *a* content. No O_2 evolution ability was found in either complex. Thornber[33] isolated complexes that appeared to be related to photosystem I and contained the bulk of the chlorophyll *a*.

Redlinger and Gantt[34] isolated two photosystem II complexes that contained 7% of the total chlorophyll *a* of the cell, whereas a single photosystem I complex had 41% of the chlorophyll *a*. The remaining 52% of the chlorophyll *a* was unbound from protein. Huang and Berns[35] and Huang et al.[36] have isolated several chlorophyll-protein complexes from the cyanobacterium *Phormidium luridum* by a nondetergent method. One of their six complexes was enriched in photosystem I activity and had approximately 49% of the total cellular chlorophyll *a*. Other complexes isolated by this method also exhibited photosystem I activity.

Guikema and Sherman[37,38] isolated chlorophyll-protein complexes from the cyanobacterium *Anacystis nidulans*. They studied the lactoperoxidase-catalyzed iodination of the membrane proteins and determined which of the thylakoid proteins were partially exposed on the membrane surface. The chlorophyll-proteins of photosystems I and II were among the iodinated proteins. Obviously, those proteins on the outer surface of the thylakoid were accessible to receiving excitation energy from the phycobilisomes.

D. Photosystem II-Phycobilisome Complexes

Following some earlier studies on phycobilisome-thylakoid isolations,[39,40] Clement-Metral and Gantt[41] have isolated a photosystem II-phycobilisome complex from the red alga *Porphyridium cruentum*. Photosystem II activity as measured by O_2 evolution was high in these complexes. Chereskin et al.[42] demonstrated the functional coupling of the phycobilisomes in these preparations when they showed that green light absorbed by the B- and b-phycoerythrins of this alga was effective in producing photosystem II activity. Clement-Metral et al.[43] further studied these functionally intact photosystem II-phycobilisome preparations. In the electron microscope, phycobilisomes were occasionally observed in clusters of three. A small particle was observed in the center of these clusters (Figure 2), and perhaps these were the photosystem II proteins. The photosystem II-phycobilisome complexes had a ratio of 90 chlorophyll *a* molecules per phycobilisome, whereas the cell had about 1200 chlorophyll *a* molecules per phycobilisome. P700 was absent from these complexes. These results were consistent with previously discussed studies that suggested that only a small amount of the total chlorophyll *a* was associated with photosystem II, although the absolute values of chlorophyll distribution varied from organism to organism.

III. FREEZE-FRACTURE PARTICLES

While the action spectra of photosystems I and II and the fluorescence studies at $-196°C$ indicated that the phycobilisomes were closely associated with photosystem II, electron microscopy revealed information on the structure of the thylakoid membrane and the phycobilisomes on the surface of the membrane. Excitation energy transfer from the phycobilisomes to the chlorophyll *a* in the thylakoids was very efficient, and it was reasonable to anticipate close contact between the final emitters from the phycobilisomes (allophycocyanins I and B) and a chlorophyll *a*-protein complex receptor on the surface of the thylakoid. The first clues to this organization of the light-harvesting apparatus were the observations by electron microscopy that some of the phycobilisomes and certain particles in the thylakoid occurred as parallel rows.[44-50] The spacing between the rows of phycobilisomes or rows of certain thylakoid particles (10-nm exoplasmic freeze-fracture [EF] particles) was very similar. Likewise, the distance between adjacent phycobilisomes in a row and adjacent EF particles in their rows were quite similar. Another type of thylakoid particle is the protoplasmic freeze-fracture (PF) particle. The protoplasmic side of the membrane is the monolayer that is in contact with the stroma, and the exoplasmic side is the half of the bilayer that is inside the thylakoid next to the intrathylakoid space. Freeze-fracture studies on cyanobacteria and red algae can yield views of phycobilisomes attached to the thylakoid similar to views obtained by classic sectioning and negative staining (Figure 3). When the fracture occurred through

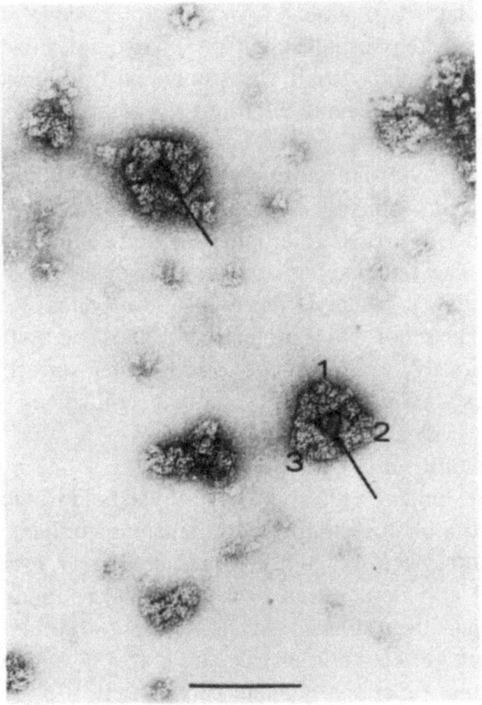

FIGURE 2. The electron microscopy of photosystem II-phycobilisome complexes. Phycobilisomes are frequently found in clusters, and one such cluster has the three phycobilisomes numbered. The arrows point to probable photosystem II particles. The bar equals 100 nm. The micrograph is courtesy of E. Gantt. (From Clement-Metral, J. D., Gantt, E., and Redlinger, T., *Arch. Biochem. Biophys.*, 238, 10, 1985. With permission.)

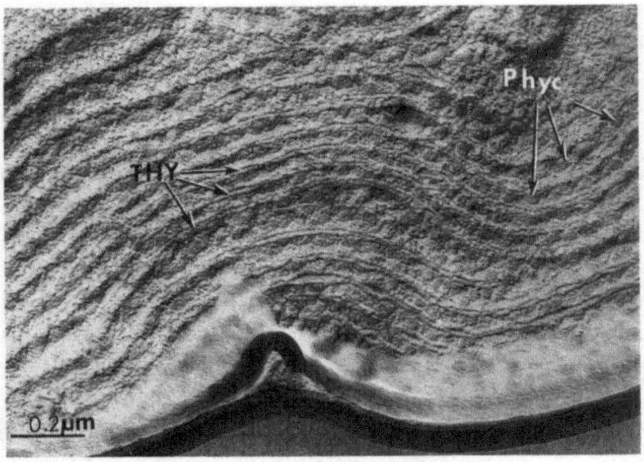

FIGURE 3. Freeze-fracture electron micrograph of the cyanobacterium *Aphanocapsa* 6308. Phycobilisomes (Phyc) are shown attached to the thylakoid (THY) membranes. The micrograph is courtesy of M. Lefort-Tran. (From Lefort-Tran, M., Cohen-Bazire, G., and Pouphile, M., *J. Ultrastruct. Res.*, 44, 199, 1973. With permission.)

FIGURE 4. Freeze-fracture electron micrograph of the red alga *Porphyridium cruentum*. Rows of EF particles are observed on the fractures of the thylakoids. Freeze-fracture entails freezing cells and cracking them with a knife. A fracture plane very readily occurs along the middle of the membrane bilayer at the interior junction of the lipids. The proteins in the membrane are then exposed and may appear on either the protoplasmic or the exoplasmic fracture side. The micrograph is courtesy of M. Neushul. (Magnification × 15,000.) (From Neushul, M., *J. Ultrastruct. Res.*, 37, 532, 1971. With permission.)

the thylakoid membrane, the particles (presumably chlorophyll *a*-protein complexes) which were integral parts of the thylakoid membrane were observed (Figures 4 to 6). The EF particles were frequently observed as long, parallel rows, and the rows were typically observed to be about 50 nanometers (nm) apart.

Lefort-Tran et al.[49] used freeze-fracture electron microscopy to study the thylakoids of various cyanobacteria and red algae. They observed phycobilisomes located in regular rows spaced 50 to 60 nm apart. EF particles were likewise arranged in rows having similar 50 to 60-nm spacings (Figure 5). They cautioned that although this was perhaps suggestive of a relationship between a phycobilisome and an EF particle, it was premature to accept fully the concept without additional data.

Lichtlé and Thomas[50] likewise observed parallel rows of both phycobilisomes and EF particles. They presented electron micrographs using thin-sectioned, negatively stained, and freeze-fracture preparations of the cyanobacterium *Oscillatoria splendida* (Figure 6). In Figure 6A, phycobilisomes were observed in the three basic views: face, edge, and top.

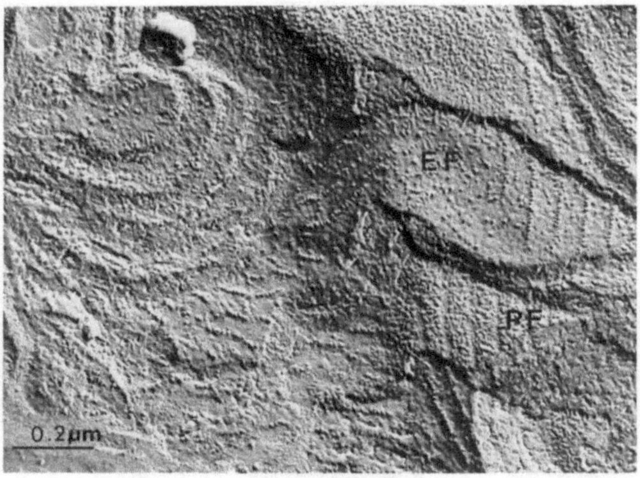

FIGURE 5. Freeze-fracture electron micrograph of *Aphanocapsa* 6308. Rows of EF particles are observed. The micrograph is courtesy of M. Lefort-Tran. (From Lefort, M., Cohen-Bazire, G., and Pouphile, M., *J. Ultrastruct. Res.*, 44, 199, 1973. With permission.)

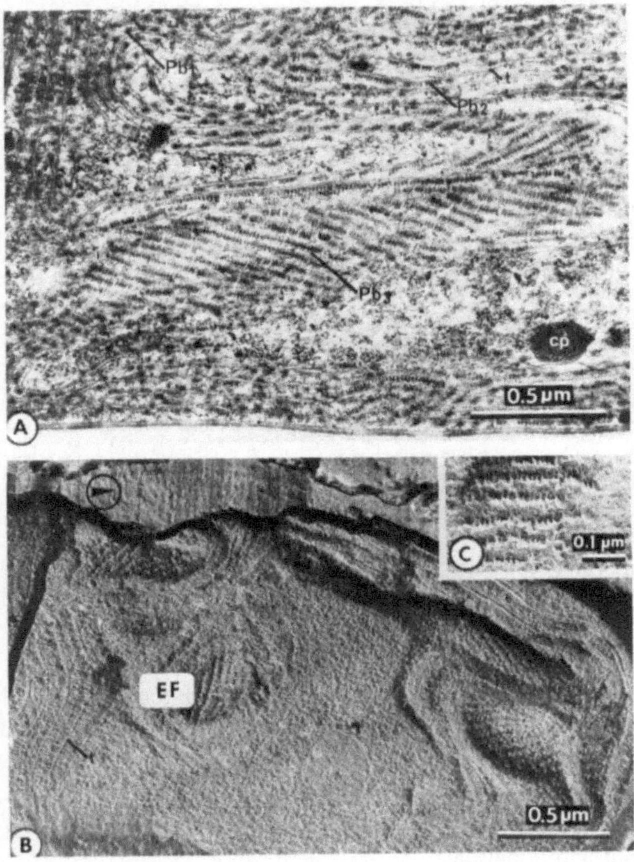

FIGURE 6. Freeze-fracture and thin-section electron micrographs of the cyanobacterium *Oscillatoria splendida.* (A) Thin-section electron micrograph showing phycobilisomes in face (Pb_1), edge (Pb_2), and top (Pb_3) views. (B) and (C) Freeze-fracture electron micrographs showing EF particles in rows. The micrographs are courtesy of J. C. Thomas. (From Lichtlé, C. and Thomas, J. C., *Phycologia*, 15, 393, 1976. With permission.)

The top view (Pb_3) showed the phycobilisomes as they were oriented on the thylakoid surface, and long rows arranged parallel to each other were clearly seen. Likewise, the freeze-fracture electron micrographs (Figures 6B and 6C) illustrated the long rows of EF particles lying parallel to each other. The row-to-row spacing was identical for phycobilisomes and EF particles.

This raised questions about the function of the EF particles and the relationship of the EF particles to phycobilisomes. Freeze-fracture studies by Mörschel and Mühlethaler[51] and Giddings et al.[52] produced views that were suggestive of a direct coupling of EF particles and phycobilisomes. Electron micrographs of the cyanelles of *Cyanophora paradoxa* exhibited the same type of phycobilisome and EF particle arrangements observed previously in other organisms.[52] Freeze-fracture electron micrographs showed that the EF particles were 9.4 nm apart (center-to-center) in long rows, and a 45-nm spacing was observed between the parallel rows. The rows of phycobilisomes were also about 45 nm apart, and the spacing between the hemi-discoidal phycobilisomes within a row was 10 nm. Freeze-fracture electron micrographs of isolated thylakoids which retained their phycobilisomes were also examined. The retention of the phycobilisomes on the membrane was accomplished by a method similar to that of Katoh and Gantt.[40] In these preparations, some of the phycobilisomes had detached, and rows of the underlying particles were revealed.[52] They concluded that the rows of phycobilisomes were directly above particles projecting out of the protoplasmic surface of the thylakoid membrane and that this was evidence for phycobilisome and EF particle association.

Simultaneously, Mörschel and Mühlethaler[51] produced very similar evidence from electron microscopy of the cyanobacterium *Mastigocladus laminosus*. As was generally observed, the EF particles were in rows 45 nm apart with a particle-to-particle spacing of 12 nm. The hemi-discoidal phycobilisomes were arranged in a similar fashion on the thylakoid surface. Some of their freeze-fracture micrographs revealed phycobilisomes and EF particles on the same thylakoid (Figure 7). These micrographs suggested a direct alignment of the two particle types.

Having obtained some support for the association of EF particles with phycobilisomes, the identity of the EF particles can be considered. Since the phycobilisomes were known to transfer excitation energy to photosystem II, the possibility of EF particles being the chlorophyll *a*-protein complexes of photosystem II was frequently discussed. Freeze-fracture micrographs of the red alga *Cyanidium caldarium*[53] and of heterocysts of the cyanobacterium *Anabaena cylindrica*[54] both supported this concept.

Wollman[53] has studied a biliprotein-less mutant and wild-type cells of *C. caldarium* in regard to the distribution of EF particles in their thylakoid membranes. Both the wild type and the biliprotein-less mutant had EF particles, but there was a 77% increase in the number in the mutant. Spectroscopy on the cells demonstrated that the photosystem II content increased by a very similar 81% in the mutant and that photosystem I showed a very much smaller increase. It was suggested that the EF particles were part of photosystem II.

Giddings and Staehelin[54] performed freeze-fracture electron microscopy on heterocysts of a cyanobacterium. Heterocysts do not have photosystem II activity. The 10-nm EF particles of the vegetative cells were lacking in the heterocysts. A 5.5-nm EF particle was found that Giddings and Staehelin suggested might be a nitrogenase complex or a residue from the 10-nm EF particles. They also concluded that the 10-nm EF particles of the vegetative cells were part of photosystem II.

Therefore a structural link exists between the phycobilisomes and the EF particles; a functional link is evident between the phycobilisomes and photosystem II, and the EF particle content can be quantitated with photosystem II activity. In certain cyanobacteria, it may well be that a single phycobilisome is situated on top of a single EF (photosystem II) particle. Giddings et al.[52] have proposed a model (Figure 8) for the structural arrangement of the

FIGURE 7. Freeze-fracture electron micrograph through a thylakoid of the cyanobacterium *Mastigocladus laminosus*. This micrograph appears to demonstrate that the EF particles are located directly under the phycobilisomes. The arrows point to the center of the phycobilisome bases. The bar equals 100 nm. This micrograph is courtesy of E. Mörschel. (From Mörschel, E. and Mühlethaler, K., *Planta*, 158, 451, 1983. With permission.)

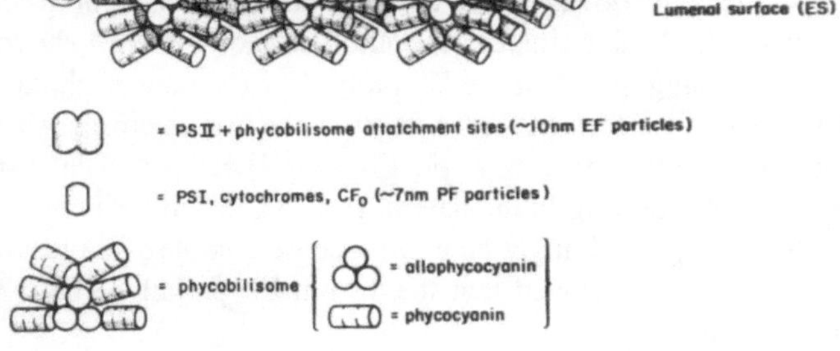

FIGURE 8. A model for a cyanobacterial thylakoid. This model is based on results from *Cyanophora paradoxa* and uses other concepts from the literature. Each EF particle is drawn showing two possible subunits as observed in electron micrographs. Rows of EF particles are shown to be located directly under rows of hemi-discoidal phycobilisomes. The CF is the subunit coupling factor. (From Giddings, T. H., Jr., Wasmann, C., and Staehelin, L. A., *Plant Physiol.*, 71, 409, 1983. With permission.)

thylakoid membranes of *Cyanophora paradoxa* that incorporated these relationships. The photosystem I complexes were assigned to the 7-nm PF particles.

Manodori et al.[55] have proposed a model for the thylakoids of *Anacystis nidulans* using data on this cyanobacterium which was found in earlier publications from various laboratories as well as their own. Each photosystem II complex had 35 chlorophyll *a* molecules plus its reaction center (P680). Each photosystem I complex had 140 chlorophyll *a* molecules plus its reaction center (P700). Manodori and Melis[56] modified this model to include excitation energy transfer from a single phycobilisome to two photosystem II complexes, each having its own P680 activity, and suggested that a reaction center could uncouple from a phycobilisome. In Chapter 6, other examples of possible phycobilisome-thylakoid uncoupling are discussed.

IV. NUMBER OF PHYCOBILISOMES PER REACTION CENTER

A major consideration in defining the thylakoid structure is determining the ratio of phycobilisomes to photosystem II reaction centers. Stevens and Myers[57] suggested that the ratio of photosystem II reaction centers to phycobilisomes was variable. They studied the wild type and a nitrosoguanidine-induced pigment mutant of the cyanobacterium *A. nidulans* and estimated 1.1 phycobilisomes per reaction center II in the wild type and 3.5 per reaction center II in a particular mutant. This mutant which showed a high reversion frequency had a slightly higher C-phycocyanin content and considerably less chlorophyll *a*. Vierling and Albert[58] discussed these ratios for *A. nidulans*. In Chapter 7, mutants of this cyanobacterium which were induced by far-red light are discussed.

Kursar and Alberte[59] studied the relationship between the phycobilisomes and reaction centers for the red alga *Neoagardhiella bailyei*. Although many of the photosynthetic properties for this alga were similar to those for *A. nidulans*, it was found to have more reaction center II complexes per phycobilisomes (4.1) than did the cyanobacterium (1.7). Kursar and Alberte[59] suggested that the larger size of the red alga phycobilisome might influence its association with the reaction centers. Ley[60] determined the number of photosystem II reaction centers coupled to a single phycobilisome by measuring the effective absorption cross-sections of the red alga *P. cruentum*. All the photosystem II reaction centers received excitation energy from phycobilisomes. Although a definitive calculation was not feasible from existing data, it was clear that several (perhaps three or four) photosystem II reaction centers could receive excitation energy from one phycobilisome. This result differed from that of Diner,[61] whose data on *Cyanidium caldarium* showed that about half of the photosystem II reaction centers were not attached to phycobilisomes. Freeze-fraction electron microscopy showed that a single EF particle might associate with a single phycobilisome (Figure 7). It is therefore possible that each EF particle may represent more than one reaction center.

Another topic concerning thylakoid structure is the ratio of photosystem II reaction centers to photosystem I reaction centers. Kawamura et al.[62] studied the quantities of the two reaction centers for three cyanobacteria. When grown in weak-intensity light, the P700 content was always greater than the number of photosystem II reaction centers as measured from O_2 yield. As the growth light intensity increased, the ratio approached unity. They suggested that the photosystem I reaction centers were the variable component in the thylakoid. Manodori et al.[55] also found comparably fewer photosystem II reaction centers in a cyanobacterium. Kursar and Alberte[59] found that the ratio of photosystem reaction centers I to II was approximately 2.3 for both a cyanobacterium and a red alga. Myers et al.[63] studied the ratio of photosystem reaction centers I to II for *A. nidulans* and its pigment mutants. For several distinct clones, the ratios were quite constant under identical growth conditions: for clones grown in gold light, the ratio was 2:1, and for those grown far-red light, the ratio was 1:1.

V. CRYPTOMONAD THYLAKOIDS

Very little is known about the thylakoids of the cryptomonads. Since cryptomonad thylakoids contain chlorophyll c_2, this is a starting point in understanding the properties of these membranes. Because of its absence in land plants, chlorophyll *c* is studied much less than chlorophyll *a* or *b*. However, it is found extensively in marine organisms and is sure to receive detailed scrutiny in the future.

Jeffrey[64-66] has studied many of the properties of this chlorophyll. It is found in two types. Some organisms — brown seaweeds, diatoms, and chrysomonads — contained both chlorophyll c_1 and c_2, while dinoflagellates and cryptomonads had only chlorophyll c_2. A purification procedure was developed for both types. Chlorophyll c_1 had visible absorption maxima at 580 and 630 nm, and its fluorescence emission maximum was at 633 nm with a small maximum at 694 nm. The 630-nm absorption band was stronger than the 580-nm band. Chlorophyll c_2 had maxima at 581 and 630 nm, with each absorption band having comparable absorbances. Its fluorescence emission maximum was at 635 nm with a smaller maximum at 696 nm. All these spectra were for chlorophylls dissolved in acetone.[64,65] The cryptomonads were the only chlorophyll *c*-containing organisms to have biliproteins. These biliproteins, however, do not form phycobilisomes (see Chapter 9 for details on cryptomonads). Jeffrey[65] found that for the cryptomonad *Chroomonas* sp. the ratio of chlorophyll *a* to c_2 was 2.5:1. Strain et al.[67] determined that chlorophyll c_1 was magnesium tetradehydropheoporphyrin a_5 monomethyl ester and chlorophyll c_2 was magnesium hexadehydropheoporphyrin a_5 monomethyl ester.

Ingram and Hiller[68] have isolated a chlorophyll a/c_2 complex from the cryptomonad *Chroomonas* sp. Using sodium dodecyl sulfate gel electrophoresis, three chlorophyll-protein complexes were obtained. Two bands contained chlorophyll *a*, and the other contained chlorophyll *a* and c_2 in a 1.4:1 molar ratio. This latter complex had absorption maxima at 672, 640, and 588 nm and a principal fluorescence emission band at 682 nm. The Soret band maximum for chlorophyll *a* was at 436 nm, and for chlorophyll c_2 it was at 458 nm. When the chlorophyll a/c_2-protein complex at -196°C was excited at 460 nm, the main emission band was at 682 nm, with a much smaller band at 640 nm. The emission at 682 nm demonstrated efficient excitation energy transfer from chlorophyll c_2 to *a*. The small 640-nm component showed that the efficiency was less than 100%. Chlorophyll *c*-protein complexes have also been studied for organisms other than the cyrptomonads.[69-76]

In Chapter 9, fluorescence and O_2 evolution studies on cryptomonads are discussed.

REFERENCES

1. **Engelmann, T. W.,** Farbe und Assimilation, *Bot. Ztg.*, 41, 1, 1883.
2. **Engelmann, T. W.,** Untersuchungen über die quantitativen Beziehungen zwischen Absorption des Lichtes und Assimilation in Pflanzenzellen, *Bot. Ztg.*, 42, 81, 1884.
3. **Emerson, R. and Lewis, C. M.,** The photosynthetic efficiency of phycocyanin in *Chroococcus*, and the problem of carotenoid participation in photosynthesis, *J. Gen. Physiol.*, 25, 579, 1942.
4. **Emerson, R. and Lewis, C. M.,** The dependence of the quantum yield of *Chlorella* photosynthesis on the wave length of light, *Am. J. Bot.*, 30, 165, 1943.
5. **Haxo, F. T. and Blinks, L. R.,** Photosynthetic action spectra of marine algae, *J. Gen. Physiol.*, 33, 389, 1950.
6. **Emerson, R., Chalmers, R., and Cederstrand, C.,** Some factors influencing the long-wave limit of photosynthesis, *Proc. Natl. Acad. Sci. U.S.A.*, 43, 133, 1957.
7. **Haxo, F. T. and Fork, D. C.,** Photosynthetically active accessory pigments of cryptomonads, *Nature*, 184, 1051, 1959.

8. **Duysens, L. N. M. and Amesz, J.**, Function and identification of two photochemical systems in photosynthesis, *Biochim. Biophys. Acta*, 64, 243, 1962.
9. **Amesz, J. and Duysens, L. N. M.**, Action spectrum, kinetics and quantum requirement of phosphopyridine nucleotide reduction and cytochrome oxidation in the blue-green alga *Anacystis nidulans*, *Biochim. Biophys. Acta*, 64, 261, 1962.
10. **Jones, L. W. and Myers, J.**, Enhancement in the blue-green alga, *Anacystis nidulans*, *Plant Physiol.*, 39, 938, 1964.
11. **Gingras, G.**, Étude comparative, chez quelques algues, de la photosynthése et de la photoréduction réalisée en présence d'hydrogéne, *Physiol. Vég.*, 4, 1, 1966.
12. **Ley, A. C. and Bulter, W. L.**, The distribution of excitation energy between photosystem I and photosystem II in *Porphyridium cruentum*, *Plant Cell Physiol.*, photosynthetic organelles issue, 33, 1977.
13. **Tel-Or, E. and Malkin, S.**, The photochemical and fluorescence properties of whole cells, spheroplasts and spheroplast particles from the blue-green alga *Phormidium luridum*, *Biochim. Biophys. Acta*, 459, 157, 1977.
14. **Wang, R. T., Stevens, C. L. R., and Myers, J.**, Action spectra for photoreactions I and II of photosynthesis in the blue-green alga *Anacystis nidulans*, *Photochem. Photobiol.*, 25, 103, 1977.
15. **Mimuro, M. and Fujita, Y.**, Estimation of chlorophyll *a* distribution in the photosynthetic pigment systems I and II of the blue-green alga *Anabaena variabilis*, *Biochim. Biophys. Acta*, 459, 376, 1977.
16. **Larkum, A. W. D. and Weyrauch, S. K.**, Photosynthetic action spectra and light-harvesting in *Griffithsia monilis* (Rhodophyta), *Photochem. Photobiol.*, 25, 65, 1977.
17. **Harnischfeger, G. and Codd, G. A.**, Liquid nitrogen fluorescence studies of the photosynthetic apparatus of blue-green algae, *Br. Phycol. J.*, 12, 225, 1977.
18. **Ried, A., Hessenberg, B., Metzler, H., and Ziegler, R.**, Distribution of excitation energy among photosystem I and photosystem II in red algae. I. Action spectra of light reactions I and II, *Biochim. Biophys. Acta*, 459, 175, 1977.
19. **Lemasson, C., Tandeau de Marsac, N., and Cohen-Bazire, G.**, Role of allophycocyanin as a light-harvesting pigment in cyanobacteria, *Proc. Natl. Acad. Sci. U.S.A.*, 70, 3130, 1973.
20. **Peterson, R. B., Dolan, E., Calvert, H. E., and Ke, B.**, Energy transfer from phycobiliproteins to photosystem I in vegetative cells and heterocysts of *Anabaena variabilis*, *Biochim. Biophys. Acta*, 634, 237, 1981.
21. **Hill, R. and Bendall, F.**, Function of the two cytochrome components in chloroplasts: a working hypothesis, *Nature*, 186, 136, 1960.
22. **Murata, N., Nishimura, M., and Takamiya, A.**, Fluorescence of chlorophyll in photosynthetic systems. III. Emission and action spectra of fluorescence — three emission bands of chlorophyll *a* and the energy transfer between two pigment systems, *Biochim. Biophys. Acta*, 126, 234, 1966.
23. **Murata, N.**, Control of excitation transfer in photosynthesis. I. Light-induced change of chlorophyll *a* fluorescence in *Porphyridium cruentum*, *Biochim. Biophys. Acta*, 172, 242, 1969.
24. **Bonaventura, C. and Myers, J.**, Fluorescence and oxygen evolution from *Chlorella pyrenoidosa*, *Biochim. Biophys. Acta*, 189, 366, 1969.
25. **Ley, A. C. and Butler, W. L.**, Energy distribution in the photochemical apparatus of *Porphyridium cruentum* in state I and state II, *Biochim. Biophys. Acta*, 592, 349, 1980.
26. **Ley, A. C. and Butler, W. L.**, Efficiency of energy transfer from photosystem II to photosystem I in *Porphyridium cruentum*, *Proc. Natl. Acad. Sci. U.S.A.*, 73, 3957, 1976.
27. **Reid, A. and Reinhardt, B.**, Distribution of excitation energy between photosystem I and photosystem II in red algae. III. Quantum requirements of the induction of a state 2-state 1 transition, *Biochim. Biophys. Acta*, 592, 76, 1980.
28. **Biggins, J., Campbell, C. L., and Bruce, D.**, Mechanism of the light state transition in photosynthesis. II. Analysis of phosphorylated polypeptides in the red alga, *Porphyridium cruentum*, *Biochim. Biophys. Acta*, 767, 138, 1984.
29. **Biggins, J.**, Mechanism of the light state transition in photosynthesis. I. Analysis of the kinetics of cytochrome *f* oxidation in state 1 and state 2 in the red alga, *Porphyridium cruentum*, *Biochim. Biophys. Acta*, 724, 111, 1983.
30. **Bruce, D., Biggins, J., Steiner, T., and Thewalt, M.**, Mechanism of the light state transition in photosynthesis. IV. Picosecond fluorescence spectroscopy of *Anacystis nidulans* and *Porphyridium cruentum* in state *1* and state 2 at 77K, *Biochim. Biophys. Acta*, 806, 237, 1985.
31. **Karukstis, K. K. and Sauer, K.**, Energy transfer and distribution in the red alga *Porphyra perforata* studied using picosecond fluorescence spectroscopy, *Biochim. Biophys. Acta*, 766, 141, 1984.
32. **Ogawa, T., Vernon, L. P., and Mollenhauer, H. H.**, Properties and structure of fractions prepared from *Anabaena variabilis* by the action of Triton X-100, *Biochim. Biophys. Acta*, 172, 216, 1969.
33. **Thornber, J. P.**, Comparison of a chlorophyll *a*-protein complex isolated from a blue-green alga with chlorophyll-protein complexes obtained from green bacteria and higher plants, *Biochim. Biophys. Acta*, 172, 230, 1969.

34. **Redlinger, T. and Gantt, E.,** Photosynthetic membranes of *Porphyridium cruentum*. An analysis of chlorophyll-protein complexes and heme-binding proteins, *Plant Physiol.*, 73, 36, 1983.
35. **Huang, C. and Berns, D. S.,** Partial characterization of six chlorophyll *a*- protein complexes isolated from a blue-green alga by a nondetergent method, *Arch. Biochem. Biophys.*, 220, 145, 1983.
36. **Huang, C., Berns, D. S., and Guarino, D. U.,** Characterization of components of P-700-chlorophyll *a*-protein complex from a blue-green alga, *Phormidium luridum, Biochim. Biophys. Acta*, 765, 21, 1984.
37. **Guikema, J. and Sherman, L.,** Protein composition and architecture of the photosynthetic membranes from the cyanobacterium, *Anacystis nidulans* R2, *Biochim. Biophys. Acta*, 681, 440, 1982.
38. **Guikema, J. A. and Sherman, L. A.,** Chlorophyll-protein organization of membranes from the cyanobacterium *Anacystis nidulans, Arch. Biochem. Biophys.*, 220, 155, 1983.
39. **Cohen-Bazire, G. and Lefort-Tran, M.,** Fixation of phycobiliproteins to photosynthetic membranes by glutaraldehyde, *Arch. Mikrobiol.*, 71, 245, 1970.
40. **Katoh, T. and Gantt, E.,** Photosynthetic vesicles with bound phycobilisomes from *Anabaena variabilis, Biochim. Biophys. Acta*, 546, 383, 1979.
41. **Clement-Metral, J. D. and Gantt, E.,** Isolation of oxygen-evolving phycobilisome-photosystem II particles from *Porphyridium cruentum, FEBS Lett.*, 156, 185, 1983.
42. **Chereskin, B. M., Clement-Metral, J. D., and Gantt, E.,** Characterization of a purified photosystem II-phycobilisome particle preparation from *Porphyridium cruentum, Plant Physiol.*, 77, 626, 1985.
43. **Clement-Metral, J. D., Gantt, E., and Redlinger, T.,** A photosystem II-phycobilisome preparation from the red alga, *Porphyridium cruentum:* oxygen evolution, ultrasturcture, and polypeptide resolution, *Arch. Biochem. Biophys.*, 238, 10, 1985.
44. **Gantt, E. and Conti, S. F.,** Phycobiliprotein localization in algae, *Brookhaven Symp. Biol.*, 19, 393, 1966.
45. **Bourdu, R. and Lefort, M.,** Structure fine, observée en cryodécapage, des lamelles photosynthétiques des Cyanophycées endosymbiotiques: *Glaucocystis nostochinearum* Itzigs, et *Cyanophora paradoxa* Korschikoff, *C.R.*, 265, 37, 1967.
46. **Guérin-Dumartrait, E., Sarda, C., and Lacourly, A.,** Sur la structure fine du chloroplaste de *Porphyridium* sp. (Lewin), *C.R.*, 270, 1977, 1970.
47. **Neushul, M.,** A freeze-etching study of the red alga *Porphyridium, Am. J. Bot.*, 57, 1231, 1970.
48. **Neushul, M.,** Uniformity of thylakoid structure in a red, a brown, and two blue-green algae, *J. Ultrastruct. Res.*, 37, 532, 1971.
49. **Lefort-Tran, M., Cohen-Bazire, G., and Pouphile, M.,** Les membranes photosynthétiques des algues à biliproteines observées apres cryodécapage, *J. Ultrastruct. Res.*, 44, 199, 1973.
50. **Lichtlé, C. and Thomas, J. C.,** Étude ultrastructurale des thylacoides des algues a phycobiliproteines, comparaison des résultats obtenus par fixation classique et cryodécapage, *Phycologia*, 15, 393, 1976.
51. **Mörschel, E. and Mühlethaler, K.,** On the linkage of exoplasmatic freeze-fracture particles to phycobilisomes, *Planta*, 158, 451, 1983.
52. **Giddings, T. H., Jr., Wasmann, C., and Staehelin, L. A.,** Structure of the thylakoids and envelope membranes of the cyanelles of *Cyanophora paradoxa, Plant Physiol.*, 71, 409, 1983.
53. **Wollman, F.-A.,** Ultrastructural comparison of *Cyanidium caldarium* wild type and III-C mutant lacking phycobilisomes, *Plant Physiol.*, 63, 375, 1979.
54. **Giddings, T. H., Jr. and Staehelin, L. A.,** Changes in thylakoid structure associated with the differentiation of heterocysts in the cyanobacterium, *Anabaena cylindrica, Biochim. Biophys. Acta*, 546, 373, 1979.
55. **Manodori, A., Alhadeff, M., Glazer, A. N., and Melis, A.,** Photochemical apparatus organization in *Synechococcus* 6301 *(Anacystis nidulans)*. Effect of pycobilisome mutation, *Arch. Microbiol.*, 139, 117, 1984.
56. **Manodori, A. and Melis, A.,** Phycobilisome-photosystem II association in *Synechococcus* 6301 (Cyanophyceae), *FEBS Lett.*, 181, 79, 1985.
57. **Stevens, C. L. R. and Myers, J.,** Characterization of pigment mutants in a bluegreen alga, *Anacystis nidulans, J. Phycol.*, 12, 99, 1976.
58. **Vierling, E. and Alberte, R. S.,** Functional organization and plasticity of the photosynthetic unit of the cyanobacterium *Anacystis nidulans, Physiol. Plant.*, 50, 93, 1980.
59. **Kursar, T. A. and Alberte, R. S.,** Photosynthetic unit organization in a red alga. Relationships between light-harvesting pigments and reaction centers, *Plant Physiol.*, 72, 409, 1983.
60. **Ley, A. C.,** Effective absorption cross-sections in *Porphyridium cruentum*. Implications for energy transfer between phycobilisomes and photosystem II reaction centers, *Plant Physiol.*, 74, 451, 1984.
61. **Diner, B. A.,** Energy transfer from the phycobilisomes to photosystem II reaction centers in wild type *Cyanidium caldarium, Plant Physiol.*, 63, 30, 1979.
62. **Kawamura, M., Mimuro., M., and Fujita, Y.,** Quantitative relationship between two reaction centers in the photosynthetic system of blue-green algae, *Plant Cell Physiol.*, 20, 697, 1979.
63. **Myers, J., Graham, J.-R., and Wang, R. T.,** Light harvesting in *Anacystis nidulans* studied in pigment mutants, *Plant Physiol.*, 66, 1144, 1980.

64. **Jeffrey, S. W.**, Properties of two spectrally different components in chlorophyll *c* preparations, *Biochim. Biophys. Acta*, 177, 456, 1969.
65. **Jeffrey, S. W.**, Preparation and some properties of crystalline chlorophyll c_1 and c_2 from marine algae, *Biochim. Biophys. Acta*, 279, 15, 1972.
66. **Jeffrey, S. W.**, The occurrence of chlorophyll c_1 and c_2 in algae, *J. Phycol.*, 12, 349, 1976.
67. **Strain, H. H., Cope, B. T., Jr., McDonald, G. N., Svec, W. A., and Katz, J. J.**, Chlorophylls c_1 and c_2, *Phytochemistry*, 10, 1109, 1971.
68. **Ingram, K. and Hiller, R. G.**, Isolation and characterization of a major chlorophyll a/c_2 light-harvesting protein from a *Chroomonas* species (Cryptohyceae), *Biochim. Biophys. Acta*, 722, 310, 1983.
69. **Kirk, J. T. O.**, Thermal dissociation of fucoxanthin-protein binding in pigment complexes from chloroplasts of *Hormosira* (Phaeophyta), *Plant Sci. Lett.*, 9, 373, 1977.
70. **Holdsworth, E. S. and Arshad, J. H.**, A manganese-copper-pigment-protein complex isolated from the photosystem II of *Phaeodactylum tricornutum*, *Arch. Biochem. Biophys.*, 183, 361, 1977.
71. **Prézelin, B. B. and Alberte, R. S.**, Photosynthetic characteristics and organization of chlorophyll in marine dinoflagellates, *Proc. Natl. Acad. Sci. U.S.A.*, 75, 1801, 1978.
72. **Boczar, B. A., Prézelin, B. B., Markwell, J. P., and Thornber, J. P.**, A chlorophyll *c*-containing pigment-protein complex from the marine dinoflagellate, *Glenodinium* sp., *FEBS Lett.*, 120, 243, 1980.
73. **Barrett, J. and Anderson, J. M.**, The P-700-chlorophyll *a*- protein complex and two major light-harvesting complexes of *Acrocarpia paniculata* and other brown seaweeds, *Biochim. Biophys. Acta*, 590, 309, 1980.
74. **Alberte, R. S., Friedman, A. L., Gustafson, D. L., Rudnick, M. S., and Lyman, H.**, Light-harvesting systems of brown algae and diatoms. Isolation and characterization of chlorophyll *a/c* and chlorophyll *a*/fucoxanthin pigment-protein complexes, *Biochim. Biophys. Acta*, 635, 304, 1981.
75. **Gugliemelli, L. A., Dutton, H. J., Jursinic, P. A., and Siegelman, H. W.**, Energy transfer in a light-harvesting carotenoid-chlorophyll *c*- chlorophyll *a*-protein of *Phaeodactylum tricornutum*, *Photochem. Photobiol.*, 33, 903, 1981.
76. **Friedman, A. L. and Alberte, R. S.**, A diatom light-harvesting pigment-protein complex. Purification and characterization, *Plant Physiol.*, 76, 483, 1984.

Chapter 9

CRYPTOMONADS

I. CELLULAR FEATURES

The cryptomonads are a groups of eukaryotic organisms found in fresh and marine waters (Figure 1). They have some features characteristic of protozoa but also have chloroplasts which contain biliprotein, chlorophyll c_2 and chlorophyll *a*. The chloroplasts have thylakoids (Figures 2 and 3) that do not exhibit phycobilisomes on their outer membrane surfaces when examined by electron microscopy.[1-3] Wehrmeyer[2] observed that there was a dense granular material in the intrathylakoidal space. Dodge[1] also observed this material and suggested that it should be tested to determine if it was the biliproteins.

Gantt et al.[3] followed the extraction of biliproteins from three cyrptomonads and demonstrated that the disappearance of the intrathylakoid material was linked with the appearance of biliprotein in aqueous extracts. This was good evidence for an intrathylakoid location of the biliproteins (Figure 4). The possible inversion of the thylakoid was considered, since the biliproteins in cyanobacteria and red algae were located on the stroma side rather than the intrathylakoid side of the membrane.

Faust and Gantt[4] tested this proposed location of the biliproteins by examining the cryptomonad *Chroomonas* sp. grown in various light intensities. Algae grown in low light intensity had 30% more phycocyanin 645 than the cells grown at high light intensities. As expected from a location of the biliproteins within the thylakoid, the width of the thylakoids — grown in low light intensity — was 30 nm, and those grown in high intensity had a narrower (20 nm) width. Subsequently, a corresponding result was obtained in a similar experiment.[5]

Lichtlé and Thomas[6] studied the *Cryptomonas* thylakoid by freeze-fracture electron microscopy. They contained thylakoid surfaces which varied in the density of particles. The surface with a low density was designated EFu, and the higher population occurred on the EFs surface. These EF (exoplasmic freeze-fracture) particles may be the chlorophyll-protein complexes of photosystem II (see Chapter 8). The EFu particles may occur in unstacked regions of the thylakoid, and the EFs in stacked regions.

Several morphological features of the cryptomonads were distinct from the other biliprotein-containing organisms; one example was the ejectosomes or trichocysts.[7] Its name was derived from the ability of the ejectosomes to be expelled out of the organism. An electron micrograph shows a series of levels of extrusion of three different ejectosomes from the same cryptomonad (Figure 5). A particular section of an ejectosome particle suggests that it is built up as a spiral, and a possible end point of the spiral is evident (Figure 6). The inset (Figure 6) shows some typical views at a lower magnification.

Despite various distinctive features, there are appreciable data that indicate an evolution of the cryptomonad chloroplast from red algal and cyanobacterial precursors (see Chapter 10).

II. BILIPROTEINS

A. Distribution

Biliproteins were first isolated from cryptomonads in 1959 by three laboratories.[8-10] The absorption spectra of these biliproteins were quite distinctive. In general, only one biliprotein was found in any cryptomonad, with the possible exception of *Sennia* sp.[9] This alga had phycoerythrin 545, but a small amount of a phycocyanin was also noted. The phycocyanin

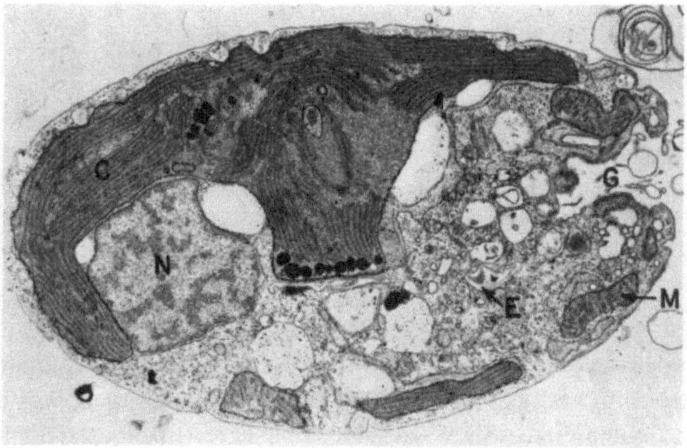

FIGURE 1. Electron micrograph of a thin section of a cryptomonad cell. The organism is *Chroomonas* sp. The symbols are M, mitochondrion; G, gullet; E, ejectosome; N, nucleus; and C, chloroplast. Although not decipherable at this magnification, the chloroplast is encompassed by four membranes. The micrograph is courtesy of M. R. Edwards. (Magnification × 11,000.)

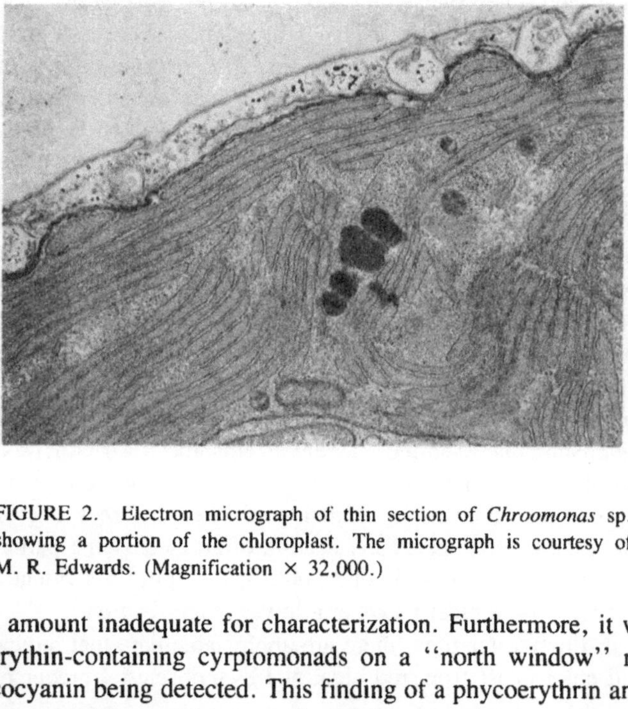

FIGURE 2. Electron micrograph of thin section of *Chroomonas* sp. showing a portion of the chloroplast. The micrograph is courtesy of M. R. Edwards. (Magnification × 32,000.)

was found in an amount inadequate for characterization. Furthermore, it was reported that growing phycoerythin-containing cyrptomonads on a "north window" resulted in small amounts of phycocyanin being detected. This finding of a phycoerythrin and a phycocyanin in the same cryptomonad has not been confirmed by any subsequent investigator, and it is usual to accept the premise that each cryptomonad produces only one biliprotein.

O'hEocha et al.[11] demonstrated that there were six spectroscopic classes of cryptomonad biliproteins: three phycoerythrins and three phycocyanins. These classes were obtained from the fairly close groupings of the longest wavelength absorption maxima of these biliproteins, and the proteins were named for the position of this characteristic maximum in nanometers. The classes were phycoerythrins 545, 555, and 566 and phycocyanins 612, 630, and 645 (Figures 7 to 9). Phycocyanin 630 was found only in a single cryptomonad *Cryptomonas cyanomagna,* but the remaining types have each been isolated from several species. The

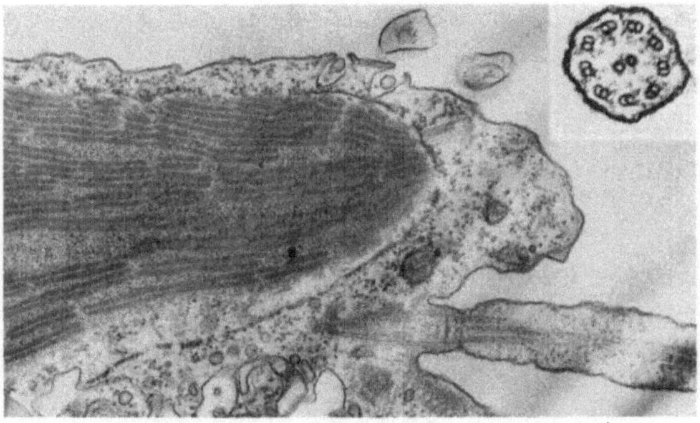

FIGURE 3. Electron micrograph of thin section of *Rhodomonas lens* showing a portion of the chloroplast. The flagella is shown in both longitudinal section and cross-section (inset). The thylakoids are shown in a characteristically paired arrangement. The micrographs are courtesy of M. R. Edwards. (Magnification × 24,000; inset × 39,000.)

FIGURE 4. Model of the thylakoids in cryptomonad. Most electron micrographs of thin sections of cryptomonads show that the thylakoids are paired.

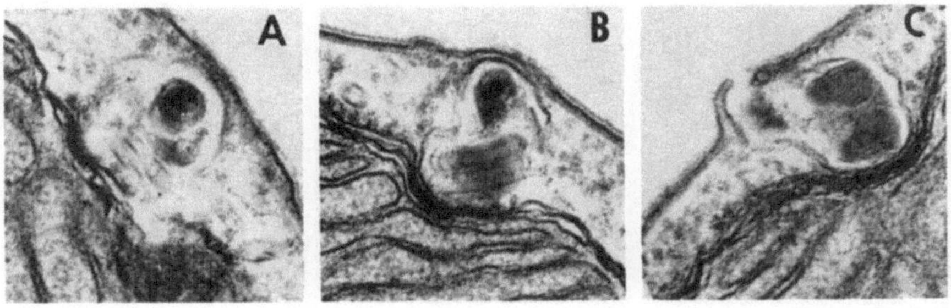

FIGURE 5. Electron micrographs of thin sections showing three stages of ejectosome release. These three views are selected from a single electron micrograph of *Chroomonas* sp., and ultimately the ejectosomes are completely extruded into the surroundings. The micrographs are courtesy of M. R. Edwards.

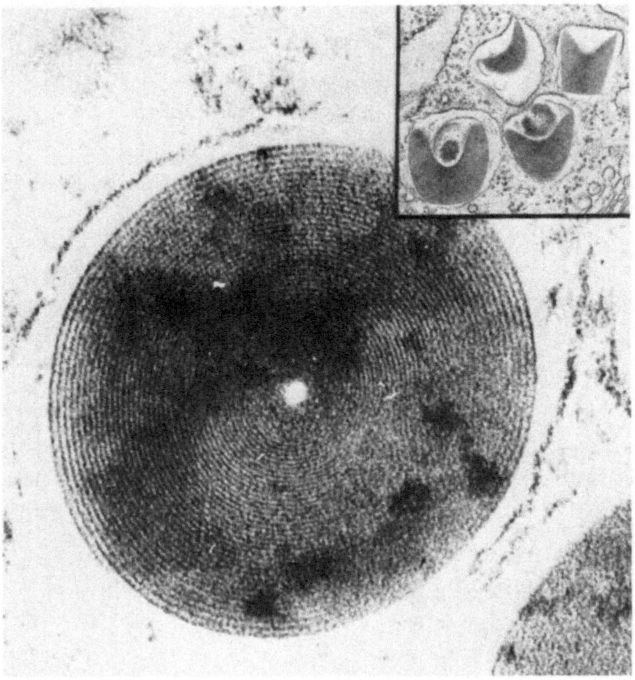

FIGURE 6. Electron micrograph of thin section of an ejectosome from *Chroomonas* sp. This micrograph suggests a spiral assembly of subunits. The inset shows a variety of typical views of ejectosomes. The magnification of the single ejectosome, × 115,000. The micrographs are courtesy of M. R. Edwards.

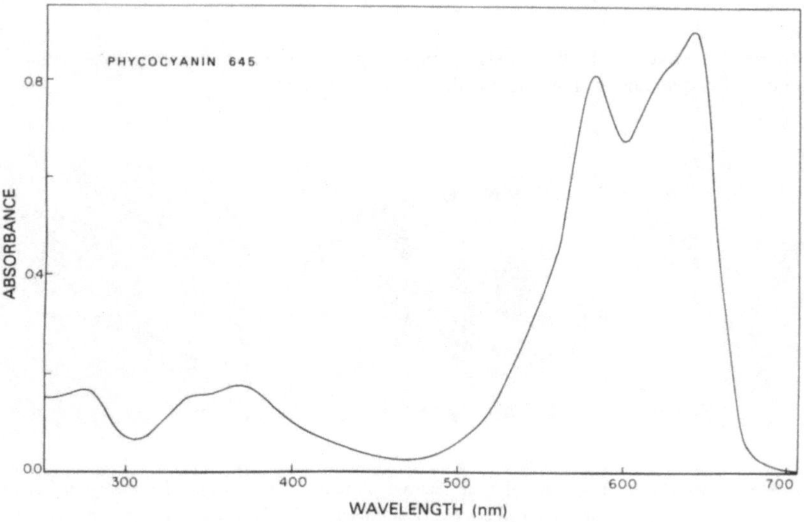

FIGURE 7. Absorption spectrum of phycocyanin 645 (*Chroomonas* sp.). The sample is in pH 6.0, 0.1-ionic strength sodium phosphate buffer at room temperature. The absorption band containing cryptoviolin is prominent at 585 nm.

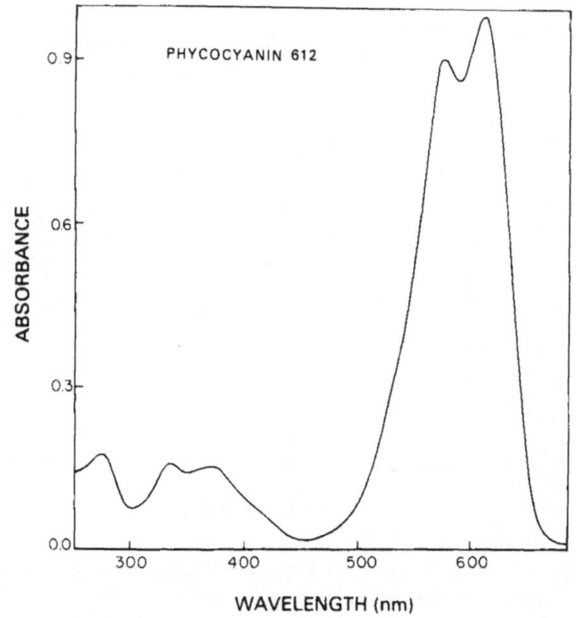

FIGURE 8. Absorption spectrum of phycocyanin 612 *(Hemiselmis virescens)*. The sample is in pH 6.0, 0.1-ionic strength sodium phosphate buffer at room temperature. The band at 575 nm contains a major cryptoviolin contribution, and the band at 612 nm is largely phycocyanobilin.

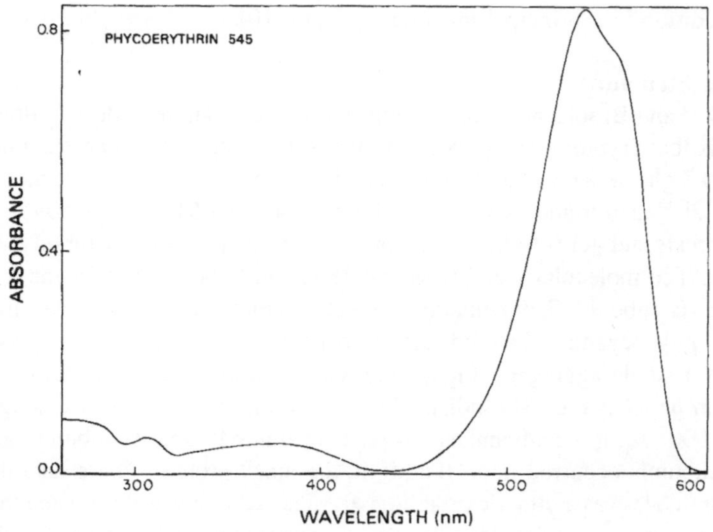

FIGURE 9. Absorption spectrum of phycoerythrin 545 *(Rhodomonas lens)*. The sample is in pH 6.0, 0.1-ionic strength sodium phosphate buffer at room temperature.

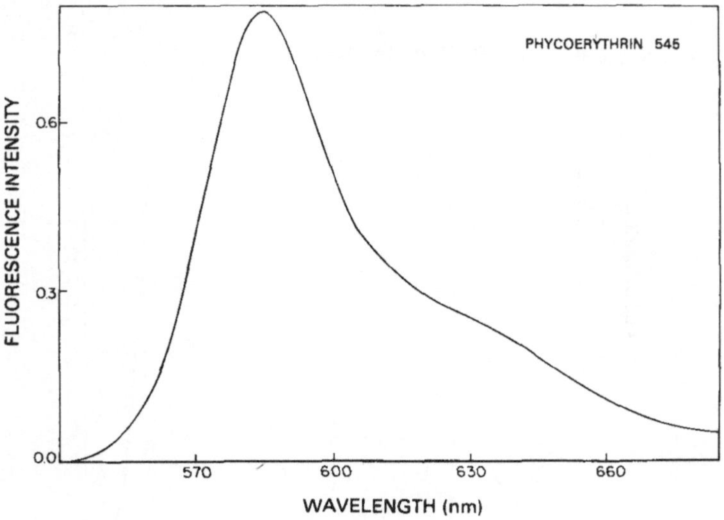

FIGURE 10. Fluorescence emission spectrum (fully corrected) of phycoerythrin 545 *(Rhodomonas lens)*. The sample has an absorption of 0.05 at 545 nm in a 1-cm light path. The emission spectrum was taken on a sample in pH 6.0, 0.1-ionic strength sodium phosphate buffer at room temperature. Excitation of various wavelengths throughout the visible absorption band gave identical emission bands. Picosecond time-resolved fluorescence kinetics studies have demonstrated, however, that the emission is not completely homogenous. A small amount of fluorescence leakage does occur from higher-energy "sensitizing" chromophores.

isolated cryptomonad biliproteins were fluorescent and typically had a long wavelength shoulder in addition to a principal maximum (Figure 10).

B. Biliprotein Structure

Glazer et al.[12] and Brooks and Gantt[13] demonstrated by sodium dodecyl sulfate (SDS) gel electrophoresis that cryptomonad phycoerythrins were composed of two subunits dissimilar in molecular weight: α with 10,000 mol wt and β with 18,000 to 19,000 mol wt.

MacColl et al.[14] also found two subunits for phycocyanin 645 *(Chroomonas* sp.) by both gel electrophoresis and gel filtration in the presence of the protein denaturant 6 *M* guanidine hydrochloride. The molecular weights were determined to be 10,000 for the α subunit and 16,000 for the β subunit. Sedimentation velocity studies on the analytical ultracentrifuge indicated that phycocyanin 645 (and later also phycoerythrin 545 and phycocyanin 612) sedimented as a single aggregate (Figure 11). Gel filtration and sedimentation equilibrium experiments on phycocyanin 645 indicated that the molecular weight of this aggregate was 49,000 to 53,000. Amino acid analysis of phycocyanin 645 and its subunits indicated that the α and β subunits occurred in a 1:1 ratio in the intact protein. These data demonstrated that phycocyanin 645 was a dimeric protein, $\alpha_2\beta_2$. The αβ unit was designated the monomer.

Further studies on the molecular weights of cryptomonad biliproteins — phycocyanins 612 and 645 and phycoerythrin 545 — demonstrated that they all had molecular weights of 54,000 to 57,000 and were best characterized as dimers.[15-17] The molecular weight determinations included gel filtration, sedimentation equilibrium ultracentrifugation (Figure 12), and osmotic pressure determinations. The stoichiometry of the two subunits was calculated by measuring the areas of the subunit bands on SDS gels after staining for protein. Assuming that the dye-binding capacity of the subunits was in proportion to their molecular weights, the results clearly showed a 1:1 molar ratio for subunits of phycoerythrin 545.[16] The sedimentation coefficient of 4.3S supported a dimeric structure.[14,16,17] After partial cross-linking

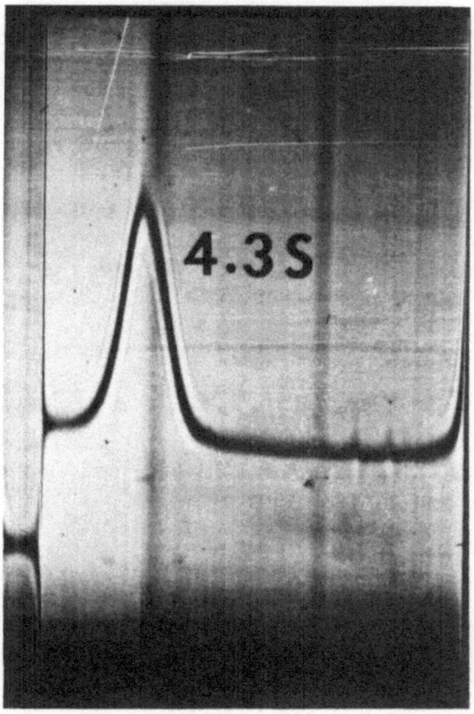

FIGURE 11. Sedimentation velocity schlieren pattern of phycoerythrin 545 *(Rhodomonas lens)*. The experiment was performed on model E analytical ultracentrifuge at 60,000 rpm. The photograph was taken 48 min after reaching full speed. The protein concentration was 11 mg/mℓ, and the solvent was pH 6.0, 1.0-ionic strength sodium phosphate buffer. The temperature was 20°C. Sedimentation coefficients ($s_{20,w}$) for the cryptomonad biliproteins are 4.3 to 4.4S in this solvent.

of cryptomonad biliproteins with the bifunctional agent dimethylsuberimidate, seven bands were obtained on SDS gels. Since eight was the maximum number of bands possible for an $\alpha_2\beta_2$ protein, this result supported the dimeric structure.[14]

Occasionally, SDS gels exhibited a splitting of the α subunit into a closely spaced doublet (Figure 13).[16,18] The α subunit was usually observed as a single band on these gels.[12-17,19] The occasional splitting, nonetheless, suggested the possibility of multiple α subunits (see Section IV).

A preliminary X-ray diffraction study was performed on phycocyanin 645 by Morisset et al.[20] Two different crystalline forms of the biliprotein were obtained.

Biliproteins yielded many bands in isoelectric focusing experiments.[12,14,18] Glazer et al.[12] reported that some of the isoelectric focusing bands were sensitive to irradiation. When bands from the isoelectric focusing of phycocyanin 645 were individually placed on an isoelectric focusing column, each refocused as the original.[14] Mörschel and Wehrmeyer[18] found that these various bands had qualitatively similar spectra, and they suggested that the bands might be due to charge heterogeneity.

C. Chromophores

O'hEocha et al.,[11] in addition to observing the tetrapyrroles, phycoerythrobilin on cryptomonad phycoerythrins and phycocyanobilin on cryptomonad phycocyanins, determined

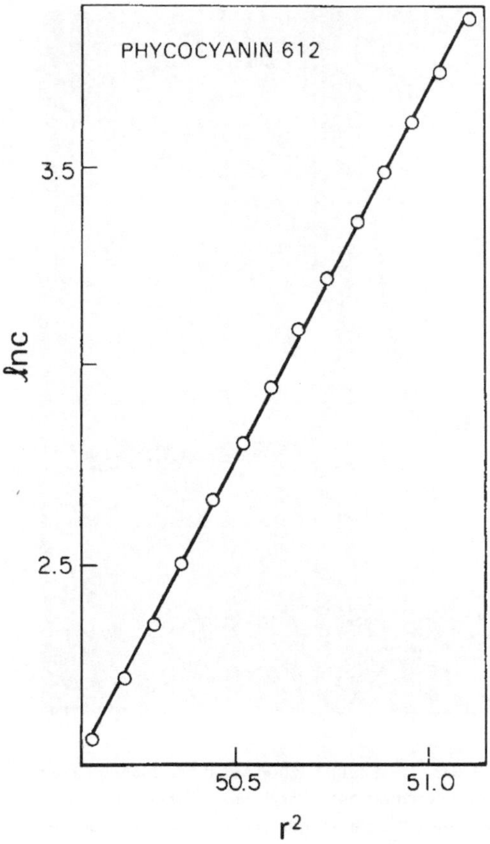

FIGURE 12. Plot of ln concentration (c) vs. the square of the distance from the center of rotation (r^2) for a sedimentation equilibrium experiment on phycocyanin 612 *(Hemiselmis virescens)*. The slope of this plot is proportional to the molecular weight of the protein. The sample had an absorbance of 0.3 at 280 nm in a 1-cm light path. The centrifugation was performed on a model E analytical ultracentrifuge at 24,000 rpm at 20°C. Measurements were taken using the photoelectric scanner at 280 nm after 24 hr of centrifugation using the meniscus depletion method.

that cryptomonad phycocyanins had a second chromophore now called cryptoviolin. Glazer and Cohen-Bazire[15] found on the subunit of phycocyanin 645 a third chromophore that had a 697-nm absorption maximum in acidic urea. Little is known concerning this bilin, and it will be referred to as the 697-nm bilin. It occurred only on this one biliprotein. Quite some time after the discovery of cryptoviolin in cryptomonad phycocyanins,[11] phycoerythrin 545 *(Rhodomonas lens)* was likewise shown to have, in addition to its phycoerythrobilin chromophore, cryptoviolin bilins on its α subunit.[21] The β subunits of phycoerythrin 545 had only phycoerythrobilins.[21] The cryptomonad biliproteins had four distinct bilins (Figure 14) but lacked phycourobilin.

Salient progress has been made on the quantitative assignments of the chromophores on phycocyanins 612 and 645.[17,22] MacColl and Guard-Friar[17] have determined the chromophore distribution for phycocyanin 612 *(Hemiselmis virescens)* and its subunits. This biliprotein had phycocyanobilin and cryptoviolin chromophores, and their spectra in acidic 8.0 *M* urea

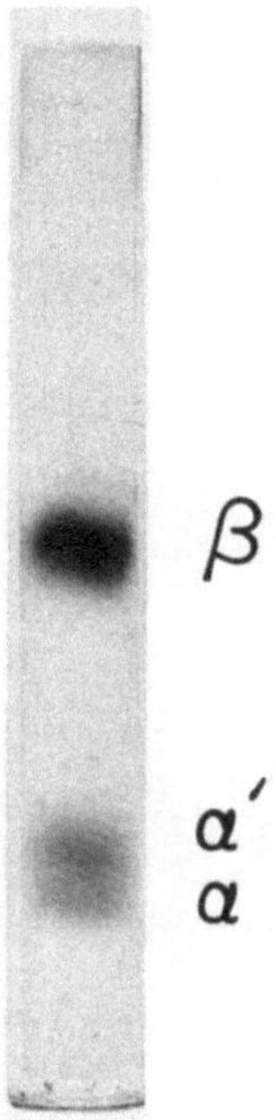

FIGURE 13. The subunits of phycoerythrin 545 *(Rhodomonas lens)* separated by SDS gel electrophoresis. The closely spaced doublet of the band at 10,000 mol wt is not observed on all gels. The higher-molecular weight band is never observed to split. The gels are 12% polyacrylamide at pH 6.4

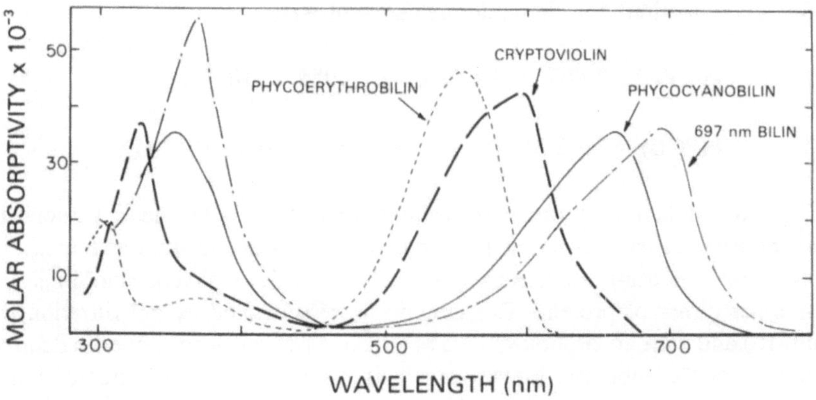

FIGURE 14. Absorption spectra in acidic urea for the four cryptomonad bilins. Spectra are plotted as molar absorptivities, and data were recorded at room temperature. The 697-nm bilin is the spectrum of the α subunit of phycocyanin 645. The cryptoviolin spectrum is from the β subunit of phycocyanin 612 with the phycocyanobilins subtracted. The phycocyanobilin and phycoerythrobilin spectra are those of C-phycocyanin and C-phycoerythrin, respectively, since these latter cyanobacterial biliproteins only have a single chromophore.

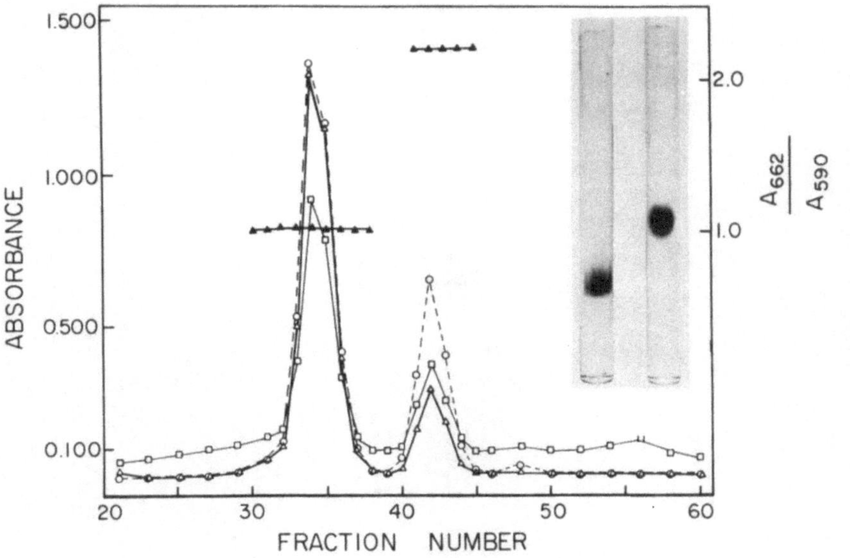

FIGURE 15.. Chromatographic separation of the α and β subunits of phycocyanin 612 *(Hemiselmis virescens)*. A gel filtration column equilibrated with 8.0 *M* urea, 0.01 *M* mercaptoethanol at pH 3.0 was used to separate the subunits. SDS gel electrophoresis demonstrates the good separation of subunits by this method. The caveat is that mercaptoethanol will affect the spectra of these biliproteins during long-term exposure. However, omission of the reducing agent from this column results in poor subunit separation.

should be relatively free of apoprotein influence and therefore dependent only on the quantity of chromophore. Using the following molar absorptivities in acidic urea, Beer's law equations were obtained for measurement of these two bilins: phycocyanobilin at 590 nm was 15,560 m^{-1} and at 662 nm it was 35,500; cryptoviolin at 590 nm was 43,000 cm^{-1} and at 662 nm it was 3334. The equations for an acidic urea solvent were

$$c(CV) = 2.407 \times 10^{-5}A_{590} - 1.055 \times 10^{-5}A_{662} \tag{1}$$

$$c(PCB) = -2.261 \times 10^{-6}A_{590} + 2.916 \times 10^{-5}A_{662} \tag{2}$$

where c(CV) is the molar concentration of cryptoviolin and c(PCB) is the molar concentration of phycocyanobilin. After dissolving phycocyanin 612 in acidic urea and measuring the visible absorption spectrum, Equations 1 and 2 yielded 5.73 phycocyanobilins and 2.07 cryptoviolins per dimer of protein. The subunits were separated by gel filtration in acidic urea (Figure 15) and their absorption spectra obtained. SDS gel electrophoresis demonstrated good separation of the subunits. Equations 1 and 2 showed that the ratios of bilins were phycocyanobilins on β to phycocyanobilins on α = 2.3; phycocyanobilins on β to cryptoviolins on β = 2.2; and phycocyanobilins on α to cryptoviolins on β = 0.98. A distribution of chromophores for phycocyanin 612 consistent with all this information was as follows: each of the two α subunits had one phycocyanobilin and each of the two β subunits had two phycocyanobilins and one cyrptoviolin.[17]

MacColl and Guard-Friar[22] have used similar techniques to evaluate the chromophore content of phycocyanin 645 *(Chroomonas* sp.). This biliprotein was known to have three spectrally distinct bilins, and the Beer's law equations were derived from the following molar absorptivities in acidic urea: phycocyanobilin at 697 nm was 12,823 M^{-1} cm^{-1} at 662 nm was 35,500, and 590 nm was 15,560; cryptoviolin at 697 nm was 0, at 662 nm

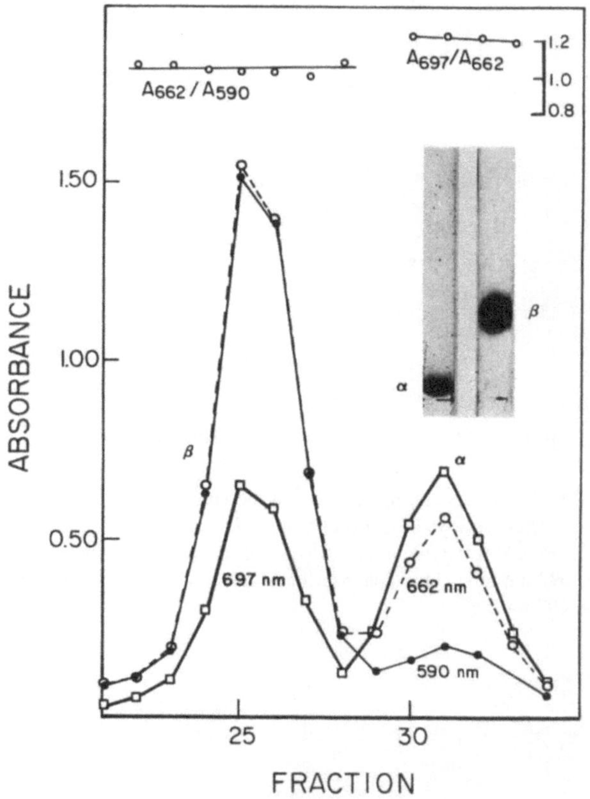

FIGURE 16. Chromatographic separation of the α and β subunits of phycocyanin 645 *(Chroomonas* sp.). A gel filtration column equilibrated with 8.0 *M* urea, 0.01 *M* mercaptoethanol at pH 3.0 was used. SDS gel electrophoresis shows that the subunits are well separated by this method.

was 3334, and at 590 nm was 43,000; and 697-nm bilin at 697 nm was 35,500, at 662 nm was 27,519, and at 590 nm was 8701 (Figure 14). The equations were

$$c(697B) = 3.927 \times 10^{-5}A_{697} - 1.469 \times 10^{-5}A_{662} + 1.139 \times 10^{-6}A_{590} \quad (3)$$

$$c(PCB) = -3.074 \times 10^{-5}A_{697} + 4.066 \times 10^{-5}A_{662} - 3.152 \times 10^{-6}A_{590} \quad (4)$$

$$c(CV) = 3.178 \times 10^{-6}A_{697} - 1.174 \times 10^{-5}A_{662} + 2.417 \times 10^{-5}A_{590} \quad (5)$$

Using these equations with the absorptivities of phycocyanin 645 dissolved in acidic urea showed that the protein had four phycocyanobilins, two cryptoviolins, and two 697-nm bilins per dimer. The two subunits were separated by gel filtration chromatography in acidic urea (Figure 16), and their chromophore distribution was obtained. The α subunits each have a single 697-nm bilin chromophore, and each β subunit has two phycocyanobilins and one cryptoviolin (Figure 17). The subunits were also separated by chromatography on a cation exchange resin. SDS gel electrophoresis demonstrated a good degree of subunit separation by these methods (Figure 18).

Guard-Friar and MacColl[23] have analyzed a pitfall associated with cryptomonad bilin quantitation. Reducing agents were always added to the acidic urea solvents used in these experiments. It was shown that mercaptoethanol changed the spectrum of phycocyanobilin

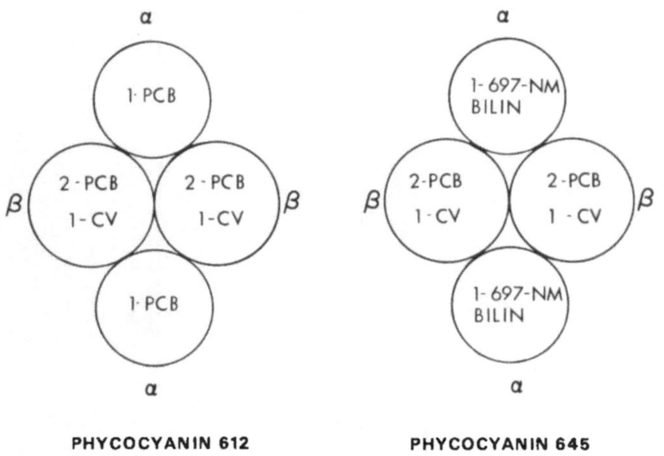

FIGURE 17. Schematic presentation of the chromophores of phycocyanins 612 *(Hemiselmis virescens)* and 645 *(Chroomonas* sp.). The chromophores are CV for cryptoviolin; PCB for phycocyanobilin; and 697-NM for the 697-nm bilin-chromophore with an absorption maximum in acidic urea at 697 nm.

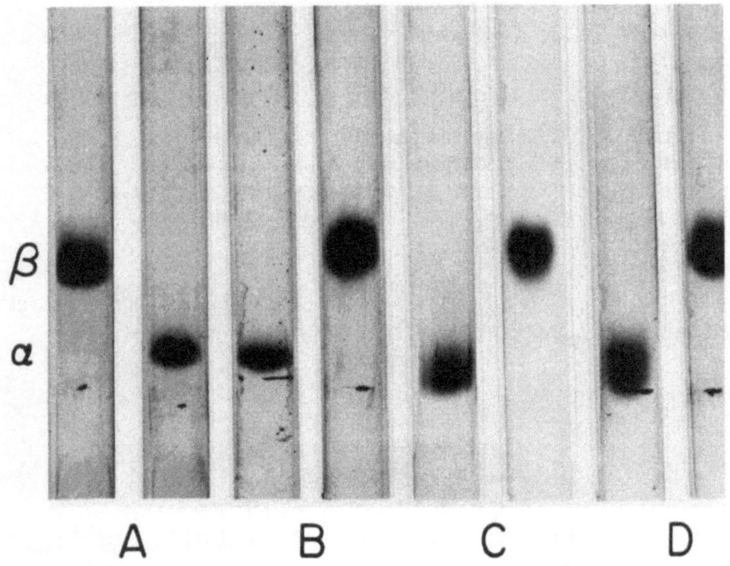

FIGURE 18. SDS gel electrophoresis on separated α and β subunits of cryptomonad biliproteins. (A) Phycocyanin 645 *(Chroomonas* sp.) subunit separated by a cation exchange resin using a urea gradient at pH 2.2, (B) the same subunits separated by a gel filtration column at pH 3.0 8.0 *M* urea; (C) the gel filtration separation for phycocyanin 612 *(Hemiselmis virescens)*; (D) the gel filtration separation for phycoerythrin 545 *(Rhodomonas lens)*.

so that the product overlapped extensively with the acid urea spectrum of cryptoviolin. If this effect was not addressed, an erroneous phycocyanobilin to cryptoviolin ratio was possible. Equations were presented to compensate for the modified phycocyanobilin spectrum, and deletion of reducing agents from the solvent was appropriate wherever practical. An additional factor that was very troublesome in these calculations was that the values of the

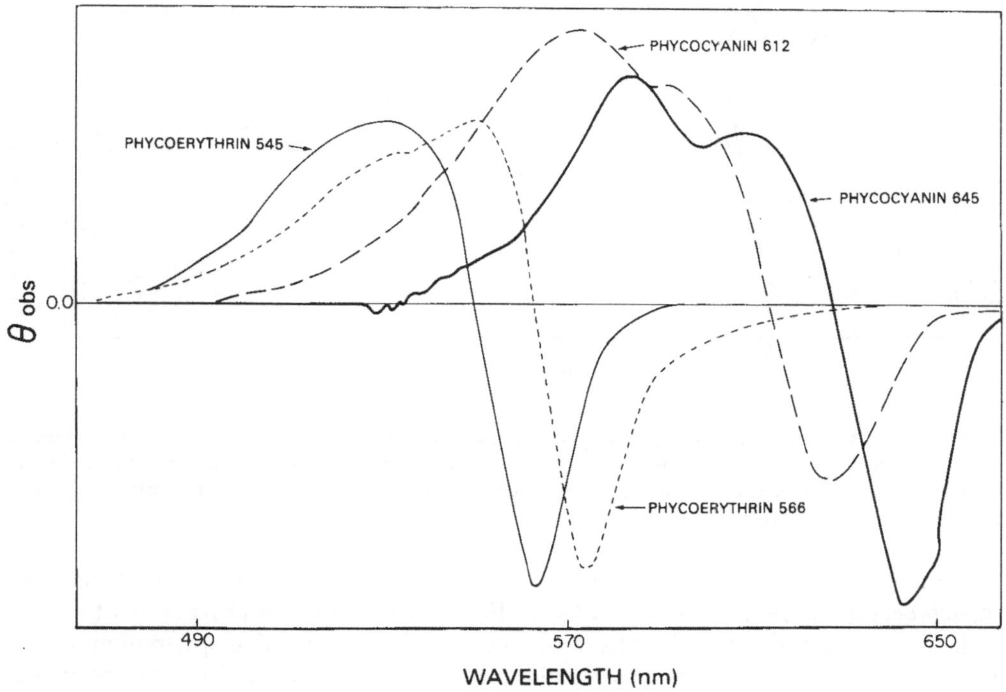

FIGURE 19. CD spectra of four cyrptomonad biliproteins. Spectra were taken at room temperature in a 1-cm light path. The samples had an absorption in a 1-cm light path of 0.6 to 0.8 at their visible maxima and were dissolved in pH 6.0, 0.1-ionic strength sodium phosphate buffer.

absorptivities were difficult to measure exactly. The 697-nm bilin was assigned absorptivities that were merely guesses based on a similarity to phycocyanoblin (also see Chapter 3).

D. Spectra

The absorption, fluorescence, and circular dichroism (CD) spectra of these biliproteins depended on both the types of chromophores and the apoprotein-chromophore or chromophore-chromophore interactions. The visible CD spectra of the various biliproteins exhibited long-wavelength negative bands and one or two shorter-wavelength positive bands (Figure 19).[14,16,17]

Jung et al.[24] have studied the CD spectrum of phycocyanin 645 *(Chroomonas* sp.) in considerable detail. They first showed that the absorption and thus the CD spectra could be deconvoluted into five bands with maxima at 652, 636, 617, 592, and 573 nm. The CD spectrum was shown to bleach when the protein was treated with potassium permanganate. The 592- and 652-nm bands exhibited identical decay rates. They pointed out that there were three chromophore types (phycocyanobilin, cryptoviolin, and 697-nm bilin) that generated the five spectral bands. Where do the two additional bands originate? The influence of the apoprotein on the chromophores or chromophore-chromophore interaction were two possibilities. Crediting the similar behaviors of two bands during the $KMnO_4$ treatment, Jung et al.[24] proposed that one of the additional bands arose from strong exciton coupling of two phycocyanobilin chromophores. Strong coupling caused spectral splitting and resulted in negative/positive CD spectra (Figure 19). They calculated that the distance between these two coupled chromophores was 0.9 nm. Each of the strongly coupled chromophores was on a different β subunit, and the bilins came into close contact during assembly of the dimeric protein structure. Differences in chromophore-apoprotein interactions were a possible explanation for the fifth spectral band.[24]

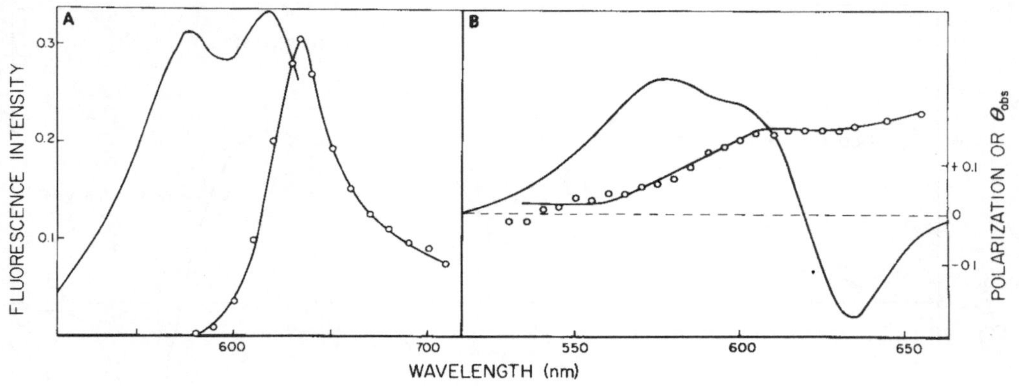

FIGURE 20. (A) Fluorescence excitation and emission spectra and (B) fluorescence polarization and CD spectra of phycocyanin 612 *(Hemiselmis virescens)*. The emission and excitation spectra are fully corrected for instrumentation effects. The fluorescence sample had absorption of 0.05 in a 1-cm light path at the visible maximum, and the spectra were all taken at room temperature.

The bilins on the isolated cryptomonad biliproteins were arranged for very efficient excitation energy transfer. MacColl and Guard-Friar[17] have shown for phycocyanin 612 that excitation at a series of wavelengths throughout the visible absorption spectrum gave rise to a group of completely superimposable fluorescence emission spectra. Since the absorption band at 585 nm contained a major contribution from cryptoviolin and the 612-nm absorption band was principally phycocyanobilin, excitation at several different wavelengths through this region will excite varying proportions of cryptoviolin and phycocyanobilin. Since the steady-state fluorescence emission spectrum was unchanged as a function of the wavelength of excitation, this demonstrated very efficient excitation energy transfer from the higher-energy cryptoviolins to the phycocyanobilins.[17] Fluorescence emission studies at the picosecond time-resolved level demonstrated (Section III.B) that a very small amount of fluorescence was actually emitted by the cryptoviolins. This small leakage of nontransferred energy allowed the measurement of the rate of excitation energy from cryptoviolin to phycocyanobilin.

The fluorescence polarization spectra of phycocyanin 612 and the other cryptomonad biliproteins had two or three distinct plateau regions throughout their visible absorption spectra.[14,16,17] The fluorescence polarization spectrum of phycocyanin 612 (Figure 20) changed in the wavelength region between cryptoviolin and phycocyanobilin absorbance and supported the idea of excitation energy transfer from cryptoviolin to phycocyanobilin.[17]

Finally, the fluorescence excitation spectrum of phycocyanin 612 (Figure 20) closely resembled its absorption spectrum (Figure 8). Since there was only a very small amount of fluorescence emission from cryptoviolin, the coincidence of the corrected excitation spectrum with the absorption spectrum in the visible region again pointed to very efficient energy transfer from cryptoviolins to lower-energy-absorbing phycocyanobilins. This transfer would cause the appearance of cryptoviolin in the excitation spectrum.[17] Fluorescence emission from phycocyanin 612 occurred from the low-energy phycocyanobilins.

In the cryptomonad phycoerythrins the cryptoviolins have a lower energy than phycoerythrobilins, and excitation would tend to flow from higher to lower energies. The relocation of cryptoviolin from the β subunit of phycocyanin where it served as a sensitizing (s) chromophore to the α subunit of phycoerythrin where it might function as a fluorescing (f) chromophore suggested that perhaps the α subunit was structurally oriented for extrinsic excitation energy transfer. The lowest-energy bilins on each isolated biliprotein were the primary f chromophores, and when part of the intact system, chromophores were responsible for interprotein energy transfer (see Chapter 6 for details on s and f chromophores). Further study of more types of cyrptomonad biliproteins is necessary to test the idea that cryptoviolin may be a ubiquitous constituent of the cryptomonad biliproteins.[21]

The specific absorptivities of purified cryptomonad biliproteins have been obtained for phycocyanin 645 and phycoerythrin 545. These values at the visible maxima and at concentrations of 1 mg/mℓ in pH 6.0, 0.1-ionic strength sodium phosphate buffer for a 1-cm light path were 11.4 for phycocyanin 645 and 12.6 for phycoerythrin 545.[14,16]

III. EXCITATION ENERGY TRANSFER

A. Intact Organisms

Haxo and Fork[10] studied O_2 evolution for the cryptomonads *R. lens* and *Cryptomonas ovata*. In both organisms the phycoerythrins showed very high effectiveness in photosynthesis. The regions of chlorophyll *a* absorbance were less effective. The biliproteins therefore must have transferred their excitation energy with high efficiency to the photosynthetic reaction centers.

Lichtlé et al.[25] have studied excitation energy transfer for the cryptomonad *C. rufescens* by analyzing the fluorescence spectra measured at −196°C. Phycoerythrin was the light-harvesting antenna for photosystem II in this alga. Harnischfeger and Herold[26] have examined excitation energy transfer between the biliprotein and chlorophyll for the cryptomonad *Chroomonas* sp. The effect of preillumination on fluorescence emission at −196°C was examined.

A void in what is known about the photosynthetic excitation energy migration scheme of the cryptomonad involved chlorophyll *c* and the biliproteins. Do they interact? If so, does energy flow from one to the other? Are they both components of photosystem II?

B. Isolated Biliproteins

Steady-state spectral studies have previously suggested (Section II.D) that excitation energy migrated among the bilins on a cryptomonad dimer. Kobayashi et al.[27] studied energy transfer of phycocyanin 645 by picosecond absorption spectroscopy. The decay curves consisted of a fast and a slow component. The component of less than 8 psec represented the rate of energy transfer from the s chromophore to the f chromophore. The slower process which was longer than 1 nsec was the emission decay of the f chromophore.

In another study using phycocyanin 645 from a different *Chroomonas* sp., Holzwarth et al.[28] examined the fluorescence kinetics which also exhibited a fast- and a slow-decay component. The wavelength-dependent 15-psec decay component was considered to represent the time necessary for energy transfer from the s to f chromophores.

Another cryptomonad biliprotein, phycocyanin 612 from *H. virescens,* was studied by picosecond fluorescence spectroscopy.[29] Hanzlik et al.[29] found a 10- to 12-psec decay component as well as long-lived fluorescence (1.0 to 1.5 nsec) and reported evidence for s and f chromophores within the biliprotein. This paper included a second approach to the energy transfer process by collecting spectral data taken at various times relative to the excitation pulse. These "intensity vs. wavelength" spectra at given times confirmed the earlier "intensity vs. time" results. They established cryptoviolin as the chromophore that decayed in the very rapid energy transfer step.

In addition to phycocyanins 612 (Figure 21) and 645, the fluorescence of phycoerythrin 545 *(R. lens)* was time resolved in the picosecond region. Guard-Friar et al.[30] found that the decay time of the two phycocyanins did not vary with excitation fluence. The lack of these exciton annihilation effects confirmed the assignment of the fast decay to s to f transfer. However, the decay of fluorescence from phycoerythrin 545 was excitation intensity dependent. At low fluences, the decay time was approximately exponential, while at very high fluences, its appearance was biphasic. Picosecond experiments for all types of biliproteins are discussed in Chapter 6 in more detail.

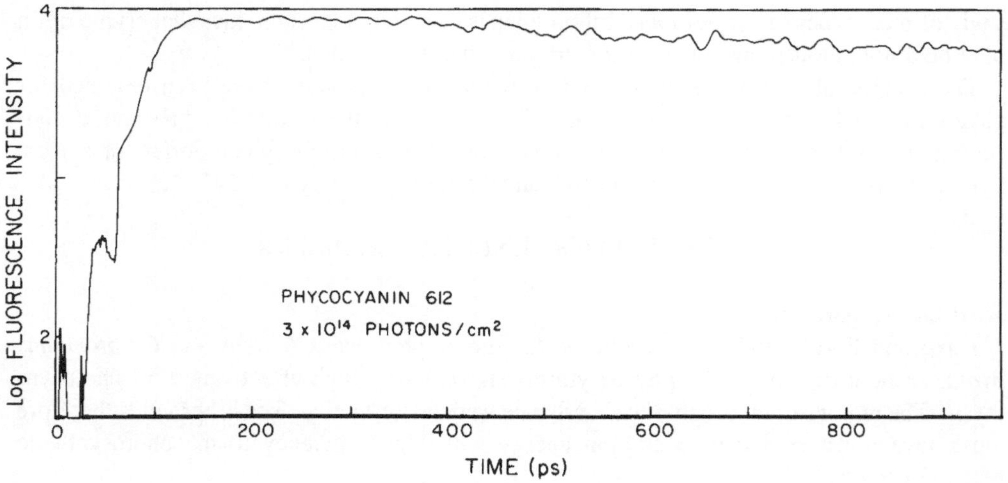

FIGURE 21. Fluorescence emission decay of phycocyanin 612 *(Hemiselmis virescens)* measured on the picosecond time scale. The biliprotein is in pH 6.0, 0.1-ionic strength sodium phosphate buffer. The sample was excited at 532 nm with an intensity of 3×10^{14} photons per square centimeter. Picosecond studies on various biliproteins are presented in Chapter 6. Using special methods, it was shown that a very small amount of very rapid emission from cryptoviolins also contributed to this fluorescence. (From Guard-Friar, D., MacColl, R., Berns, D. S., Wittmershaus, B., and Knox, R. S., *Biophys. J.*, 47, 787, 1985. With permission.)

IV. α SUBUNIT HETEROGENEITY

As discussed earlier (Section II.B), SDS gel electrophoresis occasionally showed two α subunit bands (Figure 13) for certain cryptomonad biliproteins. It might be suggested from these data that there were at least two types of α subunits. Glazer and Apell,[31] however, in a partial amino acid sequencing of a cryptomonad phycocyanin identified 24 out of the first 26 amino acid residues of the α subunit starting from the N terminus without finding a trace of heterogeneity.

Recent studies have clarified that some if not all cyrptomonads have at least two different α subunits as part of their biliprotein structure.[32-34] Offner[32] has partially sequenced the amino acids of the two α and the β subunits of phycoerythrin 545 *(R. lens)*. The α subunit residues near the N terminus were very similar, but residue 10 had valine for one α polypeptide and glutamine for the other. Offner reported an unusual amino acid, hydroxylysine, at residue 4 from the N terminus for both α subunits.

Sidler et al.[33] have partially sequenced two α subunits from both phycoerythrin 545 *(Cryptomonas maculata)* and phycocyanin 645 *(Chroomonas* sp.). The α subunits were separated by ion exchange chromatography using an ammonium acetate gradient. The samples were applied to the column in 8 *M* urea at pH 5.0. A second chromatography step was used to purify each of the α subunits. Partial N terminal Edman degradations clearly demonstrated that for both of these cryptomonad biliproteins there were two distinct genes contributing to the α subunits of each of these biliproteins. For phycoerythrin 545, two amino acid residues at positions 10 and 11 were different in sequence out to residue 26. Differences were more common from residues 26 to 42. The remainder of the sequences was not determined. The β subunits of phycocyanin 645 and phycoerythrin 545 were closely related to each other with only two differences in the first 30 residues. Immunochemistry results had previously shown that these two biliproteins were closely related (see Chapter 10).

Guard-Friar and MacColl[34] separated two α subunits from phycocyanin 645. The subunits were separated on a cation exchange resin in a urea gradient at pH 2.2. Amino acid analysis of the two α fractions showed that they had distinct amino acid compositions. For example, one of these subunits had six isoleucines and three leucines per polypeptide, and the other had three isoleucines and six leucines. In this study it was shown that both α subunits of phycocyanin 645 had identical chromophores. They both had the unusual 697-nm bilin.

The obvious and probably the correct interpretation of these data would be that the isolated dimeric protein of the cryptomonads is an $\alpha\alpha'\beta_2$ structure. There is another less direct possibility that can be suggested despite Occam's razor. This alternative would be that cryptomonad biliproteins exist as a mixture of dimers: $\alpha_2\beta_2$, $\alpha_2'\beta_2$, and $\alpha\alpha'\beta_2$. Perhaps even more than two α subunits are actually present and thus more dimeric forms. This second possibility might explain the multiple bands on the isoelectric focusing experiments (Section II.B).

REFERENCES

1. **Dodge, J. D.,** The ultrastructure of *Chroomonas mesostigmatica* Butcher (Cryptophyceae), *Arch. Mikrobiol.*, 69, 266, 1969.
2. **Wehrmeyer, W.,** Zur Feinstruktur der Chloroplasten einiger Photoautotropher Cryptophyceen, *Arch. Mikrobiol.*, 71, 367, 1970.
3. **Gantt, E., Edwards, M. R., and Provasoli, L.,** Chloroplast structure of the Cryptophyceae. Evidence for phycobiliproteins within intrathylakoidal spaces, *J. Cell Biol.*, 48, 280, 1971.
4. **Faust, M. A. and Gantt, E.,** Effect of light intensity and glycerol on the growth, pigment composition, and ultrastructure of *Chroomonas* sp., *J. Phycol.*, 9, 489, 1973.
5. **Thinh, L.-V.,** Effect of irradiance on the physiology and ultrastructure of the marine cryptomonad, *Cryptomonas* strain Lis (Cryptophyceae), *Phycologia*, 22, 7, 1983.
6. **Lichtlé, C. and Thomas, J. C.,** Étude ultrastructurale des thylacoides des algues à phycobiliproteines, comparaison des résultats obtenus par fixation classique et cryodécapage, *Phycologia*, 15, 393, 1976.
7. **Anderson, E.,** A cytological study of *Chilomonas paramecium* with particular reference to the so-called trichocysts, *J. Protozool.*, 9, 380, 1962.
8. **Allen, M. B., Dougherty, E. C., and McLaughlin, J. J. A.,** Chromoprotein pigments of some cryptomonad flagellates, *Nature*, 184, 1047, 1959.
9. **O'hEocha, C. and Raftery, M.,** Phycoerythrins and phycocyanins of cryptomonads, *Nature*, 184, 1049, 1959.
10. **Haxo, F. T. and Fork, D. C.,** Photosynthetically active accessory pigments of cyrptomonads, *Nature*, 184, 1051, 1959.
11. **O'hEocha, C., O'Carra, P., and Mitchell, D.,** Biliproteins of cryptomonad algae, *Proc. R. Ir. Acad.*, 63B, 191, 1964.
12. **Glazer, A. N., Cohen-Bazire, G., and Stanier, R. Y.,** Characterization of phycoerythrin from a *Cryptomonas* sp., *Arch. Mikrobiol.*, 80, 1, 1971.
13. **Brooks, C. and Gantt, E.,** Comparison of phycoerythrins (542, 566 nm) from cryptophycean algae, *Arch. Mikrobiol.*, 88, 193, 1973.
14. **MacColl, R., Habig, W., and Berns, D. S.,** Characterization of phycocyanin from *Chroomonas* species, *J. Biol. Chem.*, 248, 7080, 1973.
15. **Glazer, A. N. and Cohen-Bazire, G.,** A comparison of cryptophytan phycocyanins, *Arch. Microbiol.*, 104, 29, 1975.
16. **MacColl, R., Berns, D. S., and Gibbons, O.,** Characterization of cryptomonad phycoerythrin and phycocyanin, *Arch. Biochem. Biophys.*, 177, 265, 1976.
17. **MacColl, R. and Guard-Friar, D.,** Phycocyanin 612: a biochemical and photophysical study, *Biochemistry*, 22, 5568, 1983.
18. **Mörschel, E. and Wehrmeyer, W.,** Cryptomonad biliprotein: phycocyanin-645 from *Chroomonas* species, *Arch. Microbiol.*, 105, 153, 1975.
19. **Mörschel, E. and Wehrmeyer, W.,** Multiple forms of phycoerythrin-545 from *Cryptomonas maculata*, *Arch. Microbiol.*, 113, 83, 1977.
20. **Morisset, W., Wehrmeyer, W., Schirmer, T., and Bode, W.,** Crystallization and preliminary x-ray diffraction data of the cryptomonad biliprotein phycocyanin-645 from a *Chroomonas* spec., *Arch. Microbiol.*, 140, 202, 1984.

21. **MacColl, R., Guard-Friar, D., and Csatorday, K.,** Chromatographic and spectroscopic analysis of phycoerythrin 545 and its subunits, *Arch. Microbiol.*, 135, 194, 1983.
22. **MacColl, R. and Guard-Friar, D.,** Phycocyanin 645. The chromophore assay of phycocyanin 645 from the cryptomonad protozoa *Chroomonas* species, *J. Biol. Chem.*, 258, 14,327, 1983.
23. **Guard-Friar, D. and MacColl, R.,** Spectroscopic properties of tetrapyrroles on denatured biliproteins, *Arch. Biochem. Biophys.*, 230, 300, 1984.
24. **Jung, J., Song, P.-S., Paxton, R. J., Edelstein, M. S., Swanson, R., and Hazen, E. E., Jr.,** Molecular topography of the phycocyanin photoreceptor from *Chroomonas* species, *Biochemistry*, 19, 24, 1980.
25. **Lichtlé, C., Jupin, H., and Duval, J. C.,** Energy transfers from photosystem II to photosystem I in *Cryptomonas rufescens* (Cryptophyceae), *Biochim. Biophys. Acta*, 591, 104, 1980.
26. **Harnischfeger, G. and Herold, B.,** Aspects of energy-transfer between phycobilins and chlorophyll in *Chroomonas* spec. (Cryptophycea), *Ber. Dtsch. Bot. Ges.*, 94, 65, 1981.
27. **Kobayashi, T., Degenkolb, E. O., Bersohn, R., Rentzepis, P. M., MacColl, R., and Berns, D. S.,** Energy transfer among the chromophores in phycocyanins measured by picosecond kinetics, *Biochemistry*, 18, 5073, 1979.
28. **Holzwarth, A. R., Wendler, J., and Wehrmeyer, W.,** Studies on chromophore coupling in isolated phycobiliproteins. I. Picosecond fluorescence kinetics of energy transfer in phycocyanin 645 from *Chroomonas* sp., *Biochim. Biophys. Acta*, 724, 388, 1983.
29. **Hanzlik, C. A., Hancock, L. E., Knox, R. S., Guard-Friar, D., and MacColl, R.,** Picosecond fluorescence spectroscopy of the biliprotein phycocyanin 612: direct evidence for fast energy transfer, *J. Lumin.*, 34, 99, 1985.
30. **Guard-Friar, D., MacColl, R., Berns, D. S., Wittmershaus, B., and Knox, R. S.,** Picosecond fluorescence of cryptomonad biliproteins. Effects of excitation intensity on the fluorescence decay times of phycocyanin 612, phycocyanin 645, and phycoerythrin 545, *Biophys. J.*, 47, 787, 1985.
31. **Glazer, A. N. and Apell, G. S.,** A common evolutionary origin for the biliproteins of cyanobacteria, Rhodophyta, Cryptophyta, *FEMS Lett.*, 1, 113, 1977.
32. **Offner, G. D.,** Primary Structure of Red Algal and Cryptophytan Phycobiliproteins, Ph.D. thesis, Boston University, 1984.
33. **Sidler, W., Kumpf, B., Suter, F., Morisset, W., Wehrmeyer, W., and Zuber, H.,** Structural studies on cryptomonad biliprotein subunits. Two different α-subunits in *Chroomonas* phycocyanin-645 and *Cryptomonas* phycoerythrin-545, *Hoppe-Seyler's Z. Physiol. Chem.*, 366, 233, 1985.
34. **Guard-Friar, D. and MacColl, R.,** Subunit separation (α,α′,β) of cryptomonad biliproteins, *Photochem. Photobiol.*, 43, 81, 1986.

Chapter 10

BIOPHYSICAL AND IMMUNOCHEMICAL TECHNIQUES

I. DEUTERIUM ISOTOPE SUBSTITUTION AND SOLVENT ISOTOPE EFFECTS

Cyanobacteria can be successfully grown in 99.8% D_2O and can thereby become a source for fully deuterium-substituted molecules. Since C-phycocyanin was the major protein component of the cyanobacteria, it was selected for purification and elucidation. Deuterated protein dissolved in H_2O will have its nonexchangeable bonds (e.g., C–D bonds) deuterated and the exchangeable deuterium molecules replaced by hydrogens. Conversely, when dissolved in D_2O, the exchangeable positions will all have deuterium. In either solvent, when the cyanobacterium is grown in D_2O, its protein will be called deuterated phycocyanin.

The cyanobacterium *Plectonema calothricoides* was cultured in D_2O and its C-phycocyanin extracted and purified in buffered H_2O.[1] The purified protein was returned to D_2O and then dried. The deuterium content of this protein was established by two methods. The purified and dried protein was burned, and D_2O was found to be 98.4% of the water of combustion. Second, IR spectroscopy showed no detectable C–H and N–H bonds, but C–D and N–D bonds were observed. Hence the completely deuterated nature of this C-phycocyanin was established.

The characterization of this deuterated C-phycocyanin continued with amino acid analysis and thermal denaturation studies.[2] Both normal and deuterated C-phycocyanin had very similar amino acid compositions and appeared to differ chemically only in isotopic substitution. Thermal denaturation was studied by monitoring the changes in fluorescence intensity. The fluorescence changes were measured between 25 and 60°C. The curve showed three distinct regions: at first the fluorescence decreased in a linear fashion, then a very sharp drop occurred, and finally it returned to a slower decline.[2] The temperature at the onset of the sharp decline was considered the beginning of irreversible denaturation.[2,3] Berns et al.[2] showed that the deuterated C-phycocyanin began to denature at lower temperatures than the normal protein. The difference in denaturation temperature between the isotopically substituted proteins was about 5°C.

The thermal denaturation of normal and deuterated C-phycocyanin was also examined in D_2O.[4] In a phosphate buffer the deuterated protein began to denature at 45°C and the normal protein at 48°C.

Immunochemistry is a sensitive tool to examine protein structure. Antiserums to both normal and deuterated C-phycocyanin were prepared by injecting these proteins into rabbits.[5] Both Ouchterlony double diffusion and a quantitative precipitation protocol were used to assay the reactivity of deuterated and normal protein with antiserum to either deuterated or normal protein. In Ouchterlony experiments, normal and deuterated C-phycocyanin from *P. calothricoides* showed lines of mutual identity with no spurring against either type of antiserum. The lines in the agar were definitely caused by precipitation of C-phycocyanin because when irradiated with near-UV light the precipitation lines fluoresced with the characteristic red color of C-phycocyanin. The quantitative immunochemical tests were performed by reacting antigen and antibody for a sufficient period; the precipitates were centrifuged and washed, and nitrogen was determined by the Kjeldahl method. When antiserum to normal C-phycocyanin was used, identical amounts of nitrogen were measured for either the normal or the deuterated homologous antigen, and the very close correspondence of antigenic sites between the two isotopes was established. When a heterologous antigen, C-phycocyanin from a different cyanobacterium, was used, a lower amount of nitrogen was measured than found for either the normal or deuterated homologous antigen. These results supported the idea that the isotope substitution was the principal difference between the two types of C-phycocyanin and the deuterium substitution was not recognized by these antibodies.

When isolated and purified, C-phycocyanin existed as a number of different aggregates. The effect of deuteration in nonexchangeable sites has been investigated by sedimentation on the analytical ultracentrifuge as a function of pH, temperature, and ionic strength.[6] An important aggregate in these solutions sedimented at 11S. The 11S aggregate was a hexamer ($\alpha_6\beta_6$) which was the basic building block of the phycobilisome rods. The deuterated protein had significantly lower percentages of 11S aggregate around neutral pH than the normal protein.

Differential scanning calorimetry was also used to study the relative stability of C-phycocyanin isolated from normal and deuterated cultures.[7] For deuterated protein the temperature of denaturation was found to be 5 to 7°C lower than for normal protein. This differential was quite similar to that found using fluorescence techniques. During unfolding the heat capacity changes were lower for the deuterated protein.[7] The deuterated protein was also more readily denatured by urea than was the normal protein. Typical urea concentrations for denaturation were 3.6 *M* for deuterated protein and 4.8 *M* for the homologous normal protein.

D_2O as a solvent for normal or deuterated (in nonexchangeable positions) C-phycocyanin exerted striking effects on the aggregation of the protein.[8,9] At pH 7.0, the normal protein underwent a 112% increase in the amount of 11S when the H_2O was replaced with D_2O in the buffer. Under the same circumstances, the deuterated protein had 58% increase in the 11S aggregate. The suggestion was made that effect of D_2O was to increase the magnitude of hydrophobic interaction stabilizing the hexamers. The stability of aggregates larger than the 11S was increased in D_2O. For C-phycocyanin the effect on aggregation of D_2O as a solvent was the opposite of deuterium substitution in nonexchangeable positions in the protein.

The effects of nonexchangeable deuterium on isolated phycobilisomes was similar to that for C-phycocyanin.[10] Phycobilisomes from *Porphyridium cruentum* grown in 75% D_2O with deuteriums in nonexchangeable positions were less stable than the normal counterpart. Also analogous to the C-phycocyanin result was the finding that D_2O as a solvent protected phycobilisomes from dissociation.[10]

II. INTERFACE STUDIES IN MODEL SYSTEMS

A. Bilayer Lipid Membranes (BLM)

BLM were studied as a model for biological membranes. The basic piece of apparatus was a Teflon® cup 1 in. in diameter that had a 1-mm-diameter hole punched in its side. The hole was the channel of communication between a solution inside the cup and another surrounding it (Figure 1). The hole was literally painted over using a small brush. The "paint" contained lipids in organic solvents that, when applied to the hole, spontaneously spread to form a lipid bilayer. For photosynthetic research the lipid solution may be prepared from chloroplasts or a mixture of lipid, chlorophyll, and carotenoids. After the orifice was covered with a BLM, the contents of the inner and outer solututions could be changed to create asymmetry across the bilayer. Electrodes were placed in both compartments to measure the potential and capacitance across the bilayer and had the ability to apply a current across the membrane (Figure 1). When a redox potential existed across the membrane, shining light of appropriate wavelengths on the orifice created a photovoltage. The photoresponse was defined as the membrane potential during illumination minus the membrane potential in the dark. To study the interfacial properties of biliproteins, biliproteins were added to the inside of the cup, a variety of experiments were performed, and the photoresponses were recorded.[11,12]

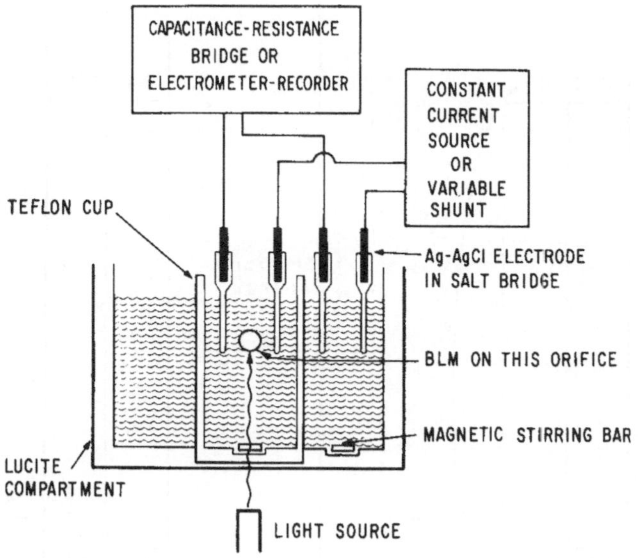

FIGURE 1. A typical experimental arrangement for a BLM experiment. The major value of the BLM setup is the ability to make very precise electrical measurements across a bilayer. (Reprinted with permission from Berns, D. S., *Photochem. Photobiol.*, 24, 117, 1976. Copyright 1976, Pergamon Press, Ltd.)

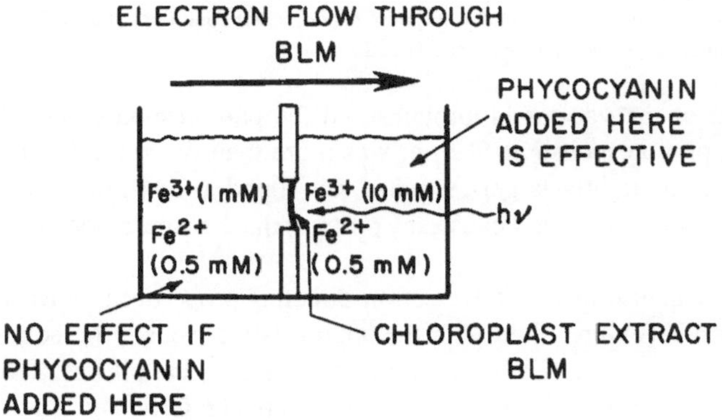

FIGURE 2. Typical BLM enhancement experiment using C-phycocyanin. In this experiment C-phycocyanin enhances the photovoltage when placed on the reducing side of the bilayer. A redox gradient is established by a ferric chloride-ferrous chloride concentration cell. The electron flow reversal experiments require a much smaller ferric ion concentration differential. The right side of the BLM is 0.7 m*M* ferric and the left is 0.5 m*M* ferric in the reversal experiment. The electron flow will still be left to right, but only a small photvoltage (2 mV) is obtained. C-Phycocyanin is then added to the left side, and its electron flow is now right to left and about 5 mV. (Reprinted with permission from Berns, D. S., *Photchem. Photobiol.*, 24, 117, 1976. Copyright 1976, Pergamon Press, Ltd.)

The bilayer, in the absence of biliproteins, showed the establishment of an electric current across the membrane when light was applied and a redox gradient existed across the bilayer. When a biliprotein was added (Figure 2) the photoresponse increased.[11,12] C-Phycocyanin, C- and B-phycoerythrin, and phycocyanin 645 all augmented the photoresponse. When the illuminating light was regulated in terms of its wavelength, a correspondence between the

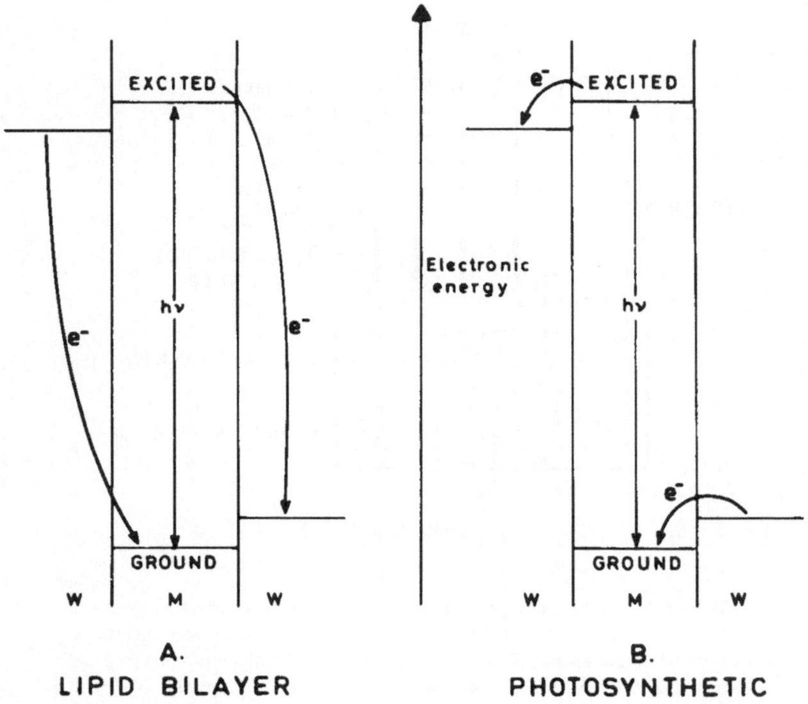

FIGURE 3. A schematic representation of the ordinary BLM experiment vs. photosynthesis. When C-phycocyanin is present under optimum experimental conditions, the BLM situation resembles, in general, the photosynthetic situation.

absorption spectra of the various biliproteins and the photoresponse was obtained. For C-phycocyanin, the photoresponse at 619 nm was more than twice the value at 544 nm, and for C-phycoerythrin a slightly larger response was found at 544 nm than at 619 nm. This biliprotein response was apparently caused by proteins that became associated with the surface of the bilayer.

The biliproteins can also reverse the electron-directing properties of a BLM.[11-13] The inner and outer compartments were filled with the appropriate ferrous/ferric solutions so that the concentration differences of ferric solution was small across the bilayer. When illuminated, an electron flow was observed toward the side of greater ferric ion concentration in the absence of C-phycocyanin. When the biliprotein was added to the side of lower ferric ion concentration, a reversal in the direction of the photoresponse was usually observed. The biliprotein thus had modified the energy barrier across the bilayer so that electron migration over the barrier was more probable. It was noted that BLM experiments without biliproteins resulted in the dissipation of a preexisting energy gradient (Figure 3). However, with a biliprotein and under the proper set of experimental conditions, the effect was to convert the BLM to a system more analogous to the functions of energy storage in photosynthesis.

C-Phycocyanin was employed in other BLM experiments.[14,15] BLM prepared from chlorophyll *a*, β-carotene, and specific lipids were affected by C-phycocyanin.[14] The biliprotein operated by directing electron transfer from the bilayer. C-phycocyanin was also studied as part of an investigation of the effects of micromolar amounts of various oxidants and reductants on BLM prepared from chloroplasts.[15]

B. Monolayer Studies on C-Phycocyanin

Monolayers are a unique system to study the behavior of proteins at aqueous-hydrophobic (membrane-like) interfaces. Information on the configurations of these proteins at interfaces is obtained. Phycobilisomes are extrinsic membrane organelles. Biliproteins may therefore have surface-active properties.

To investigate this question, C-phycocyanin was injected into the subphase of buffer which established the required bulk protein concentration, and the surface compressibility was measured at the air-water interface.[16] The surface pressure (π) as a function of protein penetration onto the surface was analyzed as follows:

$$\ln \frac{\pi_{ss} - \pi_t}{\pi_{ss} - \pi_0} = \frac{t}{\tau} \tag{1}$$

where π_{ss}, π_0, and π_t are the surface pressures under steady-state, time-zero, and any time (t), respectively, and τ is the relaxation time. When C-phycocyanin was injected into the subphase, the surface tension at the air-water interface was lowered. A plot of ln $[(\pi_{ss} - \pi_t)/(\pi_{ss} - \pi_0)]$ vs. t was characterized by two different slopes, indicating that the adsorption process was controlled by two different relaxation times. The change in surface tension for C-phycocyanin was greater than that measured for bovine serum albumin under the same conditions. Its fairly strong attraction for the interface indicated that C-phycocyanin had a proclivity toward hydrophobic associations. Monomers of C-phycocyanin were observed to be even more hydrophobic than aggregates.

Further monolayer studies should focus on the penetration of biliproteins into lipid monolayers. These studies will be directly related to the extrinsic membrane location of biliproteins.

C. Biliproteins on Solid Surfaces

Photoelectrical activity of biliproteins has been measured by forming a film of protein on a platinum electrode.[17,18] Phycoerythrin films on the electrodes were dipped into ethanol, and in the presence of atmospheric oxygen, illumination of the film resulted in a negative photopotential that was reversible in the dark. The authors pointed out that this ability to modulate electron transfer was not a necessary faculty for phycoerythrin to perform its known function of excitation energy transfer. They suggested that biliproteins might have a yet undefined role that utilized this photoelectrical capacity.[17-20]

Another example of this type of research employed a layer of biliprotein sandwiched between transparent semiconducting and metal electrodes.[21] Illumination of the cell produced a photopotential that was dark reversible.

Biliproteins and their chromophores have also been studied on various filter papers and films.[22-24] Phycoerythrin and phycocyanin have been incorporated into polyvinyl alcohol films and studied as the film was stretched.[22,23] The spectroscopic properties of biliproteins on Millipore® filters have been examined.[24]

III. THERMODYNAMICS OF C-PHYCOCYANIN INTERACTIONS

Thermodynamic measurements have been made on the interactions of C-phycocyanin with quaternary ammonium salts[25,26] and alcohols.[27] Tetraakyl ammonium salts have four alkyl groups providing these cations with the ability to enter into hydrophobic interactions. Azoniaspiroalkane halides are related to the quaternary ammonium salts, but instead of four alkyls they have two cyclic alkyl groups. The larger the alkyl, the more readily these cations enter into hydrophobic interactions, and the tetraalkyls are more hydrophobic than the azoniaspiroalkanes.

The interaction of these alkyl salts with C-phycocyanin *(Phormidium luridum)* has been studied by microcalorimetry and analytical ultracentrifugation.[25,26] At pH 6.0, C-phycocyanin prepared by ammonium sulfate fractionation was mainly composed of two aggregates: trimers ($\alpha_3\beta_3$) and hexamers ($\alpha_6\beta_6$). These aggregates can be reversibly interconverted. They sep-

Table 1
THERMODYNAMIC PARAMETERS IN THE BINDING OF QUATERNARY AMMONIUM SALTS TO TRIMERS OF C-PHYCOCYANIN

Salt	ΔH_B, kcal/mol	ΔS_B[a], eu	m[b]
$(C_4H_9)_4NBr$	49	170	1.1
$(C_3H_7)_4NBr$	26	93	1.1
$(C_2H_5)_4NBr$	−1	2	1.7
$(CH_2)_6N(CH_2)_6Br$	8	32	1.2
$(CH_2)_4N(CH_2)_4Br$	−12	−37	1.0

[a] ΔS_B is calculated from $\Delta G_B = \Delta H_B - T\,\Delta S_B$, where ΔH_B is obtained from calorimetry data and ΔG from $\Delta G_B = -RT \ln K_B$.

[b] Moles of salt bound to each mole of trimer.

arated well under centrifugation. The stability of the hexamers was in part a consequence of hydrophobic interactions between two trimers. The enthalpy (ΔH_B) changes for the salt-protein interaction were calculated from calorimetry data (Table 1). The trimer-hexamer equilibrium constants were calculated from the analytical ultracentrifuge data in the presence (K_{app}) and absence (K_D) of binding to small molecules. The binding of small molecules (X) with stoichiometry (m) was assumed to occur only to trimers. The following equilibrium expressions were used in the calculations:

$$\alpha_6\beta_6 \rightleftharpoons 2\alpha_3\beta_3 \tag{2}$$

$$K_D = \frac{[\alpha_3\beta_3]^2}{[\alpha_6\beta_6]} \tag{3}$$

Equations 2 and 3 were in the absence of X,

$$\alpha_3\beta_3 + mX \rightleftharpoons \alpha_3\beta_3 - X_m \tag{4}$$

where $\alpha_3\beta_3$-X_m was bound trimer,

$$K_B = \frac{[\alpha_3\beta_3 - X_m]}{[\alpha_3\beta_3]_u\ [X]^m} \tag{5}$$

where u indicated unbound trimer and the apparent dissociation constant (K_{app}) for hexamers in the presence of X was

$$K_{app} = \frac{[\alpha_3\beta_3]_T^2}{[\alpha_6\beta_6]} \tag{6}$$

where the concentrations of hexamer and $[\alpha_3\beta_3]_T$, the total trimer [free + bound], were those observed in the ultracentrifuge in the presence of X, since bound and free trimer sedimented at the same rate. These equilibrium expressions of K_{app} and K_D were obtained directly from experimental data, and K_B, ΔG_B, and ΔS_B were calculated from them and ΔH_B.

The salts affected the aggregation of C-phycocyanin by increasing the trimer concentration and decreasing the hexamer concentration in the following order: $(C_4H_9)_4NBr > (C_3H_7)_4NBr$

Table 2
EFFECT OF ALCOHOLS ON HEXAMER-TRIMER EQUILIBRIUM OF C-PHYCOCYANIN

Alcohol	Molar alcohol concentration	% Trimer	% Hexamer
Cyclohexanol	0.20	58	42
Butanol	0.25	54	46
Propanol	0.30	48	52
Ethanol	0.40	46	54
Methanol	0.40	46	54
Ethylene glycol	0.40	46	54

> 6:6 Br > 4:4 Br, $(C_2H_5)_4NBr$, where 6:6 Br and 4:4 Br are $(CH_2)_6N(CH_2)_6Br$ and $(CH_2)_4N(CH_2)_4$ Br, respectively. The thermodynamic parameters (Table 1) showed that the largest alkyl, tetrabutyl ammonium bromide, had the largest positive entropy and enthalpy.

The ordering of the effectiveness of the salts together with the large positive enthalpy and entropy suggested, especially for the butyl and propyl derivatives, the importance of hydrophobic forces in the binding. The large positive entropy was caused by the release of bound water from the surface of the salts and the protein. It is suggested that the salt-bound trimers cannot readily associate to form hexamers because the bulky alkyl groups sterically hinder trimer-trimer approach. Another possible factor might be that the binding of these salts introduced a positive charge onto the trimeric protein surface, and coulombic repulsion could negate the tendency for hexamer formation.

The situation with alcohol binding to C-phycocyanin trimers was also characterized by Chen[27] using Equations 2 to 6. Six alcohols — butanol, cyclohexanol, ethanol, ehtylene glycol, methanol, and propanol — were studied. These alcohols had a known ability to enter into pairwise self-interaction that indicated their relative abilities to enter into hydrophobic interactions: cyclohexanol > butanol > propanol > ethanol > methanol > ethylene glycol. Analytical ultracentrifuge results (Table 2) showed that nearly the same order of alcohols was effective in increasing trimer formation: Cyclohexanol > butanol > propanol > ethanol, methanol, and ethylene glycol. The latter three alcohols had no effect at the highest concentration (0.40 *M*) specified in the protocol. As with the ammonium salts, the best alcohols at altering the protein equilibrium had large positive entropy and enthalpy. Cyclohexanol had a ΔS_B of 177 eu and a ΔH_b of 51 kcal/mol. The interaction of the alcohols with the trimer was thus also probably a hydrophobic event. Since the alcohols were not ionic, the hydrophobic concept was better substantiated. The alcohol results tended to support the quaternary ammonium halide and azoniaspiroalkane halide hydrophobicity of binding proposals.[25-27]

IV. IMMUNOCHEMICAL STUDIES

A. Some Comments on Evolution

Life is divided into the prokaryotes (eubacteria and archaebacteria) and eukaryotes. The eukaryotes are considered to be more advanced in their evolution, and one aspect in which they differ from the prokaryotes is their membrane-bound organelles. It has been recognized for a long time that the cyanobacteria evolved early and have a simpler organization.[28] Cyanobacteria, which are still frequently referred to as blue-green algae, are prokaryotes and their thylakoid membranes are not encased in a chloroplast. The red algae and cryptomonads are both eukaryotes but differ in the number of chloroplast-limiting membranes. For the red algae, there are two membranes around the chloroplast, and for the cryptomonads there are four. It has frequently been suggested that the chloroplasts of the eukaryotes arose from symbiosis involving prokaryotes[29-34] and that the red algal chloroplasts were derived from the degeneration of a cyanobacterial inclusion in a eukaryote. The occurrence of very similar phycoilibsomes in both organisms supported such a proposal. The biliproteins of the cryptomonads do not form phycobilisomes, and their biliproteins are located in the intra-thylakoid space rather than on the stroma side of the thylakoid. Cryptomonads also have chlorophyll c_2, which is lacking in the others. Therefore the relationship among cryptomonads, red algae, and cyanobacteria is not straightforward.

In Section IV.B, the immunochemical properties of the various biliproteins that contribute to the understanding of evolution will be presented. The number of membranes surrounding their chloroplasts is also of interest and can help provide a framework for considering the biliproteins. From symbiosis, the two membranes enclosing the red algal chloroplast had an obvious origin. If indeed the original red alga arose from the endosymbiotic entrapment of a cyanobacterium, the innermost membrane of the resulting chloroplast would have arisen from the cell membrane of the cyanobacterium and the second chloroplast membrane would have originated from the plasma membrane of the eukaryotic host. As the cyanobacterium entered the host, the membrane of the host could have been pushed in, then pinched off, and thus enclosed the chloroplast in a second bilayer. How can four encompassing membranes of the cryptomonad chloroplast be explained? Symbiosis involving two eukaryotes is one possible explanation.[35-38] If the cryptomonad chloroplast began as an organism like a red alga which has a two-membrane chloroplast, the four membranes are explainable. A periplastidal region was observed by electron microscopy between the two inner and two outer cryptomonad chloroplast membranes. A particularly significant inclusion, the nucleomorph,[35] was observed in the periplastidal compartment. One nucleomorph was observed in each cell, and in this proposal it would be the remainder of the nucleus of the red algal symbiont. The periplastidal space would be, in this analysis, the remainder of the cytoplasm of the symbiont. An additional possibility was that the inner membrane of the chloroplast endoplasmic reticulum (the outer two membranes around the chloroplast) was actually the residual of the plasma membrane of the red algal symbiont. There is considerable flexibility in conceiving these endosymbiontic relationships, since once a symbiont began to degenerate to a chloroplast, any particular part of the original organism might be lost, retained, or modified in a variety of ways.

Although a red alga evolving to a cryptomonad chloroplast is a very appealing concept, it leaves the basis for the genesis of chlorophyll c_2 uncertain. A possibility was offered that an organism similar to a red alga, but having chlorophyll c_2, was the symbiont that formed the chloroplast of a cryptomonad.[37] No such organism has yet been discovered. The lack of phycobilisomes in the cryptomonads and the location of the biliproteins inside the thylakoids rather than on the outer thylakoid surface adjacent to the stroma have always suggested large variations between cryptomonad and red algal/cyanobacterial types. Chapters 2 and 9 present electron micrographs of these thylakoid arrangements. Therefore an immunochemical survey of the different biliprotein types would provide a reasonable test of the evolution of cryptomonad biliproteins from a red alga as suggested by the eukaryotic-eukaryotic endosymbiont hypothesis.

B. Immunochemistry of the Biliproteins

If the three biliprotein-containing phyla have a common line of evolution, their biliproteins might retain some level of similarity. This concept was first tested extensively by immunochemistry for the cyanobacterial and red algal proteins. Two basic rules were established by the use of Ouchterlony double diffusion techniques: (1) biliproteins within each of the three spectroscopic groups — phycocyanin, phycoerythrin, or allophycocyanin — were immunochemically related regardless of whether the source organism was eukaryotic or prokaryotic and (2) there was no cross-reaction between any two of three biliprotein spectral

AB = anti C-phycoerythrin

FIGURE 4. Ouchterlony double diffusion study of biliproteins using rabbit antiserum to C-phycoerythrin (cyanobacterial). The antigens are CPE, C-phycoerythrin; 612, phycocyanin 612 (cryptomonad); 566, phycoerythrin 566 (cryptomonad); and BPE, B-phycoerythrin (red alga). All antigens and the agar are in sodium phosphate buffer at pH 6.0. Lines of partial identity (spurring) are observed for C- and B-phycoerythrin which proves a degree of homology between the proteins from prokaryotic and eukaryotic sources. No reaction is observed between this antiserum and cryptomonad biliproteins. A list of organisms used in all (Figures 4 to 7) the immunochemical experiments include *Phormidium persicinum*, C-phycoerythrin; *Porphyridium cruentum*, B- and b-phycoerythrin mixture; *Hemiselmis virescens*, phycocyanin 612; *Cryptomonas ovata*, phycoerythrin 566; *Rhodomonas lens*, phycoerythrin 545; and *Chroomonas* sp., phycocyanin 645. The last four organisms are cryptomonads, *Phormidium persicinum* is a cyanobacterium, and *Porphyridium cruentum* is a red alga. The antiserums were also made with biliproteins from these organisms.

groups.[5,39-42] Therefore C- or R-phycocyanins from cyanobacteria and red algae were all immunochemically related, as were C-, B-, b-, and R-phycoerythrins (Figure 4) regardless of whether isolated from cyanobacteria (C-phycoerythrin) or red algae (R-, B-, or b-phycoerythrins).

Berns[41] demonstrated through a large number of examples that these prokaryotes and eukaryotes had very closely related phycocyanins and phycoerythrins. The close relation of cyanobacteria and red algae in evolution was therefore affirmed. Antiserums prepared against C-phycocyanins from both cyanobacteria and red algae were tested in Ouchterlony double diffusion experiments against several C-phycocyanins also purified from both prokaryote and eukaryote sources. For example, the antiserum against C-phycocyanin from the cyanobacterium *Plectonema calothricoides* gave good precipitation responses with C-phycocyanins from three cyanobacteria — *Phormidium luridum, Synechoccocus lividus,* and *Tolypothrix tenuis* — and with C-phycocyanins from two red algae — *Cyanidium caldarium* and *Porphyridium aerugineum.* Both eukaryotic and prokaryotic antigens gave results on Ouchterlony double diffusion experiments demonstrating partial identity with the homologous C-phycocyanin which was originally used to produce the antiserum. Reactivity was usually better against cyanobacterial sources. Similar findings were reported for several combinations of cyanobacterial and red algal phycoerythrins. For example, antiserum against the cyanobacterial C-phycoerythrin from *Phormidium persicinum* yielded responses demonstrating partial homology to the homologous C-phycoerythrin with several other cyanobacterial phycoerythrins from *T. tenuis, Nostoc muscorum* A and W, *Calothrix parietina,* and *Calothrix membranacea* and also with red algal B-phycoerythrin from *Porphyridium cruentum* and red algal R-phycoerythrin from *Ceramium rubrum.* Antiserum formed from B-phycoerythrin,

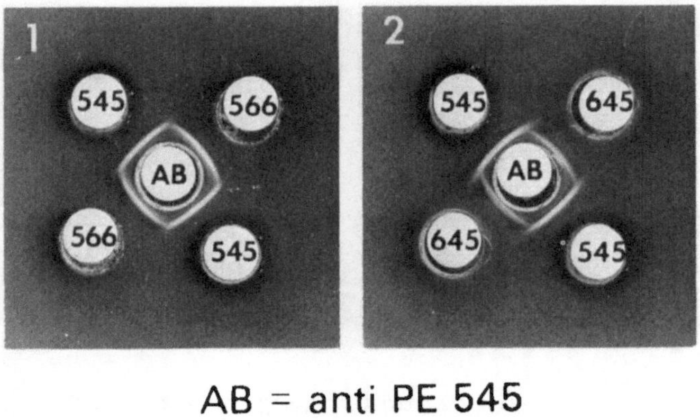

FIGURE 5. Ouchterlony double diffusion study of biliproteins using rabbit antiserum to phycoerythrin 545 (cryptomonad). Antigens are from cryptomonads: 545, phycoerythrin 545; 566, phycoerythrin 566; and 645, phycocyanin 645. Lines of partial identity between phycoerythrin 545 and the two heterologous antigens, phycoerythrin 566 and phycocyanin 645, show that the homologous antigen is related to both. The partial homology between phycocyanin and phycoerythrin is unique for the cryptomonads. All antigens and the agar are at pH 6.0

however, only gave a very minimal response with C-phycoerythrins from cyanobacteria. Nonetheless the number of heterologous cross-reactions that were positive between purified cyanobacterial and red algal biliproteins strongly endorsed the evolution of one from the other.[41]

Allophycocyanins from a number of sources were subsequently shown to be unrelated to any phycocyanin or phycoerythrin by Ouchterlony methods,[42] and the second rule for cyanobacterial and red algal biliprotein immunochemistry ascertained that phycocyanins, phycoerythrins, and allophycocyanins were each immunochemically distinct.[41,42] Quantitative immunochemical differences among C-phycoerythrins have also been studied.[43]

Cryptomonad biliproteins have also been examined extensively.[44,45] The second rule developed for immunochemical relations between cyanobacterial and red algal biliproteins did not hold. Cryptomonad phycocyanins and phycoerythrins are all very closely related to one and other (Figure 5), and these biliproteins are a much more homogenous group than the noncryptomonads.

Certain Ouchterlony plates were photographed under near-UV light as well as under normal light. The UV light produced fluorescence from the biliprotein-antibody complexes (Figure 6). The appearance of characteristic biliprotein fluorescence in the precipitin lines was early proof that the observed reactions were in fact to biliproteins. The fluorescence of biliprotein-antibiliprotein complexes will be evaluated further in Section IV.D.

When cryptomonad biliproteins were tested vs. cyanobacterial and red algal biliproteins in immunochemical studies, it was possible in certain cases to show that biliproteins from all sources were related in varying degrees.[44,45] Antiserum against cryptomonad phycoerythrin 545 was used to show lines of partial identity between phycoerythrin 545 and B-phycoerythrin (red alga) and phycoerythrin 545 and C-phycoerythrin (cyanobacterium) in Ouchterlony experiments (Figure 7). In other cases, antiserum to a cryptomonad biliprotein failed to exhibit cross-reactivity between cyanobacterial biliproteins and cryptomonad biliproteins (data not shown), but the same antiserum demonstrated a partial homology between red algal B-phycoerythrin and cryptomonad biliproteins (Figure 6). The appearance of the precipitin lines in these experiments also indicated a closer level of similarity between cryptomonad and red algal biliproteins than between cryptomonad and cyanobacterial biliproteins.

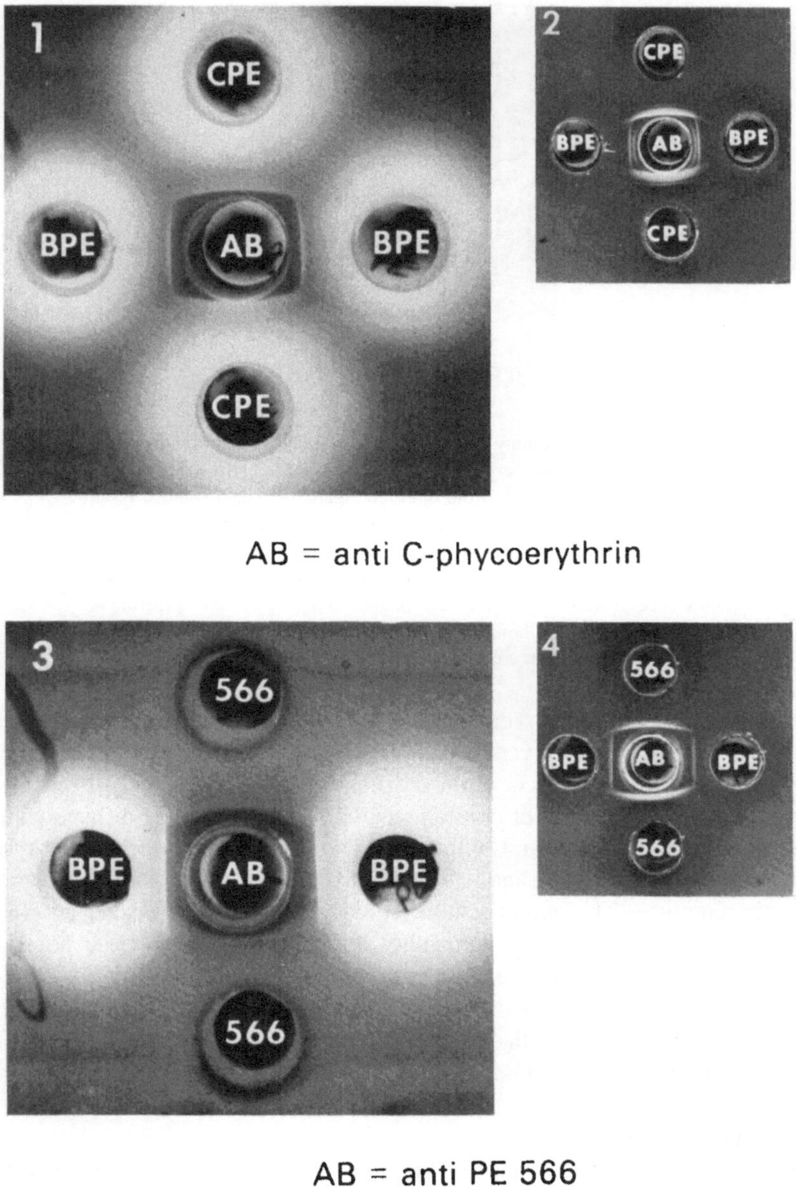

FIGURE 6. Ouchterlony double diffusion study of biliproteins using rabbit antiserums against C-phycoerythrin (cyanobacterial) and phycoerythrin 566 (cryptomonad). The two photographs at the left (1 and 3) are taken of fluorescent precipitin lines under near-UV light. The photographs at right (2 and 4) are the same Ouchterlony plates under normal lighting conditions. Antiserum to phycoerythrin 566 is used to demonstrate a partial homology between a red algal and a cryptomonad biliprotein. All antigens and the agar are at pH 6.0. The fluorescence of the cryptomonad phycoerythrin is observed to be much less intense than B- and C-phycoerythrin in these experiments.

Likewise, antiserum to B-phycoerythrin was also capable of demonstrating partial identity between B-phycoerythrin and several cryptomonad biliproteins,[44,45] but antiserums to cyanobacterial biliproteins have yet to show a reaction with any cryptomonad biliprotein (Figure 4).

The experiments performed thus far have revealed the curious fact that both cryptomonad

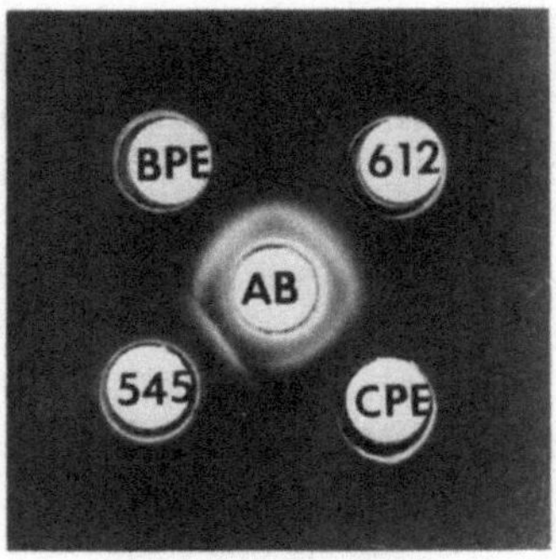

AB = anti PE 545

FIGURE 7. Ouchterlony double diffusion study of biliproteins using antiserum against phycoerythrin 545 (cryptomonad). The antigens used represented phycoerythrins from both red algal (BPE) and cyanobacterial (CPE) sources, and cryptomonad phycoerythrin (545) and phycocyanin (612) were also employed. All four different types of biliproteins showed lines of mutual partial homology. Even though some other serums did not show precipitation of certain heterologous biliproteins, this experiment clearly illustrates that homologies among cyanobacterial, red algal, and cryptomonad biliproteins are indicated through immunochemical techniques.

phycocyanins and phycoerythrins seemed to be much closer in immunochemical relatedness to phycoerythrins (red algal or cyanobacterial) than to C-phycocyanin. Therefore it is a possibility that cryptomonad phycocyanin did not evolve independently from a red algal phycocyanin, but that both types of cryptomonad biliproteins seemed to have undergone a common evolution from an ancestral biliprotein. The various immunochemical results provided strong evidence for a close relation through evolution of the prokaryotic cyanobacteria and eukaryotic red algae. The immunochemcial evidence showed a weaker cross-reactivity with the cryptomonads, in keeping with other dissimilarites shown by a comparison of cryptomonads to the other biliprotein-containing phyla. Nonetheless, cryptomonad-biliprotein immunochemistry clearly demonstrated a thread of similar evolution through the three phyla.[44,45] A possible evolution of the cryptomonad chloroplast from a red alga was strongly supported by the immunochemical data (Figure 7).

C. Amino Acid Sequencing

Partial sequences of the amino acids of biliproteins from the N terminus or of the bilin-containing regions by Edman degradation were also used to test relatedness. Harris and Berns,[46] Glazer et al.,[47] Glazer and Apell,[48] Sidler et al.,[49] and Bryant et al.[50] examined many cyanobacterial cryptomonad, and red algal biliproteins, discovering similarities in partial sequences that totally endorsed the earlier immunochemical findings. The properties of biliproteins were then very supportive of evolution schemes which suggested that the red algae and cryptomonads were related to cyanobacteria and to each other.

Although not expected from the immunochemistry, sequence homologies were also found among phycocyanins, phycoerythrins, and allophycocyanins.[46,47] All the various biliproteins thus originated from a single precursor.

Immunochemistry[51] and partial N terminal amino acid sequencing[47,52-59] have shown some varying amounts of homology between the α and β subunits of biliproteins (Table 3). The partial sequencing was conversely also very important in proving that the α and β subunits of the same biliprotein had distinct sequences in all cases.[46] The homology which was retained by the α and β subunits suggested that the two subunits appeared after gene duplication.

Table 3
PARTIAL N TERMINAL AMINO ACID SEQUENCES OF A RED ALGAL[a] (*CYANIDIUM CALDARIUM*) AND A CYANOBACTERIAL (*MASTIGOCLADUS LAMINOSUS*) C-PHYCOCYANIN

	***Mastigocladus laminosus*[54]**
β subunit	Ala Tyr Asp Val Phe *Thr* Lys Val Val Ser Glu *Ala Asp* Ser Arg *Gly Glu Phe Leu Ser Asn*
α subunit	Val Lys Thr Pro Ile *Thr* Asp Ala Ile Ala Ala *Ala Asp* Thr Gln *Gly* Arg *Phe Leu Ser Asn*
	***Cyanidium caldarium*[55]**
α subunit	*Met* Lys Thr Pro Ile Thr Glu Ala Ile *Ala Ala Ala Asn Ala Arg Gly Glu Phe Leu Ser Asx*
β subunit	*Met* Leu Asn Ala Phe Ala Lys Val Val *Ala Ala Ala Asn Ala Arg Gly Glu Phe Leu Ser Asx*

[a] The taxonomy of *Cyanidium caldarium* has been discussed frequently because of the atypical characteristics of the alga. It seems well established now as a red alga having unusual growth requirements.[60] Its unique role in the study of the bilins is considered in Chapter 7.

Amino acid sequences have been completed for entire biliproteins (Chapter 3)[61-68] and have fully proven and extended the concepts developed from partial sequencing. The α and β subunits of each biliprotein were indeed demonstrated by the entire amino acid sequences to be significantly homologous. From the entire amino acid sequences of various biliproteins, hypothetical schemes concerning biliprotein evolution have been devised.[63,65] Troxler et al.[63] suggested that a precursor to the biliproteins was similar to the β subunit of allophycocyanin. Gene duplication gave rise to two new precursors: one similar to the β subunit of allophycocyanin and the other to α. This second β-type precursor gave rise to the β subunit of allophycocyanin and both the α and β subunits of phycocyanin and perhaps phycoerythrin. The α-type precursor gave rise to the α subunit of allophycocyanin.

D. Biliproteins in Fluoroimmunoassay

Fluorescent probes of various types are commonly used in immunoassay protocols. Biliproteins as fluorescent probes have potential advantages in these assays

1. The spectral region used is relatively free of absorption and fluorescence by other biological materials.
2. The presence of multiple chromophores on biliproteins which undergo efficient excitation energy transfer produce a large wavelength differential between absorption and emission (Stokes shift), and a large Stokes shift helps avoid interference from Raman and Rayleigh scattering.
3. The chromophores are protected by the apoprotein from many types of quenching.
4. Biliproteins are very water soluble.
5. Biliproteins have a high-fluorescence quantum yield.
6. Biliproteins are readily obtained and easily purified.
7. Biliproteins are readily conjugated to other proteins.
8. A variety of biliproteins with different spectral properties are available.
9. Biliproteins are stable in a wide pH range.

Two types of systems have been investigated using biliproteins.[69] One used biliproteins as a label in "sandwich" solid-phase assays, and the second used them as either donor or acceptor in an excitation energy transfer experiment between labeled antigens and antibodies. Successful results were obtained using B-phycoerythrin linked to rabbit antihuman IgG in the solid-phase assay. However, the results were not as enhanced as expected.

Table 4
STOKES SHIFTS[a] FOR BILIPROTEINS

Biliprotein	Stokes shifts (nm)
Allophycocyanin	10
C-Phycocyanin	32
C-Phycoerythrin	12
R-Phycoerythrin	80
R-Phycocyanin	82
CU-Phycoerythrin	78
B-Phycoerythrin	80
b-Phycoerythrin	25
Phycoerythrocyanin	50
Phycocyanin 612	49
Phycocyanin 645	75
Phycoerythrin 545	40
Phycoerythrin 566	51
B-Phycoerythrin-allophycocyanin[b]	114

[a] Stokes shifts are measured as the nanometer difference between fluorescence emission maximum and the shortest wavelength absorption maximum.
[b] Synthetic complex.[70]

The observation that certain biliprotein-antibody complexes were intensely fluorescent extended back to the early study of biliproteins in Ouchterlony double diffusion experiments.[5] Antibodies to biliproteins were diffused toward wells containing either homologous or heterologous antigens. When precipitation lines were obtained, they were observed during irradiation with near-UV light. When C-phycocyanin was the antigen, a red fluorescence was observed from its precipitated complex with anti-C-phycocyanin. Likewise, a characteristic orange fluorescence was observed from phycoerythrin-antiphycoerythrin complexes (Figure 6).

The large Stokes shift of the biliproteins was important for their use in fluoroimmunoassays. To increase further this advantage, phycoerythrin and allophycocyanin have been chemically coupled.[70] Since phycoerythrin absorbed light very strongly at wavelengths much shorter than the 660-nm emission maximum of allophycocyanin, a very large Stokes shift was obtained. In the native phycobilisomes, phycoerythrin could not transfer excitation energy to allophycocyanin because phycocyanin was always interspaced between the two (Chapter 2). The phycoerythrin emission overlapped the allophycocyanin absorption so that energy transfer was possible. The complex prepared from B-phycoerythrin *(P. cruentum)* and allophycocyanin *(Anabaena variabilis)*, when excited at 500 nm, showed fluorescence emission primarily at 660 nm with only a small band at 576 nm. The 660-nm emission indicated excitation energy transfer from phycoerythrin to allophycocyanin, since allophycocyanin absorbed minimally at 500 nm. The small band at 576 nm came from phycoerythrin chromophores that were not effectively coupled to allophycocyanin receptors. Several naturally occurring biliproteins — R-phycocyanin and CU- and R-phycoerythrin — have different chromophores located on the same stable aggregate that also gave very large Stokes shifts (Table 4). In addition, biliproteins have been used to detect cell surface differences in fluorescence-activated cell-sorting analysis and flow cytometry.[71-77]

REFERENCES

1. **Berns, D. S., Crespi, H. L., and Katz, J. J.,** The isolation and characterization of a fully deuteriated protein, *J. Am. Chem. Soc.*, 84, 496, 1962.
2. **Berns, D. S., Crespi, H. L., and Katz, J. J.,** Isolation, amino acid composition and some physico-chemical properties of the protein deuterio-phycocyanin, *J. Am. Chem. Soc.*, 85, 8, 1963.
3. **Lavorel, J. and Moniot, C.,** Effect of temperature on the spectroscopic properties of phycocyanin, *J. Chim. Phys.*, 59, 1007, 1962.
4. **Berns, D. S.,** Studies of completely deuteriated proteins. II. Thermal denaturation in D_2O, *Biochemistry*, 2, 1377, 1963.
5. **Berns, D. S.,** Studies of completely deuteriated proteins. I. The immunochemistry of the deuterated protein and its hydrogen analog, *J. Am. Chem. Soc.*, 85, 1676, 1963.
6. **Scott, E. and Berns, D. S.,** Completely deuterated proteins. III. Deuteration effects on protein-protein interaction in phycocyanin, *Biochemistry*, 6, 1327, 1967.
7. **Chen, C.-H., Tow, F., and Berns, D. S.,** Solvent isotope effect on the differences in structure and stability between normal and deuterated proteins, *Biopolymers*, 23, 887, 1984.
8. **Berns, D. S., Lee, J. J., and Scott, E.,** Effect of deuterium oxide on protein aggregation in deuterio and protio phycocyanin and other proteins, *Adv. Chem. Ser.*, 84, 21, 1968.
9. **Lee, J. J. and Berns, D. S.,** Protein aggregation. The effect of deuterium oxide on large protein aggregates of C-phycocyanin, *Biochem. J.*, 110, 465, 1968.
10. **Zilinskas, B. A. and Glick, R. E.,** Noncovalent intermolecular forces in phycobilisomes of *Porphyridium cruentum*, *Plant Physiol.*, 68, 447, 1981.
11. **Ilani, A. and Berns, D. S.,** Photoresponse of chlorophyll-containing bileaflet membranes and the effect of phycocyanin as extrinsic membrane protein, *J. Membr. Biol.*, 8, 333, 1972.
12. **Chen, C.-H and Berns, D. S.,** Modification of intensity and direction of electron flow across bileaflet membranes, *Proc. Natl. Acad. Sci. U.S.A.*, 72, 3407, 1975.
13. **Ilani, A. and Berns, D. S.,** A theoretical model for electron transport through chlorophyll-containing bileaflet membranes, *Biophysik*, 9, 209, 1973.
14. **Chen, S. S. and Berns, D. S.,** Effect of plastocyanin and phycocyanin on the photosensitivity of chlorophyll-containing bilayer membranes, *J. Membr. Biol.*, 47, 113, 1979.
15. **Alexandrowicz, G. and Berns, D. S.,** Photovoltages in chloroplast extract bilayer membranes stimulated by micromolar amounts of oxidants and reductants, *Photobiochem. Photobiophys.*, 1, 353, 1980.
16. **Almog, R. and Berns, D. S.,** Surface properties of phycocyanin, *J. Colloid Interface Sci.*, 91, 448, 1983.
17. **Evstigneev, V. B. and Bekasova, O. D.,** Photoelectrochemical effect of phycoerythrin and phycoerythrobilin films, *Biofizika*, 15, 807, 1970.
18. **Evstigneev, V. B. and Bekasova, O. D.,** Connection between chlorophyll and phycoerythrin in the chromatophores of red algae, *Biofizika*, 17, 997, 1972.
19. **Evstigneev, V. B. and Bekasova, O. D.,** Photochemical properties of C-phycocyanin, *Mol. Biol.*, 2, 380, 1968.
20. **Bekasova, O. D. and Evstigneev, V. B.,** On the role of phycobilin pigments in photosynthesis, *Biofizika*, 22, 429, 1977.
21. **Frackowiak, D., Erokhina, L. G., Jadzyn, C., Shubin, L. M., and Shkuropatov, A. Y.,** Photovoltaic effect of biliproteins, their subunits and aggregates, *Photosynthetica*, 15, 36, 1981.
22. **Frackowiak, D., Fiksinski, K., and Pienkowska, H.,** Polarized absorption and emission of chlorophyllin, phycoerythrobilin and phycocyanobilin in stretched polyvinyl alcohol films, *Photobiochem. Photobiophys.*, 2, 21, 1981.
23. **Frackowiak, D., Pienkowska, H., and Szurkowski, J.,** Excitation energy transfer between phycoerythrin, phycocyanin and chlorophyllin in polyvinyl alcohol films, *Photosynthetica*, 16, 496, 1982.
24. **Rotolo, P.,** Spectral Properties of Phycobiliproteins and Chlorophyll *a* Adsorbed on Cellulose Nitrate, Ph.D. thesis, Polytechnic Institute of New York, New York, 1975.
25. **Chen, C.-H. and Berns, D. S.,** Thermodynamic studies of protein-salt interaction. Phycocyanin-tetra-butylammonium bromide and -tetraethylammonium bromide, *J. Phys. Chem.*, 81, 125, 1977.
26. **Chen, C.-H. and Berns, D. S.,** Thermodynamic studies of protein-salt interaction. II. Effects of geometrical configuration in alkyl-substituted quaternary ammonium salts on their interactions with phycocyanin, *J. Phys. Chem.*, 82, 2781, 1978.
27. **Chen, C.-H.,** Thermodynamic studies of protein-small molecule interaction. III. Solute-solute hydrophobic interaction of phycocyanin with alcohols, *J. Phys. Chem.*, 84, 2050, 1980.
28. **Tilden, J. E.,** A classification of the algae based on evolutionary development, with special reference to pigmentation, *Bot. Gaz.*, 95, 59, 1933.
29. **Raven, P. H.,** A multiple origin for plastids and mitochondria, *Science*, 169, 641, 1970.
30. **Lee, R. E.,** Origin of plastids and the phylogeny of algae, *Nature*, 237, 44, 1972.

31. **Fox, G. E., Stackebrandt, E., Hespell, R. B., Gibson, J., Maniloff, J., Dyer, T. A., Wolfe, R. S., Balch, W. E., Tanner, R. S., Magrum, L. J., Zablen, L. B., Blakemore, R., Gupta, R., Bonen, L., Lewis, B. J., Stahl, D. A., Luehrsen, K. R., Chen, K. N., and Woese, C. R.,** The phylogeny of prokaryotes, *Science*, 209, 457, 1980.
32. **Taylor, F. J. R.,** Symbionticism revisted: a discussion of the evolutionary impact of intracellular symbioses, *Proc. R. Soc. London Ser. B*, 204, 267, 1979.
33. **Margulis, L.,** *Symbiosis in Cell Evolution*, W. H. Freeman, San Francisco, 1981.
34. **Larkum, A. W. D. and Barrett, J.,** Light-harvesting processes in algae, in *Advances in Botanical Research*, Vol. 10, Woolhouse, H. W., Ed., Academic Press, London, 1983, 3.
35. **Greenwood, A. D., Griffiths, H. B., and Santore, U. J.,** Chloroplasts and cell compartments in Cryptophyceae, *Br. Phycol. J.*, 12, 119, 1977.
36. **Gibbs, S. P.,** The chloroplasts of some algal groups may have evolved from endosymbiotic eukaryotic algae, *Ann. N.Y. Acad. Sci.*, 361, 193, 1981.
37. **Whatley, J. M.,** Chloroplast evolution — ancient and modern, *Ann. N.Y. Acad. Sci.*, 361, 154, 1981.
38. **Dodge, J. D.,** The phytoflagellates: fine structure and phylogeny, in *Biochemistry and Physiology of Protozoa*, Vol. 1, 2nd ed., Levandowsky, M. and Hutner, S. H., Eds., Academic Press, New York, 1979, 7.
39. **Bogorad, L.,** Studies of phycobiliproteins, *Rec. Chem. Prog.*, 26, 1, 1965.
40. **Vaughan, M. H., Jr.,** Structural and Comparative Studies of the Algal Protein Phycoerythrin, Ph.D. thesis, Massachusetts Institute of Technology, Cambridge, 1964.
41. **Berns, D. S.,** Immunochemistry of biliproteins, *Plant Physiol.*, 42, 1569, 1967.
42. **Glazer, A. N., Cohen-Bazire, G., and Stanier, R. Y.,** Comparative immunology of algal biliproteins, *Proc. Natl. Acad. Sci. U.S.A.*, 68, 3005, 1971.
43. **Eder, J., Wagenmann, R., and Rüdiger, W.,** Immunological relationship between phycoerythrins from various blue-green algae, *Immunochemistry*, 15, 315, 1978.
44. **MacColl, R., Berns, D. S., and Gibbons, O.,** Characterization of cryptomonad phycoerythrin and phycocyanin, *Arch. Biochem. Biophys.*, 177, 265, 1976.
45. **Guard-Friar, D., Eisenberg, B., Edwards, M. R., and MacColl, R.,** Immunochemistry on cryptomonad biliproteins, *Plant Physiol.*, 80, 38, 1986.
46. **Harris, J. U. and Berns, D. S.,** Sequences of the N-terminus portions of biliproteins, *J. Mol. Evol.*, 5, 153, 1975.
47. **Glazer, A. N., Apell, G. S., Hixson, C. S., Bryant, D. A., Rimon, S., and Brown, D. M.,** Biliproteins of cyanobacteria and Rhodophyta: homologous family of photosynthetic accessory pigments, *Proc. Natl. Acad. Sci. U.S.A.*, 73, 428, 1976.
48. **Glazer, A. N. and Apell, G. S.,** A common evolutionary origin for the biliproteins of cyanobacteria, Rhodophyta and Cryptophyta, *FEMS Lett.*, 1, 113, 1977.
49. **Sidler, W., Kumpf, B., Suter, F., Morisset, W., Wehrmeyer, W., and Zuber, H.,** Structural studies on cryptomonad biliprotein subunits. Two different α-subunits in *Chroomonas* phycocyanin-645 and *Cryptomonas* phycoerythrin-545, *Hoppe-Seyler's Z. Physiol. Chem.*, 366, 233, 1985.
50. **Bryant, D. A., Hixson, C. S., and Glazer, A. N.,** Structural studies on phycobiliproteins. III. Comparison of bilin-containing peptides from the β subunits of C-phycocyanin, R-phycocyanin, and phycoerythrocyanin, *J. Biol. Chem.*, 253, 220, 1978.
51. **Takemoto, J. and Bogorad, L.,** Subunits of phycoerythrin from *Fremyella diplosiphon:* chemical and immunochemical characterization, *Biochemistry*, 14, 1211, 1975.
52. **Torjesen, P. A. and Sletten, K.,** C-Phycocyanin from *Oscillatoria agardhii*. I. Some molecular properties, *Biochim. Biophys. Acta*, 263, 258, 1972.
53. **Williams, V. P., Freidenreich, P., and Glazer, A. N.,** Homology of amino-terminal regions of C-phycocyanins from a prokaryote and a eukaryote, *Biochem. Biophys. Res. Commun.*, 59, 462, 1974.
54. **Frank, G., Zuber, H., and Lergier, W.,** How homologous are the α and β subunits of C-phycocyanin?, *Experientia*, 31, 23, 1975.
55. **Troxler, R. F., Foster, J. A., Brown, A. S., and Franzblau, C.,** The α and β subunits of *Cyanidium caldarium* phycocyanin: properties and amino acid sequences at the amino terminus, *Biochemistry*, 14, 268, 1975.
56. **Glazer, A. N. and Hixson, C. S.,** Subunit structure and chromophore composition of Rhodophytan phycoerythrins. *Porphyridium cruentum* B-phycoerythrin and b-phycoerythrin, *J. Biol. Chem.*, 252, 32, 1977.
57. **Troxler, R. F., Greenwald, L. S., and Zilinskas, B. A.,** Allophycocyanin from *Nostoc* sp. phycobilisomes. Properties and amino acid sequence at the NH_2 terminus of the α and β subunits of allophycocyanins I, II and III, *J. Biol. Chem.*, 255, 9380, 1980.
58. **Gardner, E. E., Stevens, S. E., Jr., and Fox, J. L.,** Purification and characterization of the C-phycocyanin from *Agmenellum quadruplicatum*, *Biochim. Biophys. Acta*, 624, 187, 1980.

59. **Freidenreich, P., Apell, G. S., and Glazer, A. N.,** Structural studies on phycobiliproteins. II. C-Phycocyanin: amino acid sequence of the β subunit. Specific cleavage of the α subunit, *J. Biol. Chem.*, 253, 212, 1978.
60. **Seckbach, J.,** On the fine structure of the acidophilic hot-spring alga *Cyanidium caldarium:* a taxonomic approach, *Microbios*, 5, 133, 1972.
61. **Frank, G., Sidler, W., Widmer, H., and Zuber, H.,** The complete amino acid sequence of both subunits of C-phycocyanin from the cyanobacterium *Mastigocladus laminosus, Hoppe-Seyler's Z. Physiol. Chem.*, 359, 1491, 1978.
62. **Offner, G. D., Brown-Mason, A. S., Ehrhardt, M. M., and Troxler, R. F.,** Primary structure of phycocyanin from the unicellular rhodophyte *Cyanidium caldarium*. I. Complete amino acid sequence of the α subunit, *J. Biol. Chem.*, 256, 12,167, 1981.
63. **Troxler, R. F., Ehrhardt, M. M., Brown-Mason, A. S., and Offner, G. D.,** Primary structure of phycocyanin from the unicellular rhodophyte *Cyanidium caldarium*. II. Complete amino acid sequence of the β subunit, *J. Biol. Chem.*, 256, 12,176, 1981.
64. **Sidler, W., Gysi, J., Isker, E., and Zuber, H.,** The complete amino acid sequence of both subunits of allophycocyanin, a light harvesting protein-pigment complex from the cyanobacterium *Mastigocladus laminosus, Hoppe-Seyler's Z. Physiol. Chem.*, 362, 611, 1981.
65. **Füglistaller, P., Suter, F., and Zuber, H.,** The complete amino-acid sequence of both subunits of phycoerythrocyanin from the thermophilic cyanobacterium *Mastigocladus laminosus, Hoppe-Seyler's Z. Physiol. Chem.*, 364, 691, 1983.
66. **Offner, G. D. and Troxler, R. F.,** Primary structure of allophycocyanin from the unicellular rhodophyte, *Cyanidium caldarium*. The complete amino acid sequences of the α and β subunits, *J. Biol. Chem.*, 258, 9931, 1983.
67. **de Lorimier, R., Bryant, D. A., Porter, R. D., Liu, W.-Y., Jay, E., and Stevens, S. E., Jr.,** Genes for the α and β subunits of phycocyanin, *Proc. Natl. Acad. Sci. U.S.A.*, 81, 7946, 1984.
68. **Pilot, T. J. and Fox, J. L.,** Cloning and sequencing of the genes encoding the α and β subunits of C-phycocyanin from the cyanobacterium *Agmenellum quadruplicatum, Proc. Natl. Acad. Sci. U.S.A.*, 81, 6983, 1984.
69. **Kronick, M. N. and Grossman, P. D.,** Immunoassay techniques with fluorescent phycobiliprotein conjugates, *Clin. Chem.*, 29, 1582, 1983.
70. **Glazer, A. N. and Stryer, L.,** Fluorescent tandem phycobiliprotein conjugates. Emission wavelength shifting by energy transfer, *Biophys. J.*, 43, 383, 1983.
71. **Oi, V. T., Glazer, A. N., and Stryer, L.,** Fluorescent phycobiliprotein conjugates for analysis of cells and molecules, *J. Cell Biol.*, 93, 981, 1982.
72. **Hardy, R. R., Hayakawa, K., Parks, D. R., and Herzenberg, L. A.,** Demonstration of B-cell maturation in X-linked immunodeficient mice by simultaneous three-colour immunofluorescence, *Nature*, 306, 270, 1983.
73. **Shapiro, H. M., Glazer, A. N., Christenson, L., Williams, J. M., and Strom, T. B.,** Immunofluorescence measurement in a flow cytometer using low-power helium-neon laser excitation, *Cytometry*, 4, 276, 1983.
74. **Hardy, R. R., Hayakawa, K., Parks, D. R., Herzenberg, L. A., and Herzenberg, L. A.,** Murine B cell differentiation lineages, *J. Exp. Med.*, 159, 1169, 1984.
75. **Phillips, J. H., Le, A. M., and Lanier, L. L.,** Natural killer cells activated in a human mixed lymphocyte response culture identified by expression of Leu-11 and class II histocompatibility antigens, *J. Exp. Med.*, 159, 993, 1984.
76. **Parks, D. R., Hardy, R. R., and Herzenberg, L. A.,** Three-color immunofluorescence analysis of mouse B-lymphocyte subpopulations, *Cytometry*, 5, 159, 1984.
77. **Loken, M. R. and Lanier, L. L.,** Three-color immunofluorescence analysis of Leu antigens on human peripheral blood using two lasers on a fluorescence-activated cell sorter, *Cytometry*, 5, 151, 1984.

INDEX

A

B

C

R

S

T

W

X

Z